MCBU
Molecular and Cell Biology Updates

Series Editors:

Prof. Dr. Angelo Azzi
Institut für Biochemie
und Molekularbiologie
Bühlstr. 28
CH - 3012 Bern
Switzerland

Prof. Dr. Lester Packer
Dept. of Molecular
and Cell Biology
251 Life Science Addition
Membrane Bioenergetics Group
Berkeley, CA 94720
USA

Biochemistry of Cell Membranes

A Compendium of Selected Topics

Edited by   S. Papa
                    J. M. Tager

Editorial Advisory Board

Birkhäuser Verlag
Basel · Boston · Berlin

Editors

Prof. Dr. S. Papa
Institute of Medical Biochemistry
and Chemistry
University of Bari
Piazza G. Cesare
I-70124 Bari
Italy

Prof. Dr. J. M. Tager
E. C. Slater Institute
University of Amsterdam
Plantage Muidergracht 12
NL-1018 TV Amsterdam
The Netherlands

Library of Congress Cataloging-in-Publication Data

Biochemistry of cell membranes: a compendium of selected topics
/ edited by S. Papa, J. M. Tager.
            p.   cm. – (Molecular and cell biology updates)
    Includes bibliographical references and index.
    ISBN 3-7643-5056-3 (hbk. : alk. paper) . – ISBN 0-8176-5056-3
(hbk. : alk. paper)
    1. Cell membranes.   2. Membrane proteins.   3. Membrane lipids.
4. Cellular signal transduction. I. Papa, S.   II. Tager, J. M.
III. Series.
QH601.B4725 1995

Deutsche Bibliothek Cataloging-in-Publication Data

Biochemistry of cell membranes : a compendium of selected
topics / ed. by S. Papa ... – Basel ; Boston ; Berlin :
Birkhäuser, 1995
    (Molecular and cell biology updates)
    ISBN 3-7643-5056-3 (Basel...)
    ISBN 0-8176-5056-3 (Boston)
    NE: Papa, Sergio [Hrsg.]

© 1995 Birkhäuser Verlag, PO Box 133, CH-4010 Basel, Switzerland
Cover illustration: N. Capitanio
Printed on acid-free paper produced from chlorine-free pulp. ∞
Printed in Germany
ISBN 3-7643-5056-3
ISBN 0-8176-5056-3

9 8 7 6 5 4 3 2 1

Contents

Bioenergetics: Energy transfer and membrane transport

Cellular ion homeostasis

Growth factors and adhesion molecules

Structural analysis of membrane proteins

Membranes and disease

Preface

This book consists of a series of reviews on selected topics within the rapidly and vastly expanding field of membrane biology. Its aim is to highlight the most significant and important advances that have been made in recent years in understanding the structure, dynamics, and functions of cell membranes. The progress that is being made in research in this field is due to the application of integrated experimental approaches, utilizing sophisticated and novel techniques in molecular biology, cell biology, biophysics and biochemistry. Due to the advances made, many problems have been or are being solved at the molecular level.

With the help of an Editorial Advisory Board consisting of Jean-Pierre Changeux (Paris), Paolo Comoglio (Torino), Rainer Jaenicke (Regensburg), Sten Orrenius (Stockholm), Lorenzo Pinna (Padova), Konrad Sandhoff (Bonn), and Gottfried Schatz (Basel), we have selected a number of topics in areas in which progress has been particularly rapid, and have invited internationally acknowledged experts in the field to review these topics. The areas covered in this monograph are: 1) Signal transduction; 2) Membrane traffic: Proteins and lipids; 3) Bioenergetics: Energy transfer and membrane transport; 4) Cellular ion homeostasis; 5) Growth factors and adhesion molecules; 6) Structural analysis of membrane proteins; and 7) Membranes and disease.

An important leitmotif in this monograph is the relationship between the structure of membrane proteins and their functions. This relationship is explored in reviews on the family of mitochondrial carriers, on porin, on Annexin V, and on other proteins. Protein transport through membranes forms another leitmotif and there are contributions on mitochondrial protein import, the transport of bacterial toxins, and the vesicular transport of proteins in the endocytic and secretory pathways. A third theme is the regulation of membrane flow in the vacuolar system, with contributions on endocytosis, autophagy, and the metabolism of glycosphingolipids. Other major topics include the transport of ions in relation to bioenergetics, signal transduction and, finally, mitochondrial ATP synthase. The monograph should serve as a benchmark for indicating the most important lines for future research in these areas.

We wish to acknowledge our indebtedness to the UNESCO Global Network on Molecular and Cell Biology, to the Administration of the Province of Bari and to the International Biomedical Institute (IBMI,

Bari) for their sponsorship of our activities and for providing us with financial support. We thank the University of Bari for its financial support. We also thank Dr. Helena Kirk for her assistance in compiling the book, and the staff of Birkhäuser Verlag, Basel, for their cooperation.

Sergio Papa
Joseph M. Tager   *Bari, October 1994*

Editorial Advisory Board:

J.P. Changeux
P. Comoglio
R. Jaenicke
S. Orrenius
L. Pinna
K. Sandhoff
G. Schatz

Biochemistry of Cell Membranes
ed. by S. Papa & J. M. Tager

The role of p21ras in receptor tyrosine kinase signaling

J.L. Bos, B.M.T. Burgering, G.J. Pronk, A.M.M. de Vries-Smits,
J.P. Medema, M. Peppelenbosch, R.M.F. Wolthuis and P. van Weeren

*Laboratory for Physiological Chemistry, Utrecht University, Stratenum, P.O. Box 80042,
NL-3508 TA Utrecht, The Netherlands*

Summary. p21ras is a small GTPase that functions as a molecular switch in signal transduction from activated receptor tyrosine kinases to cellular targets. Recently, progress has been made in the elucidation of this signaling pathway. After a ligand binds to most, if not all, receptor tyrosine kinases, a guanine nucleotide exchange factor is activated, which brings p21ras in the GTP-bound form. This GTP-bound form of p21ras interacts with the protein kinase raf1 and induces the activation of a kinase cascade involving MEK and ERKs. Subsequently, cellular targets are phosphorylated. This p21ras signaling pathway is modulated by protein kinase C, which activates raf1 in a p21ras independent manner and, in certain cell types, by protein kinase A, which inactivates raf1.

Introduction

p21ras is an important mediator of growth factor-induced differentiation and proliferation, and signaling pathways from growth factor receptors via p21ras to several cellular effectors have been identified. P21ras is the general name for three proteins encoded by the Hras, Kras, and Nras proto-oncogenes. These proteins differ structurally at their C-termini only, and thus far few, if any, functional differences between the proteins have been observed. p21ras is a small GTPase that cycles between an active GTP-bound form and an inactive GDP-bound form and functions as a molecular switch in signal transduction (Fig. 1). After its discovery in the early 1980s a search for the signaling pathways in which p21ras might be involved was started. The first real clues for the function of p21ras were obtained when p21ras function was studied by microinjection of neutralizing antibodies. These studies have shown that in a variety of cell types p21ras is required for growth factor-induced DNA synthesis and gene expression (Stacey et al., 1988; Stacey et al., 1987; Yu et al., 1988), as well as for the induction of differentiation, i.e. NGF-induced neurite outgrowth in PC12 cells (Bar-Sagi and Feramisco, 1985). A second approach to study the function of p21ras was the use of dominant negative (interfering) mutants of p21ras (Feig and Cooper, 1988). One commonly used interfering mutant is p21ras[asn17]. This mutant displays a reduced affinity for GTP, but normal

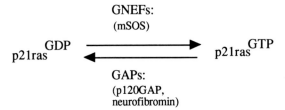

Figure 1. The p21ras cycle. p21ras cycles between an inactive GDP-bound and an active GTP-bound conformation. The cycle is regulated positively by guanine nucleotide exchange factors such as rasGRF and mSos and regulated negatively by GTPase activating proteins such as p120GAP and neurofibromin. Neurofibromin is the gene product of the neurofibromatosis type I gene.

affinity for GDP and interferes in the GDP-GTP cycling of normal p21ras (Buday and Downward, 1993a; Medema et al., 1993). Introduction of this mutant also inhibits ligand-induced proliferation and differentiation in various cell types. All these results point to a function of p21ras in growth factor receptor-mediated signal transduction.

EGF and insulin activate p21ras

Using neutralizing antibodies and inhibitory mutants of p21ras, it was shown that EGF is one of the growth factors that may require p21ras for its mitogenic signaling in fibroblasts and in PC12 cells. A possible function of p21ras in insulin signaling was indicated by the fact that insulin-induced, but not progesterone-induced maturation of *Xenopus leavis* oocytes is blocked by microinjection of neutralizing antibodies to p21ras (Deshpande and Kung, 1987; Korn et al., 1987). Convincing evidence that receptor stimulation can indeed activate p21ras was obtained by experiments showing that EGF and insulin treatment result in a shift from p21rasGDP to p21rasGTP. For insulin, these results were obtained using an NIH/3T3 cell line expressing elevated numbers of human insulin receptors (Burgering et al., 1991). In unstimulated cells approximately 15% of p21ras is in the GTP-bound state, whereas after insulin treatment 60% of p21ras is GTP-bound. Also in Swiss/3T3 cells (Burgering, unpublished observation) and in 3T3L1-adipocytes (Poras et al., 1992) insulin treatment activates p21ras. The activation of p21ras is rapid, sustains for at least 2 h and parallels the autophosphorylation of the insulin receptor β-chain (Burgering et al., 1991). EGF-induced activation of p21ras (Gibbs et al., 1990; Satoh et al., 1990) is more transient and reflects the transient nature of the activation (tyrosine phosphorylation) of the EGF receptor (Osterop et al., 1993).

Activated EGF and insulin receptors are not unique in their ability to activate p21ras and most, if not all, factors that activate receptor tyrosine kinases have been shown to activate p21ras. Also ligands that activate (non-receptor) tyrosine kinases indirectly induce the activation of p21ras. The level of p21ras activation, however, depends on the cell type and the growth factor receptor that is activated.

Mechanism of p21ras activation

Two rate-limiting steps can be distinguished in the cyclic regulation of the nucleotide content of p21ras. First, the dissociation of GDP followed by the binding of GTP, and secondly, the hydrolysis of p21ras-bound GTP. p21ras exhibits a slow intrinsic guanine nucleotide exchange and a slow intrinsic GTPase activity. However, guanine nucleotide exchange proteins regulate rapid GDP/GTP exchange, and GTPase activating proteins (GAPs) are present for the stimulation of GTP hydrolysis (Fig. 1). p21ras can be activated by either of two mechanisms: an increase in guanine nucleotide exchange activity and/or an inhibition of GAPs. For several reasons initial attention focused on the possibility that a GAP was involved in the activation of p21ras. Indeed, following PDGF stimulation p120GAP rapidly associates with the activated PDGF receptor (Kaplan et al., 1990; Kazlauskas et al., 1990) and is phosphorylated in tyrosine residues (Molloy et al., 1989). *In vitro*, PDGF receptor autophosphorylation but not p12GAP tyrosine phosphorylation, is necessary for association (Kaplan et al., 1990; Kazlauskas et al., 1990). Also, after EGF treatment (Ellis et al., 1990) and CSF-1 treatment (Reedijk et al., 1990) p120GAP becomes phosphorylated on tyrosine residues, but stable association with the respective receptors is not observed. This has led to the hypothesis that p120GAP may be an intermediate between receptor signaling and p21ras activation. Further support for this hypothesis came from studies in T-lymphocytes, where it was shown that activation of p21ras by the activated T-cell receptor correlates with a reduction of *in vitro* GAP activity (Downward et al., 1990a). After insulin treatment p120GAP is neither phosphorylated on tyrosine residues nor does it bind to the insulin receptor or to the insulin receptor substrate IRS1 (Porras et al., 1992; Pronk et al., 1992). However, p120GAP phosphorylation on tyrosine residues and association with the insulin receptor is observed after insulin treatment when the cells are pretreated with the phosphotyrosine phosphatase inhibitor phenylarsine oxide (Pronk et al., 1992). This suggests that a transient interaction between the insulin receptor and p120GAP occurs, although after insulin treatment, as well as after EGF treatment, a decrease of *in vitro* p120GAP activity was not observed (Buday and Downward, 1993b; Medema et al., 1993). Furthermore, mutant PDGF receptors that fail to bind p120GAP and fail to phosphorylate p120GAP after PDGF treatment, signal normally to p21ras, indicating that interaction of activated receptors with p120GAP is not essential for the activation of p21ras (Burgering et al., 1994; Sotoh et al., 1993; Valius and Kazlauskas, 1993).

Guanine nucleotide exchange on p21ras can be analyzed by measuring the rate of labeled nucleotide binding to p21ras in permeabilized cells. Using this approach it was found that both insulin and EGF can induce a two- to three-fold increase in nucleotide binding to p21ras (Buday and Downward, 1993b; Medema et al., 1993). Several candidate guanine nucleotide exchange factors have been isolated (Downward et al., 1990b; West et al., 1990; Wolfman and Macara, 1990) and recently the genes for two exchange factors have been cloned and characterized. A gene for rasGRF was cloned by virtue of its homology with the yeast exchange factor CDC25 (Shou et al., 1992) and, independently, by its ability to complement for a defect in CDC25 (Martegani et al., 1992). This exchange factor is predominantly expressed in brain tissue. mSos is an ubiquitously expressed protein encoded by genes homologous to the *Drosophila* SOS gene and exhibits nucleotide exchange activity *in vitro* (Bowtell et al., 1992; Chardin et al., 1993). mSos has homology with the yeast guanine nucleotide exchange factor CDC25 as well.

In the elucidation of the signaling pathway from growth factor receptor to p21ras the genetic analysis of lower eukaryotes has been crucial, particularly the analysis of vulval development in *Caenorhabditis elegans* and eye development in *Drosophila melanogaster* (Fig. 2). In both developmental processes a receptor tyrosine kinase (Let23, Sevenless) is receiving an inductive signal and a p21ras homolog (Let60, Dras) serves as an intermediate to transduce signals to downstream targets. By genetic analysis several genes have been identified that function in this pathway. One of these genes encodes Sem5, a protein containing an SH2 domain flanked by two SH3 domains (Clark et al., 1992; Sternberg and Horvitz, 1991) SH2 and SH3 domains were first identified as regions of homology between Src and Fps outside the kinase domain (Koch et al., 1991). SH2 domains are regions of about 100 amino acids that serve as a separate protein domain, which can bind to phosphotyrosine-containing peptides. The specificity of binding is determined by the first few amino acids C-terminal of the phosphotyrosine. SH3 domains are smaller and bind to proline-rich sequences. Genetic evidence places Sem5 between Let23 and Let60. A protein homologous to Sem5 has been identified by virtue of either its ability to bind to the phosphorylated EGF receptor *in vitro* (Grb2) (Lowenstein et al., 1992) or its homology to SH2 domains (Matuoka et al., 1993; Matuoka et al., 1992).

In *Drosophila melanogaster*, a gene has been identified that shows identity with a yeast exchange factor for RAS. The encoded protein (Son of Sevenless or Sos) has been placed genetically between the Sevenless receptor and Dras (Fortini et al., 1992; Simon et al., 1991).

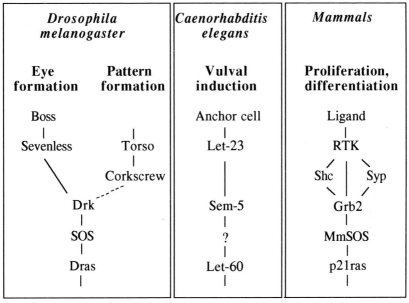

Drosophila melanogaster		Caenorhabditis elegans	Mammals
Eye formation	Pattern formation	Vulval induction	Proliferation, differentiation
Boss		Anchor cell	Ligand
│		│	│
Sevenless	Torso	Let-23	RTK
＼	│		╱ │ ＼
	Corkscrew	│	Shc │ Syp
＼-----			＼ │ ╱
Drk		Sem-5	Grb2
│		│	│
SOS		?	MmSOS
│		│	│
Dras		Let-60	p21ras
│		│	│

Figure 2. Activation of p21ras by receptor tyrosine kinases in lower and higher eukaryotes. In *Drosophila melanogaster*, Dras is involved in eye formation which is induced by the binding of the ligand Boss to the Sevenless receptor tyrosine kinase. In the formation of the tail, the Torso receptor tyrosine kinase is activated. In this pathway the Corkscrew phosphotyrosine phosphatase is involved, but details of this interaction are lacking. The SH2-SH3 domain-containing adaptor protein Drk and the guanine nucleotide exchange factor SOS mediate the signal to Dras. In *Caenorhabditis elegans*, the ras homologue Let-60 is involved in vulval development. The anchor cell activates the Let-23 receptor tyrosine kinase. The SH2-SH3 domain-containing Sem-5 protein mediates the signal to Let-60. In mammals most, if not all, receptor tyrosine kinases activate p21ras. Grb2-MmSOS associates either directly with the activated receptor, with tyrosine-phosphorylated Shc protein or with the tyrosine-phosphory-lated corkscrew homologue Syp. For further explanation see text.

Surprisingly, both the *Drosophila melanogaster* Sos protein and the mammalian counterparts mSos1 and 2, contain a proline-rich sequence at the C-terminal end of the protein. This has led to the hypothesis that Grb2 functions as an adaptor protein, which can bind to phosphoty-rosine residues of the receptor through its SH2 domain, and to the proline-rich segment of mSos through its SH3 domains. Indeed, it has been shown in *Drosophila melanogaster* that the Sem5 homolog Drk binds to both the Sevenless receptor and to Sos (Olivier et al., 1993; Simon et al., 1993). Recently, this model has been tested for the EGF receptor (Buday and Downward, 1993a; Egan et al., 1993; Gale et al., 1993; Li et al., 1993; Rozakis-Adcock et al., 1993). It was found that a Grb2-mSos complex associates with the EGF receptor after EGF treat-ment, resulting in the translocation of the mSos protein to the particu-late fraction of the cell. Interestingly, inhibition of the Grb2-EGF receptor interaction by a phosphopeptide that mimics the Grb2 binding site on the EGF receptor inhibits EGF-induced guanine nucleotide exchange in permeabilized cells (Buday and Downward, 1993a). This strongly suggests that binding of Grb2-mSos to the EGF receptor triggers the activation of p21ras. It should be noted that the association between Grb2 and mSos appears to be constitutive and may not be, or only partially, induced by EGF treatment.

An alternative pathway for the EGF receptor to recruit p21ras is through the Shc protein. Shc is a protein containing an SH2 domain and a glycine/proline-rich sequence. It binds to the activated EGF receptor and is phosphorylated on tyrosine residues upon EGF receptor activation (Pelicci et al., 1992). In addition, upon EGF treatment, Grb2 associates with Shc. Therefore, it is possible that the observed interaction between the EGF receptor and mSos is (in part) mediated by Shc. The possibility that Shc functions in p21ras signaling is supported by the observation that Shc overexpression induces p21^{ras}-dependent neurite outgrowth in PC12 cells (Rozakis-Adcock et al., 1992).

Also, after insulin treatment Shc is phosphorylated on tyrosine residues and Shc-Grb2 complexes are formed, but Shc does not associate with the insulin receptor (Kovacina and Roth, 1993; Pronk et al., 1993; Skolnik et al., 1993b; Tobe et al., 1993). In addition, Grb2 is found associated with IRS1 (Skolnik et al., 1993b; Tobe et al., 1993). Using cells that overexpress either Grb2 (Skolnik et al., 1993a) or dSOS (Baltensperger et al., 1993) it was further found that IRS1 and Grb2 form a heterotrimeric complex with mSos. However, such complexes could not be identified in fibroblasts overexpressing the insulin receptor (Pronk et al., 1994). In these cells insulin-induced complex formation between tyrosine-phosphorylated Shc and mSos was observed, suggesting that insulin may use Shc as the predominant docking protein for the Grb2-mSos complex. Taken together, complex formations involving Shc and Grb2-mSos are clearly implicated in the control of p21^{ras} guanine nucleotide status by growth factor receptor tyrosine kinases and non-receptor tyrosine kinases. Adding to this complexity is the recent observation that Syp, a phosphotyrosine phosphatase, can also bind to Grb2-mSos (Li et al., 1994). All these interactions may recruit mSos to the plasma membrane, the location of p21ras, or induce a conformational change of mSos that results in the activation of guanine nucleotide exchange activity.

In addition to complex formation, mSos is phosphorylated on serine/threonine residues after EGF or insulin treatment (Burgering et al., 1993b; Rozakis-Adcock et al., 1993). This phosphorylation occurs relatively late compared to complex formation between Shc and mSos and activation of p21ras (Burgering et al., 1993b). Therefore, the observed mSos phosphorylation appears not involved in the initial p21^{ras} activation. Furthermore, inhibition of mSos phosphorylation does not seem to affect activation of p21ras (Burgering et al., 1993b). The yeast guanine nucleotide exchange factor CDC25 is phosphorylated by protein kinase A (Gross et al., 1992). This phosphorylation releases CDC25 from the plasma membrane. Since in yeast RAS activates adenylate cyclase and thus protein kinase A, it is suggested that the phosphorylation may serve as a negative feedback in the activation of p21ras. However, also in yeast an effect of the phosphorylation of CDC25 on RAS has not yet been reported (Gross et al., 1992).

Downstream effects of growth factor-induced activation of p21ras

p21ras has been implicated in the activation of a variety of cellular proteins. The first demonstration that in mammalian cells p21ras mediates insulin-induced cellular responses came from studies using the interfering mutant p21ras[asn17]. In transient expression assays this mutant inhibits insulin-induced activation of the c-*fos*-promoter and the collagenase promoter (Medema et al., 1991). Using a recombinant vaccinia virus with p21ras[asn17] as transgene, it was shown that expression of the interfering mutant inhibits insulin-induced activation of ERK2 in A14 cells, NIH-3T3 cells transfected with the human insulin receptor (Burgering et al., 1991; de Vries-Smits et al., 1992). These results are part of a large series of reports on the signaling pathway from p21ras to downstream targets. Key enzymes in this pathway appear to be the serine/threonine kinases of the MAP2 kinase family (Blenis, 1993; Cobb et al., 1991). These enzymes phosphorylate at least *in vitro*, a variety of different cellular targets, and appear to play a crucial role in their activation. Among these targets are transcription factors, like SRF and c-*myc*, and other kinases, like MAPKAP kinase-2 (Stokoe et al., 1992) and p90rsk (Sturgill et al., 1988). In addition, ERK phosphorylates and activates phospholipase A2 (Lin et al., 1993) and may regulate through this pathway cytoskeletal rearrangements (Peppelenbosch et al., 1993). With respect to the metabolic functions of insulin, the link to p90rsk is intriguing, since p90rsk is similar to the insulin-stimulated protein kinase-1 (ISPK-1) (Sutherland et al., 1993). ISPK-1 phosphorylates a serine residue on the glycogen-binding subunit of protein phosphatase-1G (PP-1G). This phosphorylation increases the rate by which PP-1G dephosphorylates and activates glycogen synthase. Furthermore, PP-1G dephosphorylates and inactivates phosphorylase kinase. These processes underlie the activation of glycogen synthesis and inhibition of glycogen breakdown (Dent et al., 1990).

A picture of the signaling pathway between p21ras and ERKs is emerging (Fig. 3). First, ERKs are phosphorylated on tyrosine and threonine residues, which results in the activation of the enzymes. A dual specificity kinase, MEK, is responsible for the phosphorylation and

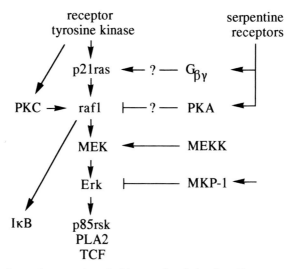

Figure 3. Agonists and antagonists of p21ras-mediated signaling. Receptor tyrosine kinases may activate raf1 via different pathways, one of which is p21ras, another protein kinase C (PKC). Activation of raf1 triggers the activation of MEK and ERKs. Certain serpentine receptors may activate raf1 through p21ras, but may inactivate raf1 through protein kinase A (PKA). MEKkinase (MEKK) may be activated by certain serpentine receptors and activates MEK, whereas the dual specific phosphatase MKP-1 inactivates ERKs. For further explanation see text.

thus for the activation. This kinase is also activated by phosphorylation and at least two different kinases have been assigned as MEK kinases, i.e., raf1 and MEKK (see Blenis for review (1993)).

raf1 is encoded by the c-raf proto-oncogene. The 70 kDa raf1 protein is phosphorylated on several serine and, in some cases, tyrosine residues after growth factor stimulation. Analogous to oncogenic p21ras, v-raf is able to transform established cell lines. Inhibition of p21ras by neutralizing antibodies or by interfering mutants of p21ras, does not abolish v-raf transformation, indicating that v-raf transforms cells independently of p21ras. An interaction between p21ras and raf1 was observed first by Moodie et al. (1993) showing that raf1 could bind to p21rasGTP containing columns and not or with reduced efficiency to columns with p21rasGDP. In addition p21ras and raf1 interact with each other in a yeast two-hybrid screen (Cooper and Kashishian, 1993; van Aelst et al., 1993; Vojtek et al., 1993), and a biochemical interaction could be demonstrated as well (Warne et al., 1993; Zhang et al., 1993). It was found that the N-terminal part of raf1 interacts with p21ras. In addition, it was shown that the raf1 binds to the effector domain of p21ras. Importantly, there is a strong correlation between effector domain mutants of p21ras that are defective in cell signaling and the failure of binding of raf1 to p21ras. *In vivo*, all signals that activate p21ras also seem to activate raf1 kinase. However, *in vitro* association of raf1 with p21ras-GTP is not sufficient to activate raf1, and another signal may be required to establish full activation of raf1 (Macdonald et al., 1993). In this respect the p21ras-raf interaction differs from the interaction of HsCDC24GTP and its kinase, which does result in the activation of the kinase. Apparently, a third component is necessary for the activation of raf1 by p21rasGTP. Recently, it has been reported that a phosphatidylcholine-specific phospholipase C (PC-PLC) functions downstream of p21ras and upstream of raf1 (Cai et al., 1993). The involvement of PC-PLC may be at the level of activation of a kinase for raf1. Perhaps this kinase is a PKC isozyme, which can be activated through PC-PLC-mediated production of diacylglycerol (DAG). Indeed, PKC activation leads to raf1 activation, possibly through direct phosphorylation of raf1 by PKC (Kolch et al., 1993; Sözeri et al., 1992). Surprisingly, as for p21ras-raf1 interaction, *in vitro* raf1 phosphorylation by PKC is also not sufficient for activation of raf1 (Macdonald et al., 1993). One model could be that PKC and p21ras cooperate in the activation of raf1. Alternatively, a third, still unknown, component could be involved in this process. Perhaps p120GAP or neurofibromin are involved in the process of raf1 activation as well.

Besides MEK phosphorylation, raf1 phosphorylates IκB (Li and Sedivy, 1993). IκB in unphosphorylated form is associated to the transcription factor NFκB in the cytoplasm. Phosphorylation of IκB results in the release of NFκB, which subsequently translocates to the nucleus, where it can bind to specific DNA sequences to regulate gene expression.

The second MEK kinase, MEKK, is probably not involved in p21ras signaling, but may mediate signals from other signaling pathways to activate ERKs, for instance, signals generated by activated serpentine receptors (Lange-Carter et al., 1993). However, some of these receptors may activate ERKs through the p21ras pathway as well (van Corven et al., 1993).

Protein kinase C

The role of protein kinase C in p21ras signaling has been a matter of debate. It has been postulated that protein kinase C is involved in growth factor-induced activation of p21ras, as well as in growth factor-induced activation of ERKs. Although many different PKC isoenzymes are present, in NIH-3T3 cells there is no evidence that PCK is involved in insulin-induced activation of ERK2. Downregulation of PKC by prolonged pretreatment with the phorbolester TPA or inhibition of PKC with H7 or sphingosine does not affect insulin-induced activation of p21ras or ERK2 (de Vries-Smits et al., 1992; Burgering, personal communication). This is in agreement with results of Blackshear and coworkers, who demonstrated that in fibroblasts insulin does not activate PKC (Blackshear et al., 1991). Activation of PKC by TPA does activate ERKs. Apparently, PKC mediates a separate pathway to activate ERK2, which is operative in certain cell types and with certain ligands only. Indeed, EGF can activate ERK2 through different pathways. One of these pathways is mediated by p21ras, the other by PKC, and a third by a calcium-dependent factor. The use of each of these pathways by EGF is cell-type dependent (Burgering et al., 1993a).

Antagonists of p21ras signaling

Growth-factor-induced activation of p21ras signaling can be antagonized by signals that elevate the levels of cAMP, thereby activating PKA (Burgering et al., 1993b; Cook and McCormick, 1993; Graves et al., 1993; Sevetson et al., 1993; Wu et al., 1993). This leads to an inhibition of growth-factor-induced activation of raf1. The inhibition is cell-type specific, i.e., it occurs in NIH/3T3 and rat1 fibroblasts, but not in Swiss-3T3 fibroblasts and PC12 cells. Surprisingly, this inhibition of raf1 reflects the inhibition of cell growth by cAMP in these cells, suggesting that inhibition of raf1 is (part of) the molecular basis of cAMP-induced inhibition of cell growth.

The inhibition of raf1 also affects TPA-induced activation of ERK2, supporting the notion that PKC activates the EK2 pathway by activating raf1. The inhibition of raf1 may be due to direct phosphorylation of ser43 by PKA, which reduces the affinity of raf1 to p21rasGTP (Wu et al., 1993). Alternatively, Rap1, which is a substrate for PKA and which can antagonize p21ras signaling, may be involved in this process (Burgering et al., 1993b). Clearly, p21ras signaling pathways are subject to both positive (PKC) and negative (PKA) interference.

Acknowledgements
The work in our laboratory is supported by the Dutch Cancer Society.

References

Baltensperger, K., Kozma, L.M., Cherniack, A.D., Klarlund, J.K., Chawla, A., Banerjee, U. and Czech, M.P. (1993) Binding of the Ras activator Son of Sevenless to Insulin Receptor Substrate-1 signaling complexes. *Science* 260: 1950–1952.
Bar-Sagi, D. and Feramisco, J.R. (1985) Microinjection of the *ras* oncogene protein into PC12 cells induces morphological differentiation. *Cell* 42: 841–848.
Blackshear, P.J., Haupt, D.M., Stumpo, D.J. (1991) Insulin activation of protein kinase C: A reassessment. *J. Biol. Chem.* 266: 10946–10952.
Blenis, J. (1993) Signal transduction via MAP kinases: proceed at your own RSK. *Proc. Natl. Acad. Sci. USA* 90: 5889–5892.
Bowtell, D., Fu, P., Simon, M. and Senior, P. (1992) Identification of murine homologues of

the *Drosophila Son* of sevenless gene: potential activators of *ras. Proc. Natl. Acad. Sci. USA* 89: 6511–6515.

Buday, L. and Downward, J. (1993a) Epidermal growth factor regulates p21[ras] through the formation of a complex of receptor, Grb2 adaptor protein, and SOS nucleotide exchange factor. *Cell* 73: 611–620.

Buday, L. and Downward, J. (1993b) Epidermal growth factor regulates the exchange rate of guanine nucleotides on p21ras in fibroblasts. *Mol. Cell. Biol.* 13: 1903–1910.

Burgering, B.M.T., Freed, E., McCormick, F. and Bos, J.L. (1994) Platelet-derived growth factor-induced p21ras-mediated signaling is independent of platelet-derived growth factor receptor interaction with GTPase-activating protein or phosphatidylinositol-3-kinase. *Cell. Growth & Diff.* 5: 341–347.

Burgering, B.M.T., de Vries-Smits, A.M.M., Medema, R.H., van Weeren, P., Tertoolen, L.G.J. and Bos, J.L. (1993a) Epidermal growth factor induces phosphorylation of extracellular signal-regulated kinase 2 via multiple pathways. *Mol. Cell. Biol.* 13: 7248–7256.

Burgering, B.M.T., Pronk, G.J., van Weeren, P.C., Chardin, P. and Bos, J.L. (1993b) cAMP antagonizes p21[ras]-directed activation of extracellular signal-regulated kinase 2 and phosphorylation of mSOS nucleotide exchange factor. *EMBO J.* 12: 4211–4220.

Burgering, B.M.T., Medema, R.H., Maassen, J.A., Van de Wetering, M.L., Van der Eb, A.J., McCormick, F. and Bos, J.L. (1991) Insulin stimulation of gene expression mediated by p21ras activation. *EMBO J.* 10: 1103–1109.

Cai, H., Erhardt, P., Troppmaier, J., Diaz-Meco, M.T., Sithanandam, G., Rapp, U.R., Moscat, J. and Cooper, G.M. (1993) Hydrolysis of phosphatidylcholine couples ras to activation of raf protein kinase during mitogenic signal transduction. *Mol. Cell. Biol.* 13: 7645–7651.

Chardin, P., Camonis, J.H., Gale, N.W., Van Aelst, I., Schlessinger, J., Wigler, J. and Bar-Sagi, D. (1993) Human Sos 1: a guanine nucleotide exchange factor for Ras that binds to Grb2. *Science* 260: 1338–1343.

Clark, S.G., Stern, M.J. and Horvitz, H.R. (1992) *C. elegans* cell-signalling gene *sem-5* encodes a protein with SH2 and SH3 domains. *Nature* 356: 340–344.

Cobb, M.H., Boulton, T.G. and Robbins, D.J. (1991) Extracellular signal-regulated kinases: ERKs in progress. *Cell Regulation* 2: 965–978.

Cook, S.J. and McCormick, F. (1993) Inhibition by cAMP of Ras-dependent activation of Raf. *Science* 262: 1069–1072.

Cooper, J.A. and Kashishian, A. (1993) In vivo binding properties of SH2 domains from GTPase-activating protein and phosphatidylinositol 3-kinase. *Mol. Cell. Biol.* 13: 1737–1745.

de Vries-Smits, A.M.M., Burgering, B.M.T., Leevers, S.J., Marshall, C.J. and Bos, J.L. (1992) Involvement of p21ras in activation of extracellular signal-regulated kinase 2. *Nature* 357: 602–604.

Dent, P., Lavoinne, A., Nakielny, S., Caudwell, F.B., Watt, P. and Cohen, P. (1990) The molecular mechanism by which insulin stimulates glycogen synthesis in mammalian skeletal muscle. *Nature* 348: 302–308.

Deshpande, A.K. and Kung, H.-F. (1987) Insulin induction of *Xenopus laevis* oocyte maturation is inhibited by monoclonal antibody against p21ras proteins. *Mol. Cell. Biol.* 1: 1285–1288.

Downward, J., Graves, J.D., Warne, P.H., Rayter, S. and Cantrell, D.A. (1990a) Stimulation of p21[ras] upon T-cell activation. *Nature* 346: 719–723.

Downward, J., Riehl, R., Wu, L. and Weinberg, R.A. (1990b) Identification of a nucleotide exchange-promoting activity for p21[ras]. *Proc. Natl. Acad. Sci. USA* 87: 5998–6002.

Egan, S. E., Giddings, B.W., Brooks, M.W., Buday, L., Sizeland, A.M. and Weinberg, R.A. (1993) Association of SOS Ras exchange protein with Grb2 is implicated in tyrosine kinase signal transduction and transformation. *Nature* 353: 45–51.

Ellis, C., Moran, M., McCormick, F. and Pawson, T. (1990) Phosphorylation of GAP and GAP-associated proteins by transforming and mitogenic tyrosine kinases. *Nature* 343: 377–381.

Feig, L.A. and Cooper, G.M. (1988) Inhibition of NIH 3T3 cell proliferation by a mutant *ras* protein with preferential affinity for GDP. *Mol. Cell. Biol.* 8: 3235–3243.

Fortini, M.E., Simon, M.A. and Rubin, G.M. (1992) Signalling by the *sevenless* protein tyrosine kinase is mimicked by Ras1 activation. *Nature* 355: 559–561.

Gale, N.W., Kaplan, S., Lowenstein, E.J., Schlessinger, J. and Bar-Sagi, D. (1993) Grb2 mediates the EGF-dependent activation of guanine nucleotide exchange on ras. *Nature* 363: 88–92.

Gibbs, J.B., Marshall, M.S., Scolnick, E.M., Dixon, R.A.F. and Vogel, U.S. (1990) Modulation of guanine nucleotides bound to ras in NIH3T3 cells by oncogenes, growth factors, and the GTPase activating protein (GAP). *J. Biol. Chem.* 265: 20437–20442.

Graves, L.M., Bornfeld, K.E., Raines, E.W., Potts, B.C., Macdonald, S.G., Ross, R. and Krebs, E.G. (1993) *Proc. Natl. Acad. Sci. USA* 90: 10300–10304.

Gross, E., Goldberg, D. and Levitzki, A. (1992) Phosphorylation of the *S. cerevisiae* Cdc25 in response to glucose results in its dissociation from Ras. *Nature* 360: 762–765.

Kaplan, D.R., Morrison, D.K., Wong, G., McCormick, F. and Williams, L.T. (1990) PDGF β-receptor stimulates tyrosine phosphorylation of GAP and association of GAP with a signaling complex. *Cell* 61: 125–133.

Kazlauskas, A., Ellis, C., Pawson, T. and Cooper, J.A. (1990) Binding of GAP to activated PDGF receptors. *Science* 247: 1578–1581.

Koch, C.A., Anderson, D., Moran, M.F., Ellis, C. and Pawson, T. (1991) SH2 and SH3 domains: Elements that control interactions of cytoplasmic signaling proteins. *Science* 252: 668–674.

Kolch, W., Heidecker, G., Kochs, G., Hummel, R., Vahldl, H., Mischak, H., Finkenzeller, G., Marmé, D. and Rapp, U.R. (1993) Protein kinase C activates RAF-1 by direct phosphorylation. *Nature* 364: 249–252.

Korn, L.J., Siebel, C.W., McCormick, F. and Roth, R.A. (1987) *Ras* p21 as a potential mediator of insulin action in *Xenopus* oocytes. *Science* 236: 840–843.

Kovacina, K.S. and Roth, R.A. (1993) Identification of Shc as a substrate of the insulin receptor kinase distinct from the GAP-associated 63 kDa tyrosine phosphoprotein. *Biochem. Biophys. Res. Commun.* 192: 1303–1311.

Lange-Carter, C.A., Pleiman, C.M., Gardner, A.M., Blumer, K.J. and Johnson, G.L. (1993) A divergence in the MAP kinase regulatory network defined by MEK kinase and Raf. *Science* 260: 315–319.

Li, N., Batzer, A., Daly, R., Yajnik, V., Skolnik, E., Chardin, P., Bar-Sagi, D., Margolis, B. and Schlessinger, J. (1993) Guanine-nucleotide-releasing factor hSOS1 binds to Grb2 and links receptor tyrosine kinases to Ras signalling. *Nature* 363: 85–88.

Li, S. and Sedivy, J.M. (1993) Raf-1 protein kinase activates the NFκB transcription factor by dissociating the cytoplasmic NFκB-IκB complex. *Proc. Natl. Acad. Sci. USA* 90: 9247–9251.

Li, W., Nishimura, R., Kashishian, A., Batzer, A.G., Kim, W.J.H., Cooper, J.A. and Schlessinger, J. (1994) A new function for a phosphotyrosine phosphatase: linking Grb2-Sos to a receptor tyrosine kinase. *Mol. Cell. Biol.* 14: 509–517.

Lin, L.-L., Wartmann, M., Lin, A.Y., Knopf, J.L., Seth, A. and Davis, R.J. (1993) cPLA is phosphorylated and activated by MAP kinase. *Cell* 72: 269–278.

Lowenstein, E.J., Daly, R.J., Batzer, A.G., Li, W., Margolis, B., Lammers, R., Ullrich, A., Skolnik, E.Y., Bar-Sagi, D. and Schlessinger, J. (1992) The SH2 and SH3 domain-containing protein GRB2 links receptor tyrosine kinases to ras signaling. *Cell* 70: 431–442.

Macdonald, S.G., Crews, G.M., Wu, L., Driller, J., Clark, R., Erikson, R.L. and McCormick, F. (1993) Reconstruction of the raf-1-MEK-ERK signal transduction pathway in vitro. *Mol. Cell. Biol.* 13: 6615–6620.

Martegani, E., Vanoni, M., Zippel, R., Coccetti, P., Brambilla, R., Ferrari, C., Sturani, E. and Alberghina, L. (1992) Cloning by functional complementation of a mouse cDNA encoding a homologue of CDC25, a *Saccharomyces cerevisiae* RAS activator. *EMBO J.* 11: 2151–2157.

Matuoka, K., Shibasaki, F., Shibata, M. and Takenawa, T. (1993) Ash/Grb2, a SH2/SH3-containing protein, couples to signaling for mitogenesis and cytoskeletal reorganization by EGF and PDGF. *EMBO J.* 12: 3467–3473.

Matuoka, K., Shibata, M., Yamakawa, A. and Takenawa, T. (1992) Cloning of ASH, a ubiquitous protein composed of one Src homology region (SH) 2 and two SH3 domains, from human and rat cDNA libraries. *Proc. Natl. Acad. Sci. USA* 89: 9015–9019.

Medema, R.H., de Vries-Smits, A.M.M., van der Zon, G.C.M., Maassen, J.A. and Bos, J.L. (1993) Ras activation by insulin and epidermal growth factor through enhanced exchange of guanine nucleotides on p21^{ras}. *Mol. Cell. Biol.* 13: 155–162.

Medema, R.H., Wubbolts, R. and Bos, J.L. (1991) Two dominant inhibitory mutants of p21^{ras} interfere with insulin-induced gene expression. *Mol. Cell. Biol.* 11: 5963–5967.

Molloy, C.J., Bottaro, D.P., Fleming, T.P., Marshall, M.S., Gibbs, J.B. and Aaronson, S.A. (1989) PDGF induction of tyrosine phosphorylation of GTPase activating protein. *Nature* 342: 711–714.

Moodie, S.A., Willumsen, B.M., Weber, M.J. and Wolfman, A. (1993) Complexes of Ras.GTP with Raf-1 and Mitogen-Activated Protein Kinase. *Science* 260: 1658–1661.

Olivier, J.P., Raabe, T., Henkemeyer, M., Dickson B., Mbamalu, G., Margolis, B., Schlessinger, J., Hafen, E. and Pawson, T. (1993) A Drosophila SH2-SH3 adaptor protein implicated in coupling the Sevenless tyrosine kinase to an activator of ras guanine nucleotide exchange, SOS. *Cell* 73: 179–191.

Osterop, A.P.R.M., Medema, R.H., van der Zon, G.C.M., Bos, J.L., Möller, W. and Maassen, J.A. (1993) Epidermal Growth Factor receptors generate Ras-GTP more efficiently than insulin receptors. *Eur. J. Biochem.* 212: 477–482.

Pelicci, G., Lanfrancone, L., Grignani, F., McGlade, J., Cavallo, F., Forni, G., Nicoletti, L., Grignani, F., Pawson, T. and Pelicci, P.G. (1992) A novel transforming protein (SHC) with an SH2 domain is implicated in mitogenic signal transduction. *Cell* 70: 93–104.

Peppelenbosch, M.P., Tertoolen, L.G.J., Hage, W.J. and De Laat, S.W. (1993) Epidermal growth factor-induced actin remodelling is regulated by 5-lipoxygenase and cyclooxygenase metabolites. *Cell* 74: 565–575.

Porras, A., Nebreda, A.R., Benito, M. and Santos, E. (1992) Activation of ras by insulin in 3T3 L1 cells does not involve GTPase-activating protein phosphorylation. *J. Biol. Chem.* 267: 21124–21131.

Pronk, G.J., de Vries-Smits, A.M.M., Buday, L., Downward, J., Maassen, J.A., Medema, R.H. and Bos, J.L. (1994) Involvement of Shc in insulin- and EGF-induced activation of p21^{ras}. *Mol. Cell. Biol.* 14: 1575–1581.

Pronk, G.J., McGlade, J., Pellici, G., Pawson, T. and Bos, J.L. (1993) Insulin-induced phosphorylation of the 46- and 52-kDa Shc proteins. *J. Biol. Chem.* 268: 5748–5753.

Pronk, G.J., Medema, R.H., Burgering, B.M.T., Clark, R., McCormick, F. and Bos, J.L. (1992) Interaction between the p21^{ras} GTPase activating protein and the insulin receptor. *J. Biol. Chem.* 267: 24058–24063.

Reedijk, M., Liu, X. and Pawson. T. (1990) Interactions of phosphatidyl kinase, GTPase activating protein (GAP) and GAP-associated proteins with the colony-stimulating factor 1 receptor. *Mol. Cell. Biol.* 10: 5601–5608.

Rozakis-Adcock, M., Fernley, R., Wade, J., Pawson, T. and Bowtell, D. (1993) The SH2 and SH3 domains of mammalian Grb2 couple the EGF receptor to the Ras activator mSOS1. *Nature* 363: 83–85.

Rozakis-Adcock, M., McGlade, J., Mbamalu, G., Pelicci, G., Daly, R., Li, W., Batzer, A., Thomas, S., Brugge, J., Pelicci, P.G., Schlessinger, J. and Pawson, T. (1992) Association of the Shc and Grb2/Sem5 SH2-containing proteins is implicated in activation of the Ras pathway by tyrosine kinases. *Nature* 360: 689–692.

Satoh, T., Endo M., Nakafuku, M., Akiyama, T., Yamamoto, T. and Kaziro, Y. (1990) Accumulation of p21^{ras}.GTP in response to stimulation with Epidermal Growth Factor and oncogene products with tyrosine kinase activity. *Proc. Natl. Acad. Sci. USA* 87: 7926–7929.

Sevetson, B.R., Kong, X. and Lawrence, J.C., Jr. (1993) Increasing cAMP attenuates activation of mitogen-activated protein kinase. *Proc. Natl. Acad. Sci. USA* 90: 10305–10309.

Shou, C., Farnsworth, C.L., Neel, B.G. and Feig, L.A. (1992) Molecular cloning of cDNAs encoding a guanine-releasing factor for ras p21. *Nature* 358: 351–354.

Simon, M.A., Dodson, G.S. and Rubin, G.M. (1993) An SH3-SH2-SH3 protein is required for p21^{ras1} activation and binds to Sevenless and SOS proteins in vitro. *Cell* 73: 169–177.

Simon, M.A., Bowtell, D.D.L., Dodson, G.S., Laverty, T.R. and Rubin, G.M. (1991) Ras1 and a putative guanine nucleotide exchange factor perform crucial steps in signaling by the sevenless protein tyrosine kinase. *Cell* 67: 701–716.

Skolnik, E.Y., Batzer, A., Li, N., Lee, C.-H., Lowenstein, E., Mohammadi, M., Margolis, B. and Schlessinger, J. (1993a) The function of Grb2 in linking the insulin receptor to Ras signaling pathways. *Science* 260: 1953–1955.

Skolnik, E.Y., Lee, C.-H., Batzer, A., Vicentini, L.M., Zhou, M., Daly, R., Myers, M.J., Backer, J.M., Ullrich, A., White, M.F. and Schlessinger, J. (1993b) The SH2/SH3 domain-containing protein Grb2 interacts with tyrosine-phosphorylated IRS1 and Shc: implications for insulin control of ras signalling. *EMBO J.* 12: 1929–1936.

Sotoh, T., Fantl, W.J., Escobedo, J.A., Williams, L.T. and Kaziro, Y. (1993) Platelets-derived growth factor receptor kinase mediates ras through different signaling pathways in different cells. *Mol. Cell. Biol.* 13: 3706–3713.

Sözeri, O., Vollmer, K., Liyanage, M., Frith, D., Kour, G., Mark III, G.E. and Stabel, S. (1992) Activation of the c-Raf protein kinase by protein kinase C phosphorylation. *Oncogene* 7: 2259–2262.

Stacey, D.W., Tsai, M.-H., Yu, C.-L. and Smith, J.K. (1988) Critical role of cellular *ras* proteins in proliferative signal transduction. *CSHSQB* 53: 871–881.

Stacey, D.W., Watson, T., Kung, H.-F. and Curran, T. (1987) Microinjection of transforming *ras* protein induced c-*fos* expression. *Mol. Cell. Biol.* 7: 523–527.

Sternberg, P.W. and Horvitz, H.R. (1991) Signal transduction during *C. elegans* vulval induction. *Trends Genet.* 7: 366–371.

Stokoe, D., Campbell, D.G., Nakielny, S., Hidaka, H., Leevers, S.J., Marshall, C.J. and Cohen, P. (1992) MAPKAP kinase-2; a novel protein kinase activated by mitogen-activated protein kinase. *EMBO J.* 11: 3985–3994.

Sturgill, T.W., Ray, L.B., Erikson, E. and Maller, J.L. (1988) Insulin-stimulated MAP-2 kinase phosphorylates and activates ribosomal S6 kinase II. *Nature* 334: 715–718.

Sutherland, C., Campbell, D.G. and Cohen, P. (1993) Identification of insulin-stimulated protein kinase-1 as the rabbit equivalent of rsk-2. *Eur. J. Biochem.* 212: 581–588.

Tobe, K., Matuoka, K., Tamemoto, H., Ueki, K., Kaburagi, Y., Asai, S., Noguchi, T., Matsuda, M., Tanaka, S., Hattori, S., Fukui, Y., Akanamu, Y., Yazaki, Y., Takanawa, T. and Kadowaki, T. (1993) Insulin stimulates association of Insulin Receptor Substrate-1 with the protein Abundant Src Homology/Growth Factor Receptor-bound protein 2. *J. Biol. Chem.* 268: 11167–11171.

Valius, M. and Kazlauskas, A. (1993) Phospholipase C-γ1 and phosphatidylinositol 3 kinase are downstream mediators of the PDGF receptor's mitogenic signal. *Cell* 73: 321–334.

van Aelst, L., Barr, M., Marcus, S., Polverino, A. and Wigler, M. (1993) Complex formation between RAS and RAF and other protein kinases. *Proc. Natl. Acad. Sci. USA* 90: 6213–6217.

van Corven, E.J., Hordijk, P.L., Medema, R.H., Bos, J.L. and Moolenaar, W.H. (1993) Pertussis toxin sensitive activation of p21^{ras} by G protein-coupled receptor agonists. *Proc. Natl. Acad. Sci. USA* 90: 1257–1261.

Vojtek, A.B., Hollenberg, S.M. and Cooper, J.A. (1993) Mammalian ras interacts directly with the serine/threonine kinase raf. *Cell* 74: 205–214.

Warne, P.H., Viciana, P.R. and Downward, J. (1993) Direct interaction of Ras and the amino-terminal region of Raf-1 in vitro. *Nature* 364: 352–355.

West, M., Kung, H. and Kamata, T. (1990) A novel membrane factor stimulates guanine nucleotide exchange reaction of *ras* proteins. *FEBS Lett.* 259: 245–248.

Wolfman, A. and Macara, I.G. (1990) A cytosolic protein catalyzes the release of GDP from p21^{ras}. *Science* 248: 57–69.

Wu, J., Dent, P., Jelinek, T., Wolfman, A., Weber, M.J. and Sturgill, T.W. (1993) Inhibition of the EGF-activated MAP kinase signaling pathway by adenosine 3′,5′-monophosphate. *Science* 262: 1065–1069.

Yu, C.-L., Tsai, M.-H. and Stacey, D.W. (1988). Cellular *ras* activity and phospholipid metabolism. *Cell* 52: 63–71.

Zhang, X., Settleman, J., Kyriakis, J.M., Takeuchi-Suzuli, E., Elledge, S.J., Marshall, M.S., Bruder, J.T., Rapp, U.R. and Avruch, J.A. (1993) Normal and oncogenic p21ras proteins-bind to the amino-terminal regulatory domain of c-raf-1. *Nature* 364: 308–313.

Biochemistry of Cell Membranes
ed. by S. Papa & J. M. Tager
© 1995 Birkhäuser Verlag Basel/Switzerland

Casein kinase-2 and cell signaling

L.A. Pinna, F. Meggio and S. Sarno

Dipartimento di Chimica Biologica and CNR, Centro per lo Studio della Fisiologia Mitocondriale, Università di Padova, via Trieste 75, I-35121 Padova, Italy

Summary. Casein kinase-2 (CK2) is a ubiquitous Ser/Thr-specific protein kinase supposed to play a relevant, albeit enigmatic, role in cellular regulation. In this review the arguments supporting the implication of CK2 in signal transduction will be summarized and the elusive mode of regulation of this pleiotropic enzyme will be discussed.

Introduction

The enzyme now termed casein kinase-2 (CK2) was one of the first protein kinases to be detected and characterized in the early 1960s, shortly after the discovery of phosphorylase kinase and before that of cAMP-dependent protein kinase (for a historical overview see Pinna et al., 1990). Despite its early discovery and the increasing circumstantial evidence that CK2 must play a pivotal role in cellular regulation (Pinna, 1990; Tuazon and Traugh, 1991) the postition(s) occupied by this ubiquitous and pleiotropic enzyme in the protein kinase network that controls cell proliferation and differentiation and the molecular mechanism that modulates its activity are still enigmatic. In particular, while protein kinases involved in signal transduction generally are silent enzymes whose activity is triggered either by second messengers or by phosphorylation (or dephosphorylation) events, CK2 is spontaneously active, being independent of any known second messenger and insensitive, at least *in vitro*, to phosphorylation either occurring through autocatalytic mechanism or promoted by other kinases.

The aim of this review is to summarize the arguments supporting the implication of CK2 in signal transduction and cellular control and to discuss the potential mechanism(s) by which CK2 can be regulated in order to fulfill its pleiotropic functions.

Overview

Arguments supporting a pleiotropic role of CK2

An intriguing outcome of studies on CK2 is the discrepancy between the abundance of circumstantial data consistent with a central role of this

Table 1. Protein substrates of casein kinase-2.

eIF3
RNA polymerase I
RNA polymerase II
Troponin-T
Glycogen synthase
PKA R-II subunit
T-substrate
mRNA particles
Androgen receptor
DNA topoisomerase I
DNA topoisomerase II
Spermin binding protein
Spectrin
Glycophorin
Pseudorabies virus-pp62
Myosin light chain
HMG 14
Inhibitor-2 of PP1
Nucleolar protein B23
Nucleolin
Calmodulin
Ornithine decarboxylase
HMG-1 (Palvimo et al., 1987)
Low density lipoprotein receptor
Fibrinogen
Insulin receptor
IGF-II receptor
Guanine nucleotide exchange factor
MAP-1B
Gap43 (= neuromodulin)
Filaggrin
Ca^{2+} channel blockers receptor
eIF4B
eIF5
HMG-like protein P1
c-Myc (Lüscher et al., 1989)
Large T Antigen (Grasser et al., 1988)
Acetyl CoA carboxylase
Clathrin light b-chain
Skeletal muscle myosin heavy chain (Murakami et al., 1988)
eIF2 β subunit
Elongation factor 1β
Hsp90
Human c-myc (Lüscher et al., 1989)
DARPP-32
CREB (Lee et al., 1990)
c-Myb (Lüscher et al., 1990)
Lipoprotein inhibitor (Girard et al., 1990)
Smooth muscle myosin heavy chain (Kelley et al., 1990)
p120 cell proliferation nucleolar protein (Valdez et al., 1990)

Brain myosin heavy chain (Murakami et al., 1990)
p53 (Meek et al., 1990)
pp35 proliferation nuclear P-protein (Malek et al., 1990)
β tubulin (Diaz et al., 1990)
Mannose-6-phosphate receptor (Meresse et al., 1990)
HPV E7 (Barbosa et al., 1990)
MYCN (Hamman et al., 1991)
SRF (Manak et al., 1991)
PKC (Taminaga et al., 1991)
Caldesmon (Wawrzynow et al., 1991)
Calsequestrin (Cala et al., 1991)
DSIP peptide (Nakamura et al., 1991)
Epididymal fast cells p22 (Diggle et al., 1991)
Nuclear DNA binding protein p210 (Sekiguchi et al., 1991)
RAB-17 (Plana et al., 1991)
P_0, P_1 and P_2 (Hasler et al., 1991)
Nef (see Marshak, 1991)
Adenovirus E1a (see Marshak, 1991)
N-myc (Hagiwara et al., 1992)
p34 cdc2 (Russo et al., 1992)
Pea lamin like protein (Li et al., 1992)
RIPP (Wu et al., 1992)
DNA ligase (Prigent et al., 1992)
mUBF (Voit et al., 1992)
c-jun (Lin et al., 1992)
Viral P-protein (Barik et al., 1992)
CK2β (Litchfield et al., 1992)
HIV1 viral protein (Henklein et al., 1992)
Nuclear Antigen 2 (Grässer et al., 1992)
Max protein (Berberich et al., 1992)
VSV protein (Takacs et al., 1992)
ZEBRA protein (Kolmer et al., 1993)
CREM (de Groot et al., 1993)
Calnexin (Cala et al., 1993)
Vitamin D receptor (Jurutka et al., 1993)
Synaptotagmin (Davletov et al., 1993)
IGFBP-1 (Jones et al., 1993)
PU-1 transcription factor (Pongubala et al., 1993)
Insulin receptor substrate-1 (Tanasijevic et al., 1993)
VZV protein (Yao et al., 1993)
PP2C (Kobayashi et al., 1993)
Dystrophin (Luise et al., 1993)
FK506 (Jin et al., 1993)
hnRNP (Mayrand et al., 1993)
30 kDa proteasome (Ludemann et al.,

Table 1. (Continued)

1993)	Prothymosin α (Barcia et al., 1993)
p98 sea urchin eggs (Ohtsuki et al., 1993)	eIF2B (Karinch et al., 1993)
p65 synaptic vesicles (Bonnett et al., 1993)	Factors Va and VIIIa (Kalafatis et al., 1993)
hnRNP A1 (Cobianchi et al., 1993)	p34 chloroplasts ribonucleoprotein
Osteopontin (Ashkar et al., 1993)	(Kanekatsu et al., 1993)

The endogenous proteins that have been reported to be phosphorylated by CK2 are chronologically listed starting from 1979, when the first physiological substrate of CK2 (eIF3) was identified by Hathaway et al., until the end of 1993. Proteins without reference are referenced in Pinna (1990). Neither phosvitin nor casein, which are thought to be artificial substrates of CK2, are considered.

enzyme in cellular regulation, and the paucity of direct information about the molecular mechanism by which CK2 could accomplish such a putative function within the cell.

Among several observations supporting the view that CK2 must play a general and relelvant role in cellular regulation are the following.

(i) Ubiquity of CK2 which was found in all eukaryotic organisms and tissues investigated so far (see Issinger, 1993).

(ii) Subcellular distribution of CK2 in several compartments: although CK2 is concentrated in nuclei (Yu et al., 1991; Krek et al., 1992) it is also present in cytosol (see Issinger, 1993) and significant CK2 activity is detectable in particulate fractions as well, such as mitochondria (Damuni and Reed, 1988), and glycogen particles (Meggio et al., 1984). CK2 targets, moreover, were detected in ribosomes and in plasma membranes (Meggio et al., 1984).

(iii) Both the catalytic (α/α') and the non catalytic (β) subunits of CK2 are highly conserved throughout evolution. Even the gene structures are highly conserved (Wirkner et al., 1984). These observations in conjunction with the lethality for cells of the disruption of the catalytic subunits, overcome by transfection and expression of the catalytic subunits (Padmanabha et al., 1990), and with the structural features found in the promoter region of the CK2β gene (Voss et al., 1991), support the concept that CK2 is a housekeeping enzyme.

(iv) The number of proteins that are known to be phosphorylated by CK2 probably surpasses that of protein targets of any other individual protein kinase considered so far. In most instances there is evidence that these proteins are also phosphorylated by CK2 in living cells. Forty-six proteins phosphorylated by CK2 were already enumerated by Pinna (1990). An up-to-date list, although still incomplete, is reported in Table 1; it includes 103 proteins implicated in a wide variety of cellular functions. Many of these proteins are involved in signal transduction, gene expression, cell growth and division. In particular a notable number of CK2 targets are nuclear proteins, including several transcription factors, a tumor suppressor protein, and enzymes involved in DNA metabolism (reviewed by Litchfield and Lüscher, 1993).

(v) CK2 activity is invariably enhanced in proliferating and transformed cells, as compared to quiescent and normal ones, respectively (reviewed by Pinna, 1990; Issinger, 1993). Interestingly, CK2 activity is increased up to 25-fold in benign prostatic hyperplasia, while its increase in prostatic carcinoma is much less pronounced (about three fold) (Yenice et al., 1993). These findings corroborate the proposal that CK2 may represent a "proliferation marker" (Schneider and Issinger, 1989). Expression of CK2 at both its mRNA and protein levels, during embriogenic development, displays characteristic oscillations and tissue distribution patterns (Mestres et al., 1994).

(vi) Various stimuli, including serum, insulin, IGF-1, EGF, bombesin, TNF, and TPA have been reported to induce transient increase of CK2 activity (reviewed by Litchfield and Lüscher, 1993). In contrast to the enchancement of CK2 observed in proliferating and transformed cells (which is accounted for by higher expression of CK2), this short-term regulation would imply some kind of reversible modulation of activity. A shortcoming of all these studies however is their failure to document the presence of hyperactive forms of CK2 with distinct molecular features. It should also be noted that the same stimulus does not always induce changes in CK2 activity (see Litchfield and Lüscher, 1993).

Regulation of CK2

In most instances CK2 holoenzyme is composed of two catalytic (α and/or α', sharing high homology) and two non catalytic β-subunits giving a very stable heterotetramer that does not dissociate unless under denaturing conditions. This latter observation in conjuction with the finding that the β-subunit is required for conferring optimal enzymatic activity to the catalytic subunits (Cochet and Chambaz, 1983; Grankowski et al., 1991; Meggio et al., 1992; Birnbaum et al., 1992) rules out any functional analogy with PKA (i.e., the only other known protein kinase sharing the same heterotetrameric structure), whose activity is triggered by the dissociation of the two regulatory subunits from the catalytic ones, driven by the physiological effector cAMP. CK2 β-subunit, on the other hand, is known to undergo autophosphorylation at its N terminal end (Ser-2/Ser-3) and phosphorylation by cdc2 in its C terminal region (Ser-209). A correlation between EGF-induced activation of CK2 and an increased ^{32}P incorporation into its β subunit has been observed in A-431 cells (Ackerman et al., 1990). It is not clear, however, whether this enhanced radiolabeling is merely due to accelerated phosphate turnover or to real hyperphosphorylation, or if the latter hypothesis is correct, whether hyperphosphorylation is the cause or just an effect of CK2 activation. Recent studies with recombinant β-subunit mutants lacking the autophosphorylation (Meggio et al., 1993; Hinrichs et al., 1993; Bodenback et al., 1994) and the cdc2 phosphorylation site (Meggio et al., 1993; Boldyreff et al., 1993) provided the clear-cut demonstration that neither of these sites can play a prominent role in the regulation of CK2 catalytic activity. In accordance with this a hyperphosphorylated form of CK2 isolated from cells arrested in mitosis exhibited normal activity (Litchfield et al., 1992).

On the other hand, mutagenesis of CK2 β-subunit within the sequence 55–64 revealed that a number of acidic residues clustered in this region are responsible for an instrinsic down-regulation of catalytic activity (Boldyreff et al., 1992, 1993). Such a negative control is normally masked by the overall beneficial effect of the β-subunit whenever CK2 activity is tested with canonical peptide substrates, but becomes dramatically evident using calmodulin as phosphorylatable substrate (Meggio et al., 1994). Thus, the β-subunits appear to display a dual function: on the one hand, they stabilize and activate the α-subunits by tightly binding to them, essentially through their C terminal domain; on the other hand, they depress the potential activity of the holoenzyme by virtue of their 55–64 acidic stretch (Boldyreff et al., 1993). The balance between these two effects is largely dependent on the nature of the phosphorylatable target. The latent activity of CK2 holoenzyme toward calmodulin can be unmasked by the addition of polybasic peptides, e.g., protamine, some histones and a peptide reproducing the 66–86 segment of the α-subunit (Sarno et al., 1993; Meggio et al., 1994), but not by polyamines, like spermine, which conversly reverse the stimulatory effect of polybasic peptides, probably through a mechanism of futile competition (Sarno et al., 1993, 1994). The same acidic sequence of the β-subunit which is responsible for downregulation of basal activity also accounts for the suppression of activity toward calmodulin and the positive response toward polycationic peptides (Meggio et al., 1994). Intriguingly, this 55–64 acidic cluster is also required for efficient autophosphorylation of the β-subunit at its N terminal end (Boldyreff et al., 1994). This reveals a spatial contiguity between these two regions and may suggest a role of the autophosphorylated Ser-2/Ser-3 residues in reinforcing the downregulation imposed by the 55–64 acidic region.

Collectively taken, these data disclose a plausible mechanism by which CK2 activity could be variably modulated depending on the nature of the substrate effectors, as schematically summarized in Figure 1. Two crucial points however remain to be elucidated: (i) how could a mechanism like that of Figure 1 put CK2 under the control signals

calmodulin   **canonical substrate**

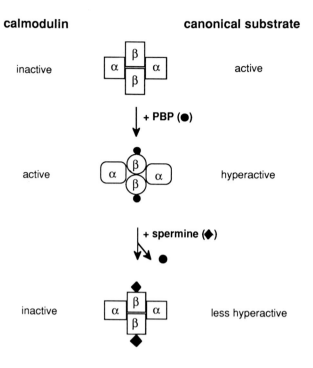

PBP= polybasic peptide (e. g. histone H4)

Figure 1. Hypothetical mechanism by which CK2 activity and specificity can be modulated by different classes of polycationic compounds. Calmodulin is representative of the protein substrates whose phosphorylation is entirely dependent on the addition of polybasic peptides (see Pinna, 1990, 1991). "Canonical substrate" refers to the majority of CK2 targets, whose phosphorylation occurs also under basal conditions, being variably stimulated by polybasic peptides.

initiated by extracellular stimuli? (ii) Which is the mechanism of regulation with those CK2 targets whose phosphorylation is insensitive to polycationic effectors, thus presumably unaffected by the intrinsic downregulation imposed by the β-subunits? In order to address the former question a better preliminary insight should be gained about the physiological compounds whose putative regulation of CK2 is mimicked *in vitro* by polybasic peptides and antagonized by spermine. As far as the latter question is concerned the possibility should be explored that, as discussed elsewhere (Pinna, 1991), a number of CK2 targets may undergo constitutive phosphorylation, the regulation being entirely committed in this case to the protein phosphatase(s) reversing the effect of CK2.

Site specificity of CK2

Some hints about the implication of CK2 in cellular functions could be also provided by its unusual site specificity. In contrast to most Ser/Thr protein kinases, CK2 recognizes acidic, rather than basic residues as specificity determinants, its minimum consensus consisting of a negatively charged side chain at position +3 (Marin et al., 1986; Marchiori et al., 1988). Efficient phosphorylation by CK2 however normally requires the presence of multiple acidic residues (see Pinna, 1990), a feature which is detrimental for most of the other Ser/Thr protein kinases. In contrast, the presence of basic residues and the motif Ser/Thr-Pro, which are specifically recognized by second messenger and cyclin dependent protein kinases, respectively, are perceived as negative determinants by CK2 (Pinna, 1990; Marin et al., 1992); consequently, the site specificity of CK2 is not only distinct from but even incompatible with that of basophilic and Pro-directed protein kinases (e.g., Pinna et al., 1984; Marin et al., 1992).

Bearing in mind that also S6 kinases and MAP kinases are basophilic (Flotow and Thomas, 1992; Donella-Deana et al., 1993) and Pro-directed (Davis, 1993), respectively, it can be argued that the unique site specificity of CK2 represents a natural hindrance against the possibility that the same residues are simultaneously affected by CK2 and the other Ser/Thr protein kinases implicated in signal transduction. A teleological corollary of this inference would be that incidental phosphorylation of CK2 sites by more or less any other Ser/Thr protein kinase, and vice versa, is especially detrimental to cellular functions. This would in turn imply that the role of CK2 in signaling is distinct from, if not opposite to that of the other protein kinases. Based on this assumption any effort to insert CK2 into one of the canonical protein kinase cascades described up to now would be unsuccessful.

Conclusions

The conclusion that CK2 must play a crucial and pleiotropic role as mediator of cellular regulation is corroborated by a wide variety of coincidental observations, notably: its ubiquity; its high conservation throughout evolution; the lethal effect of disruption of its genes; the miriad of its targets, several of which are involved in gene expression and cell proliferation; the dramatic increase of CK2 in proliferating cells; transient enhancements of CK2 activity upon treatment of cells with mitogens.

On the other hand, little information is available about the direct participation of CK2 in definite signal transduction pathway(s) and about its physiological regulation. CK2 appears to be insensitive to either second messengers or phosphorylation/dephosphorylation events which are known to control many other growth-related protein kinases. The unique site specificity of CK2, dictated by acidic residues and incompatible with the specificity determinants of second messengers dependent protein kinases as well as of cyclin dependent and mitogen activated (MAP) kinases may reflect an at least partial segregation of CK2 from the canonical pathways initiated by hormones and mitogenic stimuli. Pertinent to this may be also the observation that mitogens which invariably promote a dramatic increase of MAP kinases activity only occasionally induce a rise of CK2, whose basal activity is quite remarkable in non-stimulated cells as well.

The recent finding that the non catalytic β-subunit of CK2 possesses an acidic region responsible for an intrinsic downregulation of catalytic activity, which is especially effective with some physiological targets (notably with calmodulin) discloses new perspectives about the mechanism by which CK2 could be modulated. Such a downregulation can be reversed *in vitro* by a number of polycationic peptides while it is restored by the simultaneous addition of spermine. These observations envisage a potential device by which two categories of polycationic compounds could tune CK2 activity.

Acknowledgements
This work was supported by grants to FM from MURST and CNR (grant 93.00341.CT04) and to LAP from CNR (grant 93.02055.CT14, Target Project on Biotechnology and Bioinstrumentation and ACRO) and from AIRC.

References

Ackerman, P., Glover, C.V.C. and Osheroff, N. (1990) Stimulation of casein kinase-II by epidermal growth factor: relationship between the physiological activity of the kinase and the phosphorylation state of its β-subunit. *Proc. Natl. Acad. Sci. USA* 87: 821–825.

Ashkar, S., Teplow, D.B., Glimcher, M.J. and Saavedra, R.A. (1993) *In vitro* phosphorylation of mouse osteopontin expressed in *E. coli. Biochem. Biophys. Res. Commun.* 191: 126–133.

Barbosa, M., Edmonds, C., Fisher, C., Schiller, J., Lowry, D. and Vausden, K. (1990) The region of the HPV E7 oncoprotein homologous to adenovirus E1a and SV40 large T antigen contains separate domains for Rb binding and casein kinase II phosphorylation. *EMBO J.* 9: 153–160.

Barcia, M.G., Castro, J.M., Jullien, C.D., Gonzales, C.G. and Freire, M. (1992) Prothymosin α is phosphorylated by casein kinase-2. *FEBS Lett.* 312: 152–156.

Barik, S. and Banerjee, A.K. (1992) Phosphorylation by cellular casein kinase II is essential for transcriptional activity of vesicular stomatitis virus phosphoprotein P. *Proc. Natl. Acad. Sci. USA* 89: 6570–6574.

Berberich, S.J. and Cole, M.D. (1992) Casein kinase II inhibits the DNA binding activity of Max homodimers, but not Myc/Max heterodimers. *Genes Dev.* 6: 166–176.

Bodenbach, L., Fauss, J., Robitzki, A., Krehan, A., Lorenz, P., Lozeman, F.J. and Pyerin, W. (1994) Recombinant human casein kinase II. A stydy with the complete set of subunits (α, α' and β), site-directed autophosphorylation mutants and a bicistronically expressed holoenzyme. *Eur. J. Biochem.* 220: 263–283.

Birnbaum, M.J., Wu, J., O'Reilly, D.R., Rivera-Marrero, C.A., Hanna, D.E., Miller, L.K., and Glover, C.V.C. (1992) Expression and purification of the α and β subunits of Drosophila casein kinase II using a Baculo vector. *Protein Expression and Purification* 3: 142–150.

Boldyreff, B., Meggio, F., Pinna, L.A. and Issinger, O.-G. (1992) Casein kinase-2 structure-function relationship: creation of a set of mutants of the β subunit that variably surrogate the wildtype β subunit. *Biochem. Biophys. Res. Commun.* 188: 228–234.

Boldyreff, B., Meggio, F., Pinna, L.A. and Issinger, O.-G. (1993) Reconstitution of normal and hyperactivated forms of casein kinase-2 by variably mutated β-subunits. *Biochemistry* 32: 12672–12677.

Boldyreff, B., Meggio, F., Pinna, L. A. and Issinger, O.-G. (1994) Efficient autophosphorylation and phosphorylation of the β-subunit by casein kinase-2 require the integrity of an acidic cluster 50 residues downstream from the phosphoacceptor site. *J. Biol. Chem.* 269: 4827–4832.

Bennett, M.K., Miller, K.G. and Scheller, R.H. (1993) Casein kinase II phosphorylates the synaptic vesicle protein p65. *J. Neurosci.* 13: 1701–1708.

Cala, S.E. and Jones, L.R. (1991) Phosphorylation of cardiac and skeletal muscle calsequestrin isoforms by casein kinase II. Demonstration of a cluster of unique rapidly phosphorylated sites in cardiac calsequestrin. *J. Biol. Chem.* 266: 391–398.

Cala, S.E., Ulbright, C., Kelly, J.S. and Jones, L.R. (1993) Purification of a 90-kDa protein (Band VII) from cardiac sarcoplasmic reticulum. Identification as calnexin and localization of casein kinase II phosphorylation. *J. Biol. Chem.* 268: 2969–2975.

Cobianchi, F., Calvio, C., Stoppini, M., Buvoli, M. and Riva, S. (1993) Phosphorylation of human hnRNP protein AP1 abrogates *in vitro* strand annealing activity. *Nucleic Acid Res.* 21: 949–955.

Cochet, C. and Chambaz, E.M. (1983) Oligomeric structure and catalytic activity of G type casein kinases. *J. Biol. Chem.* 258: 1403–1406.

Damuni, Z. and Reed, L.J. (1988) Purification and properties of a protamine kinase and a type II casein kinase from bovine kidney mitochondria. *Arch. Biochem. Biophys.* 262: 574–584.

Davis, R.J. (1993) The mitogen-activated protein kinase signal transduction pathway. *J. Biol. Chem.* 268: 14553–14557.

Davletov, B., Sontag, J.-M., Hata, Y., Petrenko, A.G., Fyske, E.M., Jahn, R. and Sündhof, T.C. (1993) Phosphorylation of synaptotagmin I by casein kinase II. *J. Biol. Chem.* 268: 6816–6822.

de Groot, R.P., den Hertog, J., Vandenheede, J.R., Goris, J. and Sassone-Corsi, P. (1993) Multiple and cooperative phosphorylation events regulate the CREM activator function. *EMBO J.* 12: 3903–3912.

Diaz-Nido, J., Serrano, L., Lopez Otin, C., Vandekerckhove, J. and Avila, J. (1990) Phosphorylation of a neuronal-specific β-tubulin isotype. *J. Biol. Chem.* 265: 13949–13954.

Diggle, T.A., Schmitz-Pfeiffer, C., Borthwick, A.C., Welsh, G.I. and Denton, R.M. (1991) Evidence that insulin activates casein kinase II in rat epididymal fat-cells and that this may result in the increased phosphorylation of an acidic soluble 22 kDa protein. *Biochem. J.* 279: 545–551.

Donella Deana, A., Lavoinne, A., Marin, O., Pinna, L.A. and Cohen, P. (1993) An analysis of the substrate specificity of insulin-stimulated protein kinase-I, a mammalian homologue of S6 kinase-II. *Biochim. Biophys. Acta* 1178: 189–193.

Flotow, H. and Thomas, G. (1992) Substrate recognition determinants of the mitogen-activated 70K S6 kinase from rat liver. *J. Biol. Chem.* 267: 3074–3078.

Girard, T.J., McCourt, D., Novotny, W.F., MacPhail, L.A., Likert, K.H. and Broze, G.J. (1990) Endogenous phosphorylation of the lipoprotein-associated coagulation inhibitor at serine-2. *Biochem. J.* 270: 621–625.

Grankowski, N., Boldyreff, B. and Issinger, O.-G. (1991) Isolated and characterization of recombinant human CKII subunits α and β from bacteria. *Eur. J. Biochem.* 198: 25–30.

Grässer, F.A. Scheidtmann, K.H., Tuazon, P.T., Traugh, J.A. and Walter, G. (1988) *In vitro* phosphorylation of SV40 large T Antigen. *Virology* 165: 13–22.

Grässer, F.A. Göttel, S., Haiss, P., Boldyreff, B., Issinger, O.-G. and Müller-Lantzsch (1992) Phosphorylation of the Epstein-Barr virus nuclear antigen 2. *Biochem. Biophys. Res. Commun.* 186: 1694–1701.

Hagiwara, T., Nakaya, K., Nakamura, Y., Nakajima, H., Nishimura, S. and Taya, Y. (1992) Specific phosphorylation of the acidic central region of the N-Myc protein by casein kinase II. *Eur. J. Biochem.* 209: 945–950.

Hamann, U., Wenzel, A., Frank, R. and Schwab, M. (1991) The MYCN protein of human neuroblastoma cells is phosphorylated by casein kinase II in the central region and at serine 367. *Oncogene* 6: 1745–1751.

Hasler, P., Brot, N., Weissbach, H., Parnassa, A.P. and Elkon, K.B. (1991) Ribosomal proteins P0, P1 and P2 are phosphorylated by casein kinase II at their conserved carboxyl termini. *J. Biol. Chem.* 266: 13815–13820.

Hinrichs, M.V., Jedlicki, A., Tellez, R., Pongor, S., Gatica, M., Allende, C. and Allende, J.E. (1993) Activity of recombinant α and β subunits of casein kinase II from *Xenophus laevis*. *Biochemistry* 32: 7310–7316.

Issinger, O.-G. (1993) Casein kinase: pleiotropic mediators of cellular regulation. *Pharmac. Ther.* 59: 1–30.

Jin, Y.J. and Burakoff, S.J. (1993) The 25-kDa FK506-binding protein is localized in the nucleus and associated with casein kinase II and nucleolin. *Proc. Natl. Acad. Sci. USA* 90: 7769–7773.

Jones, J.I., Busby, W.H., Jr., Wright, G., Smith, C.E., Kimack, N.M. and Clemmons, D.R. (1993) Identification of the sites of phosphorylation in insulin like growth factor binding protein-1. Regulation of its affinity by phosphorylation at serine 101. *J. Biol. Chem.* 268: 1125–1131.

Jurutka, P.W., Hsieh, J.-C., MacDonald, P.N., Terpening, P.M., Haussler, C.A., Haussler, M.R. and Whitefield, G.K. (1993) Phosphorylation of serine 208 in the human VitD receptor. The predominant aminoacid phosphorylated by casein kinase II *in vitro* and identification of a significant phosphorylation site in intact cells. *J. Biol. Chem.* 268: 6791–6799.

Kalafatis, M., Roud, M.D., Jenny, R.J., Ehrlich, Y.H. and Mann, K.G. (1993) Phosphorylation of factor Va and factor VIIIa by activated platelets. *Blood* 81: 704–719.

Kanekatsu, M., Munakata, H., Furuzono, K. and Ohtsuki, K. (1993) Biochemical characterization of a 34 kDa ribonucleoprotein (p34) purified from the spinach chloroplast fraction as an effective phosphate acceptor for casein kinase II. *FEBS Lett.* 335: 176–180.

Karinch, A.M., Kimball, S.R., Vary, T.C. and Jefferson, L.S. (1993) Regulation of eukaryotic initiation factor-2B activity in muscle of diabetic rats. *Am. J. Physiol.* 264: E101–108.

Kelley, C.A. and Adelstein, R.S. (1990) The 204-kDa smooth muscle myosin heavy chain is phosphorylated in intact cells by casein kinase II on a serine near the carboxyl terminus. *J. Biol. Chem.* 265: 17876–17882.

Kobayashi, T., Kanno, S., Terasawa, T., Murakami, T., Ohnishi, M., Ohtsuki, K., Hiraga, A. and Tamura, S. (1993) Phosphorylation of Mg^{2+} dependent protein phosphatase alpha (type 2C α) by casein kinase II. *Biochem. Biophys. Res. Commun.* 195: 484–489.

Kolman, J. L., Taylor, N., Marshak, D.R. and Miller, G. (1993) Serine-173 of the Epstein-Barr virus ZEBRA protein is required for DNA binding and is target for casein kinase II phosphorylation. *Proc. Natl. Acad. Sci. USA* 90: 10115–10119.

Krek, W., Maridor, G. and Nigg, E.A. (1992) Casein kinase II is a predominantly nuclear enzyme. *J. Cell Biol.* 116: 43–55.

Lee, C.Q. Yun, Y., Hoeffler, J.P. and Habener, J.F. (1990) Cyclic-AMP responsive transcriptional activation of CREB-327 involves interdependent phosphorylated subdomains. *EMBO J.* 9: 4455–4465.

Li, H. and Roux, S.J. (1992) Casein kinase II protein kinase is bound to lamina-matrix and phosphorylates lamin-like protein in isolated pea nuclei. *Proc. Natl. Acad. Sci. USA* 89: 8434–8438.

Lin, A., Frost, J., Deng, T., Smeal, T., Al-Alawi, N., Kikkawa, U., Hunter, T., Brenner, D. and Karin, M. (1992) Casein kinase-II is a negative regulator of c-Jun DNA binding and AP-1 activity. *Cell* 70: 777–789.

Litchfield, D.W. and Lüscher, B., Lozeman, F.J., Eisenman, R.N. and Krebs. E.G. (1992) Phosphorylation of casein kinase II by $p34^{cdc2}$ *in vitro* and at mitosis. *J. Biol. Chem.* 267: 13943–13951.

Litchfield, D.W. and Lüscher, B. (1993) Casein kinase II in signal transduction and cell regulation. *Mol. Cell. Biochem.* 127/128: 187–199.

Ludemann, R., Lerca, K.M. and Etlinger, J.D. (1993) Copurification of casein kinase II with 20S proteasome and phosphorylation of a 30 kDa proteasome subunit. *J. Biol. Chem.* 268: 17413–17417.

Luise, M., Presotto, C., Senter, L., Betto, R., Ceoldo, S., Furlan, S., Salvatori, S., Sabbadini, R.A. and Salviati, G. (1993) Dystrophin is phosphorylated by endogenous protein kinases. *Biochem. J.* 293: 243–247.

Lüscher, B., Kuenzel, E., Krebs, E.G. and Eisenman, R.N. (1989) Myc oncoproteins are phosphorylated by casein kinase II. *EMBO J.* 8: 1111–1119.

Lüscher, B., Christenson, E., Litchfield, D.W., Krebs, E.G. and Eisenman, R.N. (1990) Myb DNA binding inhibited by phosphorylation at a site deleted during oncogenic activation. *Nature* 344: 517–521.

Malek, S.N., Katumuluwa, A.I. and Pasternack, G.R. (1990) Identification and preliminary characterization of two related proliferation-associated nuclear phosphoproteins. *J. Biol. Chem.* 265: 13400–13409.

Manak, R. and Prywes, R. (1991) Mutation of serum response factor phosphorylation sites and the mechanism by which its DNA-binding activity is increased by casein kinase II. *Mol. Cell. Biol.* 11: 3652–3659.

Marin, O., Meggio, F., Marchiori, F., Borin, G. and Pinna, L.A. (1986) Site specificity of casein kinase-2 (TS) from rat liver cytosol. A study with model peptide substrates. *Eur. J. Biochem.* 160: 239–244.

Marin, O., Meggio, F., Draetta, G. and Pinna, L.A. (1992) The consensus sequence for cdc2 kinase and for casein kinase-2 are mutually incompatible. *FEBS Lett.* 301: 111–114.

Marchiori, F., Meggio, F., Marin, O., Borin, G., Calderan, A., Ruzza, P. and Pinna, L.A. (1988) Synthetic peptide substrates for casein kinase-2. Assessment of minimum structural requirements for phosphorylation. *Biochim. Biophys. Acta* 971: 332–338.

Marshak, D.R. and Carrol, D. (1991) Synthetic peptide substrates for casein kinase II. *Methods in Enzymol.* 200: 134–156.

Mayrand, S.H., Dwen, P. and Pedersen, T. (1993) Serine/Threonine phosphorylation regulates binding of C hnRNP proteins to pre-mRNA. *Proc. Natl. Acad. Sci. USA* 90: 7764–7768.

Meek, D., Simon, S., Kikkawa, U. and Eckhart, W. (1990) The p53 tumor suppressor protein is phosphorylated at serine 389 by casein kinase II. *EMBO J.* 9: 3253–3260.

Meggio, F., Boldyreff, B., Marin, O., Pinna, L.A. and Issinger, O.-G. (1992) Role of β-subunit of casein kinase-2 on the stability and specificity of the recombinant reconstituted holoenzyme. *Eur. J. Biochem.* 204: 293–297.

Meggio, F., Boldyreff, B., Issinger, O.-G. and Pinna, L.A. (1993) The autophosphorylation and p34cdc2 phosphorylation sites of casein kinase-2 β-subunit are not essential for reconstituting the fully-active heterotetrameric holoenzyme. *Biochim. Biophys. Acta* 1164: 223–225.

Meggio, F., Boldyreff, B., Issinger, O.-G. and Pinna, L.A. (1994) Casein kinase-2 down-regulation and activation by polybasic peptides are mediated by acidic residues in the 55–64 region of the β-subunit. A study with calmodulin as phosphorylatable substrate. *Biochemistry* 33: 4336–4342.

Meggio, F., Brunati, A.M., Donella Deana, A. and Pinna, L.A. (1984) Detection of type-2 casein kinase and its endogenous substrates in the components of the microsomal fraction of rat liver. *Eur. J. Biochem.* 138: 379–385.

Meresse, S., Ludwig, T., Frank, R. and Hoflack, B. (1990) Phosphorylation of the cytoplasmic domain of the bovine cation-independent mannose-6-phosphate receptor. Serines 2421 and 2492 are the targets of a casein kinase II associated to the Golgi-derived HAI adaptor complex. *J. Biol. Chem.* 265: 18833–18842.

Mestres, P., Boldyreff, B., Ebensperger, C., Hameister, H. and Issinger, O.G. (1994) Expression of casein kinase-2 during mouse embryogenesis. *Acta Anat.* 149: 13–20.

Murakami, N., Kumon, A., Matsumura, S., Hara, S. and Ikenaka, T. (1988) Phosphorylation of the heavy chain of skeletal muscle myosin by casein kinase II: localization of the phosphorylation site to the amino terminus. *J. Biochem.* 103: 209–211.

Murakami, N., Healy-Louie, G. and Elzinga, M. (1990) Aminoacid sequence around the serine phosphorylated by casein kinase II in brain myosin heavy chain. *J. Biol. Chem.* 265: 1041–1047.

Nakamura, A. and Shiomi, H. (1991) Phosphorylation of delta sleep-inducing peptide (DSIP) by casein kinase II *in vitro*. *Peptides* 12: 1375–1377.

Ohtsuki, K., Matsumoto, M., Saito, H. and Kato, T. (1993) Characterization of casein kinase II, and of p98 as one of its effective phosphate acceptors in Sea Urchin eggs. *J. Biochem.* 113: 334–342.

Padmanabha, R., Chen-Wu, J., Hanna, D.E. and Glover, C.V.C. (1990) Isolation, sequencing and disruption of the yeast CKA2 gene: casein kinase II is essential for viability in *Saccharomyces cerevisiae*. *Molec. Cell. Biol.* 10: 4089–4099.

Palvimo, J. and Linnala-Kankkunen (1989) Identification of sites on chromosomal protein HMG1 phosphorylated by casein kinase II. *FEBS Lett.* 257: 101–104.

Pinna, L.A. (1990) Casein kinase 2: an *eminence grise* in cellular regulation? *Biochim. Biophys. Acta* 1054: 267–284.

Pinna, L.A. (1991) "Independent" protein kinases: a challenge to canons. *In*: L.M.G., Jr. Heilmeyer, Jr. (ed.): *Cellular Regulation and Protein Phosphorylation*. NATO ASI Series, Vol. 56, Springer-Verlag, Berlin, Heidelberg, pp. 179–193.

Pinna, L.A., Meggio, F. and Marchiori, F. (1990) Type-2 casein kinases: general properties and substrate specificity. *In*: B.E. Kemp (ed.): *Peptides and Protein Phosphorylation*. CRC Press, Inc., Boca Raton, Florida, pp. 145–169.

Pinna, L.A., Meggio, F., Marchiori, F. and Borin, G. (1984) Opposite and mutually incompatible structural requirements of type-2 casein kinase and cAMP-dependent protein kinase as visualized with synthetic peptide substrates. *FEBS Lett.* 171: 211–214.

Plana, M., Itarte, E., Eritja, R., Goday, A., Pages, M. and Martinez, M.C. (1991) Phosphorylation of maize RAB-17 protein by casein kinase 2. *J. Biol. Chem.* 266: 22510–22514.

Pongubala, J.M., Van-Beveren, C., Nagulapalli, S., Klemsz, M.J., McKercher, S.R., Maki, R.R.A. and Atchison, M.L. (1993) Effect of PU-1 phosphorylation on interaction with NF-EM5 and transcriptional activation. *Science* 259: 1622–1625.

Prigent, C., Lasko, D.D., Kodama, K., Woodgett, J.R. and Lindahl, T. (1992) Activation of mammalian DNA ligase through phosphorylation by casein kinase II. *EMBO J.* 11: 2925–2933.

Russo, G.L., Vandenberg, M.T., Yu, I.J., Seuk Bae, Y., Franza, B. R., Jr., and Marshak, D.R. (1992) Casein kinase II phosphorylates p34^{cdc2} kinase in G1 phase of the HeLa cell division cycle. *J. Biol. Chem.* 267: 20317–20325.

Sarno, S., Marin, O., Meggio, F. and Pinna, L.A. (1993) Polyamines as negative regulators of casein kinase-2: The phosphorylation of calmodulin triggered by polylysine and by the α[66–86] peptide is prevented by spermine. *Biochem. Biophys. Res. Commun.* 194: 83–90.

Sarno, S., Meggio, F. and Pinna, L.A. (1994) Regulation of casein kinase-2 (CK2) by polyamines and other polycationic effectors. *In*: C.M. Caldarera, C. Clo and M.S. Moruzzi (eds): *Polyamines: Biological and Clinical Aspects*, CLUEB, Bologna, Italy, pp. 105–109.

Schneider, H.R. and Issinger, O.-G. (1989) CKII, a multifunctional protein kinase and its role during proliferation. *Biotec. Europe* 6: 82–88.

Schubert, U., Schneider, T., Henklein, P., Hoffmann, K., Berthold, E., Hauser, H., Pauli, G. and Portsmann, T. (1992) Human-immunodeficiency-virus-type-1-encoded Vpu protein is phosphorylated by casein kinase II. *Eur. J. Biochem.* 204: 875–883.

Sekiguchi, T., Nohiro, Y., Nakamura, Y., Hisamoto, N. and Nishimoto, T. (1991) The human CCG1 gene, essential for progression of the G1 phase, encodes a 210 kDa nuclear DNA-binding protein. *Mol. Cell. Biol.* 11: 3317–3325.

Takacs, A.M., Barik, S., Das, T. and Banerjee, A.K. (1992) Phosphorylation of specific serine residues within the acidic domain of the phosphoprotein of vesicular stomatitis virus regulates transcription *in vitro*. *J. Virol.* 66: 5842–5848.

Taminaga, M., Kitagawa, Y., Tanaka, S. and Kishimoto, A. (1991) Phosphorylation of type II (β) protein kinase C by casein kinase II. *J Biochem.* 110: 655–660.

Tanasijevic, M.J., Myers, M.G. Jr., Thoma, R.S., Crimmins, D.L., White, M.F. and Sacks, D.B. (1993) Phosphorylation of the insulin receptor substrate IRS-1 by casein kinase II. *J. Biol. Chem.* 268: 18157–18166.

Tuazon, P.T. and Traugh, J.A. (1991) Casein kinase I and II – Multipotential protein kinases: structure, function and regulation. *Adv. Second Messenger Phosphoprotein Res.* 23: 123–164.

Valdez, B.C., Busch, R.K. and Busch, H. (1990) Phosphorylation of the human cell proliferation-associated nucleolar protein p120. *Biochem. Biophys. Res. Commun.* 173: 423–430.

Voit, R., Schnapp, A., Kuhn, A., Rosenbauer, H., Hirschmann, P., Stunnenberg, H.G. and Grummt, I. (1992) The nucleolar transcription factor mBUF is phosphorylated by casein kinase II in the C-terminal hyperacidic tail which is essential for transactivation. *EMBO J.* 11: 2211–2218.

Voss, H., Wirkner, U., Jakobi, R., Hewitt, N.A., Schwager, C., Zimmermann, J., Ansorge, W. and Pyerin, W. (1991) The structure of the gene encoding human casein kinase II subunit beta. *J. Biol. Chem.* 266: 13706–13711.

Wawrzynow, A., Collins, J.M., Bogatcheva, N.V., Vorotnikov, A. and Gusev, N.B. (1991) Identification of the sites phosphorylated by casein kinase II in smooth muscle caldesmon. *FEBS Lett.* 289: 213–216.

Wirkner, U., Voss, H., Lichter, P. Ansorge, W. and Pyerin, W. (1994) The human gene (CSNK2A1) coding for the casein kinase II subunit α is located on chromosome 20 and contains tandemly arranged *Alu* repeats. *Genomics* 19: 257–280.

Wu, C.B., Pelech, S.L. and Veis, A. (1992) The *in vitro* phosphorylation of the native rat incisor dentin phosphoryns. *J. Biol. Chem.* 267: 16588–16594.

Yao, Z., Jackson, W. and Grose, C. (1993) Identification of the phosphorylation sequence in the cytoplasmic tail of the VSV-virus Fc receptor glycoprotein gpI. *J. Virol.* 67: 4464–4473.

Yenice, S., Davis, A.T., Goueli, S.A., Akdas, A., Limas, C. and Ahmed K. (1994) Nuclear casein kinase-2 (CK2) activity in human normal, benign hyperplastic, and cancerous prostate. *The Prostate* 24: 11–16.

Yu, J.L., Spector, D.L., Base, Y.-S. and Marshak, D.R. (1991) Immunocytochemical localization of casein kinase II during interphase and mitosis. *J. Cell Biol.* 114: 1217–1232.

Biochemistry of Cell Membranes
ed. by S. Papa & J. M. Tager
© 1995 Birkhäuser Verlag Basel/Switzerland

Molecular structure of G protein-coupled P_2 purinoceptors

T.E. Webb and E.A. Barnard

Molecular Neurobiology Unit, Royal Free Hospital School of Medicine, Rowland Hill Street, London NW3 2PF, UK

Summary. In this chapter the recent cloning of cDNAs encoding G protein-coupled purinoceptors for ATP and other nucleotides is reviewed. This has revealed that they form a new family within this receptor superfamily.

Introduction

The action of purine compounds on the cardiovascular system was first observed in 1929 (Drury and Szent-Györgyi, 1929). In the following decades evidence accumulated that extracellular ATP had numerous functions within the body, which are achieved through specific purinoceptors, and include a role as a neurotransmitter and cotransmitter within the autonomic nervous system (Burnstock, 1972, 1976). The division of purinoceptors into two types, P_1 and P_2, was suggested, based on the relative potency of ATP and its derivatives and the selective antagonist effect of the methylxanthines (Burnstock, 1978). P_1 purinoceptors preferentially bind adenosine over adenine nucleotides and are blocked by methylxanthine derivatives such as theophylline. P_2 purinoceptors preferentially bind ATP and other nucleotides and are not antagonised by methylxanthines.

The P_2 purinoceptor family was subsequently divided into two subtypes, P_{2X} and P_{2Y} (Burnstock and Kennedy, 1985). The P_{2X} purinoceptors are members of the transmitter-gated ion channel family, mediating fast transmission in both the peripheral and central nervous system (Bean, 1992; Edwards, 1993). The ATP analogues α,β-methylene ATP (α,β-meATP) and β,γ-methylene ATP (β,γ-meATP) act as selective agonists at this purinoceptor type which is generally associated with the contraction of visceral and vascular smooth muscle in the periphery (Burnstock and Kennedy, 1985). Agonist binding to the P_{2Y} purinoceptor is not in GTPγS sensitive, indicating that this receptor is coupled to G proteins (Boyer et al., 1989) and thus would be expected to possess the same structural feature as other guanine nucleotide binding protein-coupled receptors (GPCRs). In addition, as found in other receptor families within the GPCR superfamily, it could be expected that a number of distinct GPC purinoceptors would exist for ATP and related ligands.

Pharmacological and biochemical characterisation of the GPC P_2 purinoceptors does indicate such a heterogeneity in ligand specificity, with these receptors hypothesized to be composed of four distinct pharmacological entities on the basis of the rank order of potency of congeners of ATP:

A. The P_{2Y} purinoceptor, as already mentioned, was the first GPC P_2 purinoceptor to be identified and has a potency order of 2-methylthio ATP (2-MeSATP) $\gg$ ATP = ADP $\gg$ α,β-meATP, β,γ-meATP,UTP. This receptor has been found in a variety of tissues including the smooth muscle of guinea pig teania coli (Burnstock and Kennedy, 1985), endothelial cells of the bovine aorta (Allsup and Boarder, 1990) and human airway epithelia (Mason et al., 1991) where they have been implicated in smooth muscle relaxation, the relaxation of arterial vasculature and transepithelial ion transport, respectively. P_{2Y} purinoceptors have also been found on astrocytes where they are suggested to mediate stellation in response to extracellular ATP (Neary and Norenberg, 1992).
B. The P_{2T} purinoceptor, found on platelets and cells with the capacity for megakaryotic differentiation, has ADP as its preferred ligand and utilizes ATP acting as an apparent antagonist (Gordon, 1986; Murgo and Sistare, 1992). ADP appears to activate several platelet responses including shape change, adherence, aggregation, secretion and inhibition of cAMP accumulation (Gordon, 1986).
C. The P_{2U} or nucleotide receptor possesses the potency order ATP = UTP > ADP > α,β-meATP,2-MeSATP (O'Connor et al., 1991). This receptor has been found on neutrophils, pituitary cells, aortic smooth muscle cells and hepatocytes, where it may play a role in wound healing, cell growth and the maintenance of vascular tone (O'Connor et al., 1991; Keppens, 1993).
D. The P_{2D} purinoceptor is proposed to bind dinucleotide polyphosphates (Hilderman et al., 1991).

A further P_2 purinoceptor, the P_{2Z} subtype, binds ATP in its tetra-anionic form (Tatham et al., 1988). This purinoceptor is thought to play a role in the inflammatory response, with its activation leading to an increased permeability in mast cells and macrophages (Muria et al., 1993), and has been implicated in ATP-mediated apotosis (Pizzo et al., 1992). However, the nature of this purinoceptor is at present unclear as agonist application leads to the opening of a relatively non-selective large membrane pore that appears to be distinct from the ligand-gated ion channels.

Evidence has acculumated that the P_2 purinoceptor population is composed of further subtypes than those mentioned above. A review by

Abbracchio and Burnstock (1994) discusses this issue and they propose that, at present, seven GPC P_2 purinoceptors can be identified. While the generation of more selective agonists and selective potent antagonists will aid the characterisation of further types and subtypes within the P_2 purinoceptor family, the cloning of the different genes for these receptors will settle any controversy. As the name P_{2Y} was first associated with a GPC P_2 purinoceptor the recombinant receptors should be named by agreement (Fredholm et al., 1994) sequentially P_{2y1}, P_{2y2} etc., to distinguish them from the pharmacologically defined receptor subtypes. When in due course a member of the latter series can be equated to an identified recombinant receptor, then the designation of the latter should be used alone for it.

Recombinant P_2 purinoceptors

Two different cloning approaches have recently yielded sequences of GPC P_2 purinoceptors. We have used a degenerate PCR based approach (Webb and Bateson, 1992) to generate a probe for library screening (Webb et al., 1991). Degenerate oligonucleotide primers were designed, based on the conservation of amino acid sequences within the transmembrane domains TM2 and TM6 of recombinant GPCRs, and were used for polymerase chain reaction amplification of other members of the superfamily. As we have previously found the chick brain to be a highly abundant source of high affinity, specific ATP binding sits (Bmax = 37 pmoles/mg protein; Simon et al., 1995), a sequence isolated from the polymerase chain reaction described above was used to screen a late-embryonic chick whole-brain cDNA (complementary DNA) library. This led to the isolation of three cDNA clones encoding proteins which all possess the typical hydropathy profile of a GPCR. cRNA (complementary RNA) transcribed from these clones were then tested electrophysiologically in the *Xenopus laevis* oocyte expression system for responsiveness to the application of adenosine nucleotides (Webb et al., 1993). Of the three clones two were found to be responsive, one to ATP (Fig. 1; Webb et al., 1993) and the other to ADP.

The application of ATP to oocytes expressing the receptor responsive to ATP led to the generation of the typical slowly-developing, calcium-activated inward chloride current of the oocyte. This current was dose dependent and could be successfully blocked by application of the general P_2 purinoceptor antagonist Suramin and could also be blocked by Reactive Blue 2, which in the concentration range used in a selective P_{2Y} antagonist (Fig. 1; Burnstock, 1991). The relative rank order of potency of ligands tested was 2-MeSATP $\geq$ ATP > ADP $\gg$ α,β-MeATP,β,γ-MeATP,UTP (Fig. 1; Webb et al, 1993). This potency order is not in agreement with those accepted for the P_2 purinoceptors

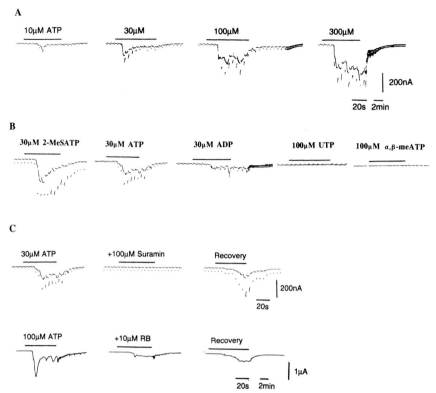

Figure 1. Responses of P_{2y1} cRNA-injected *Xenopus* oocytes: (A) Dose-dependence of membrane currents evoked by ATP (10–300 μM). (B) Agonist selectivity of the expressed P_{2y1} receptor. Note that ADP is less potent than ATP. UTP and α,β-meATP were both inactive. (C) Suramin and Reactive Blue 2 (RB) antagonized the responses to ATP, in the latter case incomplete recovery was seen. In all cases the holding potential was −40 mV. The transient downward deflections monitor the input conductance following hypepolarizing voltage steps (−10 mV, applied every 5 s for 1 s). (Taken from Webb et al., 1993.)

as known physiologically from peripheral sites. In particular the reduced potency of ADP clearly distinguishes this recombinant purinoceptor from the P_{2Y} purinoceptor characterised pharmacologically in
peripheral tissues. Furthermore, the distribution of the mRNA encoding
the recombinant purinoceptor was distinct from that proposed for a
classical P_{2Y} purinoceptor. Northern blot analysis revealed that the
mRNA encoding the purinoceptor transcript was present in brain, spinal
cord, spleen, gastrointestinal tract and leg muscle of the adult chicken
(Webb et al., 1993). However, no hybridization was seen in the heart,
kidney or liver, tissues where the presence of a P_{2Y} purinoceptor has
been reported in mammals (Zhao and Dhalla, 1990; Pfeilschifter, 1990;
Dixon et al., 1990). As the first recombinant member of the GPC P_2

purinocetor family, this receptor species was named P$_{2y1}$. We have studies the distribution pattern of the P$_{2y1}$ transcript within the brain by *in situ* hybridization. The mRNA has a widespread but specific distribution within the chick brain and is highly expressed in various nuclei of the telecephalon, diencephalon and mesencephalon as well as in the cerebellum, where it is present in the external granule, Purkinje and internal granule cell layers (Webb et al., 1994).

Concurrently, Lustig and colleagues used a functional expression system to isolate a further member of the GPC P$_2$ purinoceptor family (Lustig et al., 1993). In this case cRNA transcribed from an NG108-15 neuroblastomna × glioma hybrid cell cDNA library was injected into *Xenopus* oocytes and the pool that responded to both ATP and UTP was subdivided until a single clone was isolated which retained this responsiveness. The ligand selectivity of this clone was determined in this same expression system and was found to be ATP = UTP > ATPγS ≫ 2MeSATP = βγmeATP,ADP,αβmeATP. This potency order is in accord with that found previously for a P$_{2U}$ receptor; we have designated this P$_{2U}$-subtype receptor P$_{2y2}$. As expected for a P$_{2U}$ receptor subtype, Northern blot analysis revealed that this receptor species was widely distributed in mouse tissues and was detected in spleen, testes, kidney, liver, lung, heart and brain (Lustig et al., 1993).

The second P$_2$ purinoceptor cDNA that we isolated (P$_{2y3}$ in our scheme) was a candidate for the P$_{2T}$ subtype as it showed a preference for ADP upon expression as a cRNA in the *Xenopus* oocyte expression system. However, the relative rank order of potency of the ligands tested was similar but not identical to that expected of a P$_{2T}$ purinoceptor (B. King, personal communication). Northern blot analysis of the distribution of this purinoceptor transcript revealed its presence in chicken brain, spinal cord and various peripheral tissues (T. Webb, unpublished observations). Taken together these data may indicate that this recombinant receptor encodes a new subtype of P$_2$ purinoceptor. Interestingly, an ADP binding purinoceptor has been pharmacologically identified in cultures of rat brain capillary endothelium cells (Frelin et al., 1993).

Subsequent studies of the P$_{2y1}$ purinoceptor, transiently expressed in COS-7 cells, have confirmed that this purinoceptor has high affinity for ATP (K$_I$ = 50 nM). In this system the recombinant purinoceptor has been shown to couple to phosphatidylinositol-dependent phospholipase C (Simon et al., 1995); presumably this receptor is acting through either G$_{\alpha11}$ or G$_{\alpha q}$, which are both expressed in COS-7 cells (Wu et al., 1992). The P$_{2y2}$ purinoceptor has been stably expressed in the cell line K-562 and its signal transduction system has also been studied. Application of ATP led to an increase in intracellular calcium levels, the production of which could be partially inhibited by pretreatment with pertussis toxin (Erb et al., 1993).

Discussion

Structure of P_{2y1}, P_{2y2} and P_{2y3} purinoceptors

The three recombinant purinoceptors display the quintessential structural feature of a GPCR – the presence of seven stretches of mainly hydrophobic residues which are proposed to form transmembrane regions. In accordance with the model proposed for other GPCRs (Dohlman et al., 1987), their amino-terminal region is predicted to be extracellular while their carboxyl tail is proposed to be cytoplasmic in location (Fig. 2). Furthermore, these three sequences also possess other features common to receptors within this superfamily. For example, they have consensus sequences for asparagine-linked glycosylation in their amino-terminal region and sites for phosphorylation by PKC and

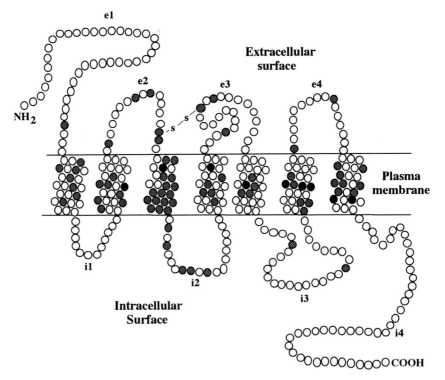

Figure 2. Schematic diagram of the sequence of the P_{2y1} purinoceptor and its similarity with the P_{2y2} and P_{2y3} purinoceptors. The P_{2y1} purinoceptor amino acid sequence is shown. The residues conserved between the three receptor sequences are indicated by shaded circles, those which are variant are denoted by open circles. Those residues which have been implicated in GPCR function, as discussed in the text, are identified by black circles. Two conserved cysteine residues found in the e2 and e3 loops are proposed to form a disulphide bridge (-s-s-). (Taken with permission from Barnard et al., 1994.)

PKA in their intracellular loops and cytoplasmic tail. Their sizes, 329 to 373 amino acids, place them amongst the smallest of the GPCRs. Comparison of these purinoceptors with other members of the GPCR superfamily indicates that these sequences fall within the main grouping in this superfamily and exhibit features exemplified by the adrenergic receptors rather than those found in the secretin receptor group or the metabotrophic glutamate receptors (Barnard, 1992). Such conserved features include an aspartate residue in TM2, proline residues in TM4, 5, 6 and 7, and aromatic residues in TM6 (Fig. 2; Hilbert et al., 1993). It should be noted, however, that the asparate residue in TM3, which has been proposed to act as a counter-ion for the amino group of the biogenic amines (Strader et al., 1988), is absent from these receptors for acidic ligands, being replaced by an aromatic residue in the P$_2$ purinoceptors discussed here.

Each purinoceptor sequence shares in the region of 40% amino acid sequence identity with the other two sequences. This degree of sequence identity is comparable to that seen for other receptors in a single family within the GPCR superfamily. For example, the human NK1 and NK2 receptors share 47% identity (Gerard et al., 1991) and the human 5-HT1e serotonin receptor shares 39% and 47% identity with the 5-HT1a and 5-HT1d sequences respectively (McAllister et al., 1992). Within the adrenergic receptor-like family the most closely related receptor sequences, which nevertheless share amino acid sequence identities of only 22–27% with the P$_2$ purinoceptors, are those for thrombin (Vu et al., 1991), angiotensin II (Sasaki et al., 1991), delta opioid (Evans et al., 1992), interleukin 8 (Murphy and Tiffany, 1991), platelet activating factor (Honda et al., 1992), and the unidentified receptor sequence RDC1 (Libert et al., 1989). Thus it can clearly be seen that these three recombinant P$_2$ purinoceptors form a distinct family of their own within the GPCR superfamily (Fig. 3). Interestingly, an unidentified receptor sequence, 6H1, isolated from activated T cells (Kaplan et al., 1993) shares a degree of sequence identity with P$_{2y3}$ which may indicate that this is a related receptor (Fig. 3).

A lower degree of sequence identity is seen when the recombinant P$_2$ sequences are compared with those of the P$_1$ purinoceptors (Fig. 3). The P$_{2y1}$ purinoceptor shares 21% to 17% identity with the receptors for adenosine and cAMP while the P$_{2y2}$ and P$_{2y3}$ sequences have less than 14% with these sequences. The lack of sequence identity shared between the recombinant P$_2$ and P$_1$ purinoceptors is at first surprising, given that the endogenous ligands for both contain a nucleoside. However, the inactivity of both adenosine and AMP at the P$_2$ purinoceptors and of ADP and ATP at P$_1$ purinoceptors must be taken into consideration. Secondly, a conserved histidine residue in TM7 of the adenosine receptors, that has been implicated in ligand binding to the A$_1$ adenosine receptor (Olah et al., 1992), is absent from the P$_2$ purinoceptors and is

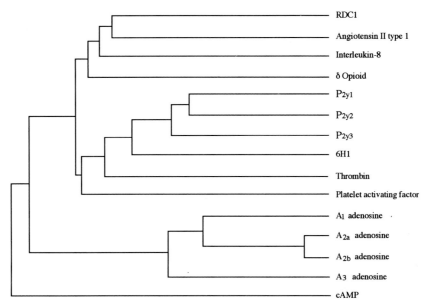

Figure 3. Multiple sequence alignment of the three P_2 purinoceptors with other GPCRs. The sequences of the recombinant cAMP and rat adenosine receptors are included as well as the most related GPCRs. These sequences were obtained from the Protein Identification Resource database (from which information regarding the original references can be obtained). Sequences were aligned and compared using the PC/Gene 6.5 CLUSTAL program (Higgins and Sharp, 1989) which weighs all sequence homology features; the lengths of lines between junctions are proportional to percentage sequence difference.

again indicative of a distinct mechanism of ligand binding to this latter receptor type. The isolation of cDNAs encoding the P_{2y1}, P_{2y2} and P_{2y3} purinoceptors has confirmed that they are indeed entirely distinct from the adenosine receptors.

Conclusions

It has now become clear that GPCRs for ATP and other nucleotides do exist. Now that these purinoceptor cDNAs have been isolated, the contribution of individual amino acids to the formation of the ligand binding site and G protein coupling can be investigated by site-directed mutagenesis and the construction of chimeric receptors.

Within the brain P_2 purinoceptors have previously been associated with glial cells (Bruner and Murphy, 1990). The abundant presence and, further, the widespread distribution of the P_{2y1} purinoceptor within the brain may indicate that this species is involved in a specific form of metabotropic neurotransmission.

Acknowledgements
J. Simon provided invaluable help and advice during this study. We also thank P. Leff for this interest in this work. This work was supported by Fisons plc.

References

Abbracchio, M.P. and Burnstock, G. (1994) Purinoceptors: Are there families of P_{2X} and P_{2Y} purinoceptors? *Pharmac. Ther.* 64: 445–453.

Allsup, D.J. and Boarder, M.R. (1990) Comparison of P_2 purinergic receptors of aortic endothelial cells with those of adrenal medulla: Evidence for heterogeneity of receptor subtype and of inositol phosphate response. *Mol. Pharmacol.* 38: 84–91.

Barnard, E.A. (1992) Receptor classes and the transmitter-gated ion channels. *Trends Biochem. Sci.* 17: 368–374.

Barnard, E.A., Burnstock, G. and Webb, T.E. (1994) The G protein-coupled receptors for ATP and other nucleotides: A new receptor family. *Trends Pharmacol. Sci.* 15: 67–71.

Bean, B. (1992) Pharmacology and electrophysiology of ATP-activated ion channels. *Trends Pharmacol. Sci.* 13: 87–90.

Boyer, J.L., Downes, C.P. and Harden, T.K. (1989) Kinetics of activation of phospholipase C by P_{2Y} purinergic receptor agonists and guanine nucleotides. *J. Biol. Chem.* 264: 884–890.

Bruner, G. and Murphy, S. (1990) ATP evoked arachidonic acid mobilisation in astrocytes is via a P_{2Y} purinergic receptor. *J. Neurochem.* 55: 1569–1575.

Burnstock, G. (1972) Purinergic nerves. *Pharmacol. Rev.* 24: 509–581.

Burnstock, G. (1976) Do some nerve cells release more than one transmitter? *Neuroscience* 1: 239–248.

Burnstock, G. (1978) A basis for distinguishing two types of purinergic receptor. *In:* R.W. Straub and L. Bolis (eds): *Cell Membrane Receptors for Drugs and Hormones: A Multidisciplinary Approach.* Raven Press, New York, pp. 107–118.

Burnstock, G. (1991) Distribution and roles of purinoceptor subtypes. *Nucleosides and Nucleotides* 10: 917–930.

Burnstock, G. and Kennedy, C. (1985) Review: Is there a basis for distinguishing two types of P_2 purinoceptor? *Gen. Pharmacol.* 5: 433–440.

Dixon, C.J., Woods, N.M., Cuthbertson, K.S.R. and Bobbold, P.H. (1990) Evidence for two Ca^{2+}-mobilizing purinoceptors on rat hepatocytes. *Biochem. J.* 269: 499–502.

Dohlman H.G., Bouvier, M., Benovic, J.L., Caron, M.G. and Lefkowitz, R.J. (1987) The multiple membrane spanning topography of the β_2-adrenergic receptor. *J. Biol. Chem.* 262: 14282–14288.

Drury, A.N. and Szent-Györgyi, A. (1929) The physiological activity of adenine compounds with special reference to their action upon the mammalian heart. *J. Physiol.* 68: 213–237.

Edwards, F.A. and Gibbs, A.J. (1993) ATP – a fast neurotransmitter. *FEBS Lett.* 325: 86–89.

Erb, L., Lustig, K.B., Sullivan, D.M., Turner, J.T. and Weisman, G.A. (1993) Functional expression and photoaffinity labelling of a cloned P_{2U} purinergic receptor. *Proc. Natl. Acad. Sci. USA* 90: 10449–10453.

Evans, C.J., Keith, D.E., Morrison, H., Magendzo, K. and Edwards, R.H. (1992) Cloning of a delta opioid receptor by functional expression. *Science* 258: 1952–1955.

Fredholm, B.B., Abbracchio, M.P., Burnstock, G., Daly, J.W., Harden, T.K., Jacobson, K.A., Leff, P. and Williams, M. (1994) Nomenclature and classification of purinoceptors. *Pharmacological Reviews* 46: 143–156.

Frelin, C., Breittmayer, J.P. and Vigne, P. (1993) ADP induces inositol phosphate-independent intracellular Ca^{2+} mobilization in brain capillary endothelial cells. *J. Biol. Chem.* 268: 8787–8792.

Gerard, N.P., Garraway, L.A., Eddy, R.L., Jr, Shows, T.B., Iijima, H., Paquet, J. and Gerard, C. (1991) Human substance P receptor (NK-1): Organisation of the gene, chromsome localisation, and functional expression of the cDNA clones. *Biochemistry* 30: 10640–10646.

Gordon, J.L. (1986) Extracellular ATP: Effects, sources and fates. *Biochem. J.* 233: 309–319.

Hibert, M.F., Trumpp-Kallmeyer, S., Hoflack, J. and Bruinvels, A. (1993) This is not a G protein-coupled receptor. *Trends Pharmacol. Sci.* 14: 7–12.

Higgins, D.G. and Sharp, P.M. (1989) Fast and sensitive multiple sequence alignments on a microcomputer. *CABIOS* 5: 151–153.

Hilderman, R.H., Martin, M., Zimmerman, J.K. and Pivorum, E.B. (1991) Identification of a unique membrane receptor for adenosine 5′,5‴-P1P4-tetraphosphate. *J. Biol. Chem.* 266: 6915–6918.

Honda, Z., Nakamura, M., Miki, I., Minami, M., Watanabe, T., Seyama, Y., Okado, H., Toh, H., Ito, K., Miyamoto, T. and Shimizu, T. (1991) Cloning by functional expression of platelet-activating factor receptor from guinea big lung. *Nature* 349: 342–346.

Kaplan, M.K., Smith, D.I. and Sundick, R.S. (1993) Identification of a G protein coupled receptor induced in activated T cells. *J. Immunol.* 151: 628–638.

Keppens, S. (1993) The complex interaction of ATP and UTP with isolated hepatocytes. How many receptors? *Gen. Pharmacol.* 24: 283–289.

Libert, F., Parmentier, M., Lefort, A., Dinsart, C., Van Sande, J., Maenhaut, C., Simons, M., Dumont, J.E. and Vassart, G. (1989) Selective amplification and cloning of four new members of the G protein-coupled receptor family. *Science* 244: 569–572.

Lustig, K.D., Shiau, A.K., Brake, A.J. and Julius, D. (1993) Expression cloning of an ATP receptor from mouse neuroblastoma cells. *Proc. Natl. Acad. Sci. USA* 90: 5113–5117.

Mason, S.J., Paradiso, A.M. and Boucher, R.C. (1991) Regulation of transepithelial ion transport and intracellular calcium by extracellular ATP in human normal and cystic fibrosis airway epithelium. *Br J. Pharmacol.* 103: 1649–1656.

McAllister, G., Charlesworth, A., Snodin, C., Beer, M.S., Noble, A.J., Middlemiss, D.N., Iversen, L.L. and Whiting, P. (1992) Molecular cloning of a serotonin receptor from human brain (5HT 1E); A fifth 5H51-like subtype. *Proc. Natl. Acad. Sci. USA* 89: 5517–5521.

Murgia, M., Hanau, S., Pizzo, P., Rippa, M. and Di Virgilio, F. (1993) Oxidized ATP – An irreversible inhibitor of the macrophage purinergic P_{2Z} receptor. *J. Biol. Chem.* 21: 8199–8203.

Murgo, A.J. and Sistare, F.D. (1992) K562 leukemia cells express P_{2T} (adenosine diphosphate) purinergic receptors. *J. Pharm. Expt. Thr.* 261: 580–585.

Murphy, P.M. and Tiffany, H.L. (1991) Cloning of complementary DNA encoding a functional human interleukin-8 receptor. *Science* 253: 1280–1283.

Neary, J.T. and Norenberg, M.D. (1992) Signalling by extracellular ATP: Physiological and pathological considerations in neuronal-astrocytic interactions. *In:* A.C.H. Yu, M.D. Hertz, E. Norenberg, E. Syková and S.G. Waxman (eds): *Progress in Brain Research*, Vol. 94, Elsevier Science Publishers, Amsterdam, pp. 145–151.

O'Connor, S.E., Dainty, I.A. and Leff, P. (1991) Further subclassification of ATP receptors based on agonist studies. *Trends Pharmacol. Sci.* 12: 137–141.

Olah, M.E., Ren, H., Ostrowski, J., Jacobson, K.A. and Stiles, G.L. (1992) Cloning, expression and characterisation of the unique bovine A_1 adenosine receptor. *J. Biol. Chem.* 267: 10764–10770.

Pfeilschifter, J. (1990) Comparison of extracellular ATP and UTP signalling in rat renal mesengial cells. No indications for the involvement of distinct purino- and pyrimidinoceptors. *Biochem. J.* 272: 469-472.

Pizzo, P., Murgia, M., Zambon, A., Zanovello, P., Bronte, V., Pietrobon, D. and Di Virgilio, F. (1992) Role of P_{2Z} purinergic receptors in ATP-mediated killing of tumor necrosis TNF-sensitive and TNF-resistant L929 fibroblasts. *J. Immunol.* 149: 3372–3378.

Sasaki, K., Yamono, Y., Bardhan, S., Iwai, N., Murray, J.J., Hasegawa, M., Matsuda, Y. and Inagami, T. (1991) Cloning and expression of a cDNA encoding a bovine adrenal angiotensin II type-1 receptor. *Nature* 351: 230–233.

Simon, J., Webb, T.E. and Barnard, E.A. (1995) Characterisation of a P_{2Y} purinoceptor in the brain. *Pharmacol. Toxicol.* 76; *in press.*

Simon, J., Webb, T.E., King, B.J., Burnstock, G. and Barnard, E.A. (1994) Pharmacological characterisation of a recombinant P_{2Y} purinoceptor; *submitted.*

Strader, C.D., Sigal, I.S., Candelore, M.R., Hill, W.S. and Dixon, R.A.F. (1988) Conserved aspartic acid residues 79 and 113 of the β-adrenergic receptor have different roles in receptor function. *J. Biol. Chem.* 263: 10267–10271.

Tatham, P.E.R., Cusack, N.J. and Gomperts, B.D. (1988) Characterisation of the ATP^{4-} receptor that mediates permeabilization of rat mast cells. *Eur. J. Pharmacol.* 147: 13–21.

Vu, T.K., Hung, D.T., Wheaton, V.I. and Coughlin, S.R. (1991) Molecular cloning of a functional thrombin receptor reveals a novel proteolytic mechanism of receptor activation. *Cell* 64: 1057–1068.

Webb, T.E. and Bateson, A.N. (1992) The use of degenerate oligonucleotides for polymerase chain-reaction-based isolation of related DNA sequences. *In:* A. Longstaff and P. Revest (eds): *Methods in Molecular Biology, Vol. 13: Protocols in Molecular Neurobiology.* Humana press, Totowa, New Jersey, pp. 67–77.

Webb, T.E., Bateson, A.N. and Barnard, E.A. (1991) Paper presented at the 17th EMBO Annual Symposium, Heidelberg, Germany, 16 September 1991, Abstract number 208.

Webb, T.E., Simon, J., Krishek, B.J., Bateson, A.N., Smart, T.G., King, B.J., Burnstock, G. and Barnard, E.A. (1993) Cloning and functional expression of a cDNA encoding a brain G protein-coupled ATP receptor. *FEBS Lett.* 324: 219–225.

Webb, T.E., Simon, J., Bateson, A.N. and Barnard, E.A. (1994) Transient expression of the recombinant chick brain P$_{2y1}$ purinoceptor and localization of the corresponding mRNA. *Cell. Mol. Biol.* 40: 437–442.

Wu, D., Katz, A., Lee, C. and Simon, M.I. (1992) Activation of pospholipase C by α_1-adrenergic receptors is mediated by the α subunits of Gq family. *J. Biol. Chem.* 267: 25798–25802.

Zhao, D. and Dhaklla, N.S. (1990) [^{35}S]ATPγS binding sites in the purified heart sacrolemma membranes. *Am. J. Physiol.* 25: C185–C188.

Biochemistry of Cell Membranes
ed. by S. Papa & J. M. Tager
© 1995 Birkhäuser Verlag Basel/Switzerland

RAS proteins and control of the cell cycle in *Saccharomyces cerevisiae*

O. Fasano†

Dipartimento di Biologia Cellulare e dello Sviluppo, Università di Palermo, via Archiraff 22, I-90123 Palermo, Italy

The *RAS1* and *RAS2* genes of *Saccharomyces cerevisiae*

Genes related to the mammalian H-, K-, and N-*ras* oncogenes were identified in *S. cerevisiae* by DNA hybridization techniques (for reviews, see Tamanoi, 1988; Gibbs and Marshall, 1989; Broach and Deschenes, 1990). According to the rules of yeast genetics (dominant genes are indicated by three capital letters followed by a number), the yeast genes were denominated *RAS1* and *RAS2* (collectively referred to as *RAS*). The corresponding RAS1 and RAS2 proteins were 309 and 322 amino acids long, respectively. The sequence similarity between the human and yeast proteins was very high, reaching 90% identity at the level of the N-terminal 80 amino acids. As a consequence, perfect sequence conservation was observed at the level of those amino acids that are critical for the oncogenic activation of mammalian RAS proteins (12, 13, 61, human coordinates).

To determine the physiological function of a protein in an organism, it is helpful to observe what happens when the function of the protein is either inhibited by microinjection of an antibody, or by disruption of the corresponding gene. The latter strategy is widely used in *S. cerevisiae*, because of the powerful gene replacement techniques available in this organism. It is important to note that the disruption of the *RAS* genes could not have been easily done in vertebrates. In fact, this would have required the elimination of the chromosomal H-, K-, and N-*ras* genes, a difficult task considering the low frequency of homologous recombination in high eukaryotes. By constructing yeast cells in which either the *RAS1* or the *RAS2* genes was disrupted, Kataoka et al. (1984) determined that the loss of either *RAS1* or *RAS2* had almost no effect on growth. However, the disruption of both genes resulted in loss of viability. Interestingly, the lethality of *ras1-ras2*-strains could be suppressed by the expression of the human Ha-*ras* gene product under the

---

†Deceased on 27 September 1994.

control of a yeast promoter (Kataoka et al., 1985a). This showed that the similarity between yeast and human proteins was both structural and functional.

Role of RAS proteins in the cell cycle

Kataoka et al. (1984) also used the gene replacement technique for the engineering of yeast cells expressing a RAS2 protein that was constitutively activated by a single amino acid change (i.e., a Gly to Val change at position 19, corresponding to the oncogenic position 12 in the human H-*ras* p21 protein). These cells could not properly arrest in G1, following deprivation of nutrients (Toda et al., 1985). This phenotype has some analogy with that of tumor cells, whose essential feature appears to be an inability to arrest in G1 or Go, following those signals that are capable of inducing a resting state in normal cells. This analogy is interesting, but it should be kept in mind that conditions triggering growth arrest in higher eukaryotes are different from just deprivation of nutrients. The latter is in fact a physiological signal for inducing a resting state in a unicellular organism like yeast, but not in multicellular organisms. Despite these differences, components or intracellular signaling pathways in yeast and vertebrates appear to be structurally related. The effect of RAS hyperactivation on the cell cycle suggests that the structural conservation of some components of the signaling machinery might also reflect a conserved function.

The observation of uncontrolled progression through G1 after activation of RAS suggested that an impaired RAS function would produce the opposite phenotype, that is G1 arrest even in good nutritional conditions. To answer this question, De Vendittis et al. (1986a) constructed yeast cells in which the chromosomal *RAS1* gene was disrupted and the chromosomal *RAS2* gene was replaced by a mutated allele expressing a RAS2 protein that was functional at 30 °C, but not at 37 °C. Upon upward temperature shift, the cells were found predominantly as unbudded, G1-arrested cells. The conclusion was that the *RAS* gene products positively regulate the progression of yeast through the G1 phase of the cell cycle. In retrospect, the lethality resulting from the disruption of both *RAS* genes (Kataoka et al., 1984) could be reinterpreted as growth arrest resulting from the inability to progress through the cell cycle.

Temperature-sensitive yeast mutants that arrest at specific stages of the cell cycle upon upward temperature shift are known as *cdc* mutants (Hartwell, 1974). Each of the known *cdc* mutants carries a genetic lesion in a gene encoding a protein whose function is at some point essential during the cell cycle. Since, to a certain extent, *RAS1* can substitute for *RAS2* and *vice versa*, it is not surprising that *RAS* genes were not originally discovered as cell-cycle-specific genes. Only after the disruption of the *RAS* gene (*RAS1*) has it been possible to show that genetic lesions of the *RAS2* gene leading to temperature-sensitivity cause cell-cycle-specific effects (De Vendittis et al., 1986a). Since yeast genes encoding key metabolic or regulatory elements are often duplicated, it is possible that more cell-cycle regulatory elements have escaped detection.

Beside their effect on growth, mutations which either activate or impair the function of the yeast *RAS* gene products affect metabolic parameters, like glycogen levels, as well as adaptive responses such as sporulation and thermotolerance (Kataoka et al., 1984; Toda et al., 1985). These results suggest that *RAS* genes are involved in the coordination of several cellular responses, including cell division and cell metabolism.

Positive regulatory elements of RAS

Signals are propagated through intracellular signaling networks by positive regulatory elements. The *CDC25* gene product, a guanine nucleotide exchange factor (GEF) acts as a positive factor for *RAS* by stimulating the rate of regeneration of the RAS-GTP complex from the GDP-bound form (Jones et al., 1991). In agreement with a role as a positive stimulator of RAS *in vivo*, the inactivation of *CDC25* in yeast cells leads to G1-specific arrest of the cell cycle (Camonis et al., 1986). CDC25 function is probably regulated by nutrient availability (for glucose regulation of CDC25, see below).

Enzymes involved in RAS proteins targeting to plasma membranes can be considered positive regulators of RAS, acting on longer time scales than GEFs. Processing, which is critical for the biological activity of RAS, involves isoprenoid modification of the C-terminal Cys-Xaa-Xaa-Leu box and includes the following steps: (i) farnesylation (and/or geranylgeranyl transferase prenylation in exceptional cases) of Cys319 (Evans Trueblood et al., 1993; Goodman et al., 1990); (ii) proteolytic cleavage of the last three amino acids (Deschenes et al., 1989; Fujiyama and Tamanoi, 1990); (iii) methyl-esterification of the C-terminal cysteine (Deschenes et al., 1989). In addition, acylation by palmitate (Fujiyama and Tamanoi, 1986; Powere et al., 1986) probably occurs at a separate cysteine. The *DPR1/ RAM1/STE16* and *RAM2* genes encode the two components (α and β) of the farnesyltransferase (Goodman et al., 1988; He et al., 1991). *RAM2* and *CDC43/CAL1* encode the components of geranylgeranyl transferase I (Finegold et al., 1991; Mayer et al., 1992). *STE14* encodes the methyl-transferase (Hrycyna et al., 1991).

Negative regulatory elements of RAS

Negative elements appear to be required in signaling networks in order to terminate signals. RAS proteins are subject to negative control by

IRA1 and *IRA2*, the yeast homologs of the mammalian neurofibro-matosis gene *NF1* (Tanaka et al., 1989; 1990a,b; Ballester et al., 1990; Martin et al., 1990; Xu et al., 1990). The IRA1 and IRA2 proteins promote the conversion of RAS-GTP to inactivate RAS-GDP by stimulating the RAS GTPase activity (Tanaka et al., 1990b).

Adenyl cyclase: The RAS target in yeast

The target for RAS appears to be different in yeast and vertebrates. In the latter, the RAS target is the Raf protein kinase (for a review, see Moodie and Wolfman, 1994). In yeast, it is well established that membrane-bound adenylyl cyclase, encoded by the gene *CYR1*, is a major, and possibly unique target for RAS (Toda et al., 1985; De Vendittis et al., 1986a).

Reconstitution of the system in bacteria suggests that RAS2 interacts directly with a C-terminal region of CYR1 (Uno et al., 1987). The 2026 amino acid-long CYR1 protein also includes a leucine-rich repeat region, that is structurally related to the mammalian *rsp-1* gene. *Rsp1* was isolated on the basis of its ability to interfere with RAS-induced oncogenic transformation (Cutler et al., 1992). The leucine-rich repeat region has been suggested to play a role in RAS2-adenylyl cyclase interaction (Colicelli et al., 1990). An adenylyl cyclase-associated protein, denominated CAP (Field et al., 1988), even though not essential for RAS-dependent stimulation, has also been shown to be required for maximal response of CYR1 to RAS2 (Wang et al., 1993).

A possible mechanism of stimulation of adenylyl cyclase by RAS has been proposed on the basis of genetic and biochemical data. The essential features of the model are: (i) regulatory regions of adenylyl cyclase, including a region located between the leucine-rich repeat region and the catalytic C-domain (Feger et al., 1991), negatively and reversibly regulate a C-terminal catalytic domain in the absence of RAS (Kataoka et al., 1985b); (ii) the negative effect of the regulatory domains on the catalytic domain are relieved by the presence of the RAS-GTP, but not of the RAS-GDP complex (De Vendittis et al., 1986a; Feger et al., 1991). In line with this model, antibodies directed against a region located between the leucine-rich repeat region and the catalytic C-domain showed a RAS-like effect by efficiently stimulating the adenylyl cyclase catalytic activity (Suzuki et al., 1993).

Additional elements of the cAMP pathway

Regulatory elements that are thought to function upstream of the RAS-adenylyl cyclase pathway are encoded by *MSI1* and *RPI1* (Ruggieri et al., 1989; Kim and Powers, 1991). Elements downstream from RAS and adenylyl cyclase include the regulatory and catalytic subunits of the cAMP-dependent protein kinase (*TPK1*, *TPK2*, *TPK3*, and *BCY1*, Toda et al., 1987a,b), two phosphodiesterases (*PDE1* and *PDE2*, Sass et al., 1986; Nikawa et al., 1987), *MSI3* and *MKS1* (Shirayama et al., 1993; Matsuura and Anraku, 1993). Elements that interact at a yet undefined level with the RAS-adenylyl cyclase pathway include the protein kinases *SCH9*, *YAK1*, and *SNF1* (Toda et al., 1988; Thompson-Jaeger et al., 1991; Di Blasi et al., 1993; Hartley et al., 1994), the *TSF1* and *SRK1* gene products (Robinson and Tatchell, 1991; Wilson et al., 1991) and possibly the *RSR1* gene involved in the determination of cell polarity (Ruggieri et al., 1992). Transcriptional elements that are controlled by cyclic AMP (see for example Klein and Struhl, 1994) are beyond the scope of this short review.

The regulatory nature of some elements of the RAS/adenylyl cyclase pathway is confirmed by the fact that the elimination of a positive element (by gene disruption) can be compensated by the elimination of a negative element. For example, *ras1- ras2-* yeast strains are unviable, while *ras1- ras2- bcy1-* strains show detectable growth (Toda et al., 1985). This observation provides a methodology that is useful for the identification of additional elements of the pathway. In fact, it can be predicted that the lethality resulting from the inactivation of a given positive element will be rescued either by the constitutive activation of downstream positive elements, or by the inactivation of a downstream negative regulatory element. The tools of genetics and molecular biology have been used to isolate and characterize the genes involved.

Glucose-induced activation of RAS

Input on both positive or negative effectors of RAS could in principle be used to modulate RAS-dependent responses. However, genetic evidence for the involvement of RAS in mitogenic stimulation is not paralleled at the biochemical level by an extensive knowledge about the molecular identity of the corresponding mitogens. The only observation in this direction comes from Thevelein's group (Mbonyi et al., 1988). These authors were interested in the mechanism whereby addition of glucose to derepressed yeast cells resulted in a transient stimulation of adenylyl cyclase activity. They and other groups found that the presence of both *RAS* and *CDC25* appeared to be required for this effect (reviewed by Broach and Deschenes, 1990; Thevelein, 1991). However, induction of the glucose-dependent cAMP signal is neither an essential nor an exclusive function for RAS. The first conclusion is supported by the observation that *ras1- ras2-*, *cyr1-* strains, whose viability is ensured by additional suppressor mutations, can grow on glucose (our unpub-

lished results). This does not imply that the RAS- and adenylyl cyclase-
dependent rise of cAMP induced by glucose is of marginal physiological
relevance. Rather, it might be interpreted as an indication for glucose
being a nutrient of such importance that several signaling pathways are
probably used to ensure its correct utilization. That RAS has other
functions beside the induction of the glucose-dependent cAMP signal is
shown by the observation that RAS proteins are absolutely required for
growth using glycerol as a carbon source (Fasano et al., 1988).

Future work will probably concentrate on the identification of other
cellular elements linking specific molecular signals to *RAS*. This is also
expected to clarify whether or not some elements (like RAS and
adenylyl cyclase) mediate the transmission of signals of different origin,
while other cellular components, for example receptors, could be endowed with specificity for a given signal.

The function of RAS proteins is regulated by the alternating binding of GTP or GDP

The molecular mechanism of conversion of RAS from an inactive to an
active conformation (GPD- *versus* GTP-bound form) is a variation on
the theme of how chemical energy (conversion of GTP to GDP) can be
converted into a vectorial process. For an introduction to this subject
see Kaziro (1978). The initial observation that human RAS proteins
shared structural similarity with GTP-binding proteins involved in
protein synthesis (like EF-Tu, Leberman and Egner, 1984) and signal
transduction (like the Gs component of adenylyl cyclase, Gilman,
1984), together with the high affinity for guanine nucleotides, suggested
that the RAS-bound GTP was a conformational effector rather than a
substrate in the usual sense (Fasano et al., 1984). To investigate this
point, the yeast RAS2 protein was purified using an *E. coli* expression
system. Yeast membranes from *ras1- ras2-* strains were used as a source
of adenylyl cyclase to perform *in vitro* complementation assays with
purified RAS2 (Broek et al., 1985; De Vendittis et al., 1986a). The
relevant conclusions of the biochemical analysis can be summarized as
follows:

- The Ras2 protein in its GTP-bound, but not in its GDP-bound form,
  was capable of stimulating adenylyl cyclase, the RAS target in yeast
  (De Vendittis et al., 1986a). Furthermore, the Gpp(NH)p-bound
  form was as effective as the GTP-bound form. Since Gpp(NH)p is
  not hydrolyzed to GDP in the time scale of the experiment, the
  GTPase activity of the protein can be considered as a turnoff mechanism for the conversion of the active to the inactive form of RAS.
- The regeneration of the active RAS-GTP from the inactive RAS-
  GDP form takes place by a nucleotide exchange reaction, in which

the rate-limiting step is the dissociation of GDP (De Vendittis et al., 1986b). Positive regulators of RAS, as for example CDC25, are able to increase the rate of this reaction *in vitro*, thus stimulating the form of the active RAS2-GTP complex (Créchet et al., 1990; Jones et al., 1991).

– The conversion from the active to the inactive form of RAS takes place through a GTPase reaction, that is stimulated by negative regulators of RAS function such as *IRA1* and *IRA2* (Tanaka et al., 1989, 1990b). The replacement of amino acids in the region 12–13 or 59–63 (human coordinates) by other residues activates the oncogenic function of the protein in higher eucaryotes (Fasano et al., 1984). This is due to an impaired GTPase activity of the protein that is rendered unresponsive to negative regulators. Therefore, RAS cannot attain the inactive (GDP-bound) state. In yeast cells the same mutations cause inability to properly arrest in the G1 phase of the cell cycle (reviewed by Tamanoi, 1988; Gibbs and Marshall, 1989; Broach and Deschenes, 1990).

How do the GTP- and GDP-bound forms of RAS recognize different molecular effectors?

It is interesting to mention that antibiotics acting on elongation factor Tu, the first GTP/GDP binding protein to be studied in great detail, can display specificity for either the GTP- or for both the GTP- and GDP-bound forms of this protein (Anborgh and Parmeggiani, 1991, 1993). This suggests that it could be possible to search for antibiotics acting selectively on oncogenically activated forms of RAS, or perhaps to engineer appropriate molecules with these characteristics. A detailed knowledge of the dynamics of the RAS activation/deactivation cycle at the molecular level could facilitate this approach. Insight into this mechanism has been provided by the crystallographic data obtained from the human H-ras p21 (Schlichting et al., 1990), as well as by the biochemical analysis of human and yeast RAS mutants that were unable to attain the active state. The relevant conclusions can be summarized as follows:

– The switch I region that includes loop L2 (Fig. 1) is important for the interaction of RAS with its target, since amino acid substitutions in this region reduced the ability of RAS to stimulate adenylyl cyclase (Sigal et al., 1986; Fasano et al., 1988). The structure of loop L2 is reorganized upon replacement of GDP by GTP (Schlichting et al., 1990; Pai, 1991). It is possible that this loop is capable of interacting with adenylyl cyclase when properly positioned by GTP, and not by GDP. The same loop is also important for interaction with the IRA1

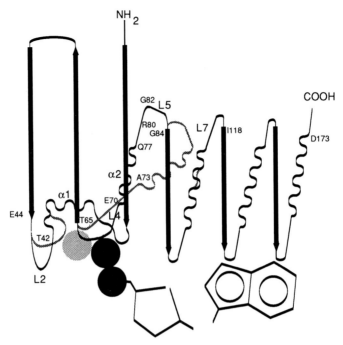

Figure 1. Cartoon view of conformational changes in the GDP- and GTP-bound forms of RAS. The topological order of secondary structure of the RAS2 G-domain is based on the crystallographic analysis of the highly homologous Ha-*ras* p21 (see text). Strands of β-sheet are shown as arrows, helices as wavy lines, and mobile loops as black (GDP-bound) or grey lines (GTP-bound conformation). The drawing is not in scale, and only relevant topological relationships are shown using yeast coordinates.

and IRA2 proteins (and with GAP in higher eucaryotes, Tanaka et al., 1992).

– The switch II region that includes loop L4-helix α2-loop L5 residues, is critical for proper interaction of GDP-bound RAS with GDP exchange factors, as shown by the inhibitory effect of amino acid substitutions within the RAS switch II region on RAS/GEFs interaction (Mistou et al., 1992; Verrotti et al., 1992; Mosteller et al., 1994). The switch II region undergoes profound nucleotide-induced conformational changes. In fact, upon replacement of GDP for GTP, helix α2 axis rotates and tilts by 65°, while winding at the C-end and unwinding at the N-end (Fig. 1; Stouten et al., 1993). Glycines 82 and 84 (loop L5, yeast coordinates) ensure backbone flexibility during the conformational change (Fasano et al., 1988; Kavounis et al., 1991; Miller et al., 1992; Stouten et al., 1993; Quilliam et al., 1994). It has been proposed that GEFs could recognize the GDP-bound form of RAS and stabilize a transition state conformation endowed with low affinity for GDP (Verotti et al., 1992; Mirisola et al., 1994). This would cause rapid release of GDP and allow the regeneration of the active RAS-GTP complex. Amino acid residues within helix α3-loop L7 also appear to play a role in RAS-GEF interaction (Willumsen et al., 1991; Segal et al., 1993).

To summarize, it appears that the presence of an additional phosphate group in GTP versus GDP is capable of stabilizing a new conformation of specific regions of RAS (loop L2, and the L4-α2-L5 region). The GTP-induced conformation of RAS differs from the GDP-induced one, not only in close proximity of the γ-phosphate, but also at distance. Therefore, it is not surprising that other cellular components might "see" or "not see" some features of RAS, depending on the bound nucleotide.

Conclusions

We do not know to what extent the similarity between the yeast and mammalian cell cycle will allow the extrapolation of the yeast model to vertebrates. Extracellular signals regulating progression through the cell cycle are clearly different. However, many intracellular components are very similar. It is possible that this network of intracellular elements constitutes an extreme form of "delocalized" and dynamic organelle, whose function is to integrate different signals from outside to generate coordinated responses inside the cell. It remains to be seen whether or not the understanding of the principles governing these interactions in yeast will help in solving the problem of oncogenesis in higher eucaryotes.

Acknowledgements
I thank Dr. E. De Vendittis for discussions and for critical reading of the manuscript.

References

Anborgh, P.H. and Parmeggiani, A. (1991) New antibiotic that acts specifically on the GTP-bound form of elongation factor Tu. *EMBO J.* 10: 779–784.

Anborgh, P.H. and Parmeggiani, A. (1993) Probing the reactivity of the GTP- and GDP-bound conformations of elongation factor Tu in complex with the antibiotic GE2270 A. *J. Biol. Chem.* 268: 24622–24628.

Ballester, R., Marchuk, D., Boguski, M., Saulino, A., Letcher, R., Wigler, M. and Collins, F. (1990) The *NF1* locus encodes a protein functionally related to mammalian GAP and yeast *IRA* proteins. *Cell* 63: 851–859.

Broach, J.R. and Deschenes, R.J. (1990) The function of *RAS* genes in *Saccharomyces cerevisiae*. *Advan. Cancer Res.* 54: 79–139.

Broek, D., Saimy, N., Fasano, O., Fujiyama, A., Tamanoi, F., Northup, J. and Wigler, M. (1985) Differential activation of yeast adenylate cyclcase by wild-type and mutant *RAS* proteins. *Cell* 41: 763–769.

Camonis, J.H., Kalekine, M., Gondre, B., Garreau, H., Boy-Marcotte, E. and Jacquet, M. (1986) Characterization, cloning and sequence analysis of the *CDC25* gene which controls the cyclic AMP level of *Saccharomyces cerevisiae. EMBO J.* 5: 375–380.

Colicelli, J., Field, J., Ballester, R., Chester, N., Young, D. and Wigler, M. (1990) Mutational mappings of RAS-responsive domains of the *Saccharomyces cerevisiae* adenylyl cyclase. *Mol. Cell. Biol.* 10: 2539–2543.

Crechet, J.B., Poullet, P., Mistou, M.Y., Parmeggiani, A., Camonis, J., Boy-Marcotte, E., Damak, F. and Jacquet, M. (1990) Enhancement of the GDP-GTP exchange of RAS proteins by the carboxy-terminal domain of SCD25. *Science* 248: 866–868.

Cutler, M.L., Bassin, R.H., Zanoni, L. and Talbot, N. (1992) Isolation of *rsp-1*, a novel cDNA capable of suppressing v-Ras transformation. *Mol. Cell. Biol.* 12: 3750–3756.

Deschenes, R.J., Stimmel, J.B., Clarke, S., Stock, J. and Broach, J.R. (1989) RAS2 Protein of *Saccharomyces cerevisiae* is methyl-esterified at its carboxyl terminus. *J. Biol. Chem.* 264: 11865–11873.

De Vendittis, E., Vitelli, A., Zahn, R. and Fasano, O. (1986a) Suppression of defective *RAS1* and *RAS2* functions in yeast by an adenylate cyclase activated by a single amino acid change. *EMBO J.* 5: 3657–3663.

De Vendittis, E., Zahn, R. and Fasano, O. (1986b) Regeneration of the GTP-bound from the GDP-bound form of human and yeast *ras* proteins by nucleotide exchange. Stimulatory effect of organic and inorganic polyphosphates. *Eur. J. Biochem.* 161: 473–478.

Di Blasi, F., Carra, E., De Vendittis, E., Masturzo, P., Burderi, E., Lambrinoudaki, I., Mirisola, M.G., Seidita, G. and Fasano, O. (1993) The *SCH9* protein kinase mRNA contains a long 5′ leader with a small open reading frame. *Yeast* 9: 21–32.

Evans Trueblood, C., Ohya, Y. and Rine, J. (1993) Genetic evidence for in vivo cross-specificity of the CaaX-Box protein prenyltransferases farnesyltransferase and geranylgeranyl-transferase-I in *Saccharomyces cerevisiae. Mol. Cell. Biol.* 13: 4260–4275.

Fasano, O., Aldrich, T., Tamanoi, F., Taparowsky, E., Furth, M. and Wigler, M. (1984) Analysis of the transforming potential of the human H-*ras* gene by random mutagenesis. *Proc. Natl. Acad. Sci. USA* 81: 4006–4012.

Fasano, O., Crechet, J.B., De Vendittis, E., Zahn, R., Feger, G., Vitelli, A. and Parmeggiani, A. (1988) Yeast mutants temperature-sensitive for growth after random mutagenesis of the chromosomal *RAS2* gene and deletion of the *RAS1* gene. *EMBO J.* 7: 3375–3383.

Feger, G., De Vendittis, E., Vitelli, A., Masturzo, P., Zahn, R., Verrotti, A.C., Kavounis, C., Pal, G.P. and Fasano, O. (1991) Identification of regulatory residues of the yeast adenylyl cyclase. *EMBO J.* 10: 349–359.

Field, J., Nikawa, J.-I., Broek, D., MacDonald, B., Rogers, L., Wilson, I.A., Lerner, R.A. and Wigler, M. (1988) Purification of a RAS-responsive adenylyl cyclase complex from *Saccharomyces cerevisiae* by use of an epitope addition method. *Mol. Cell. Biol.* 8: 2159–2165.

Finegold, A.A., Johnson, D.I., Farnsworth, C.C., Gelb, M.H., Renée Judd, S., Glomset, J.A. and Tamanoi, F. (1991) Protein geranylgeranyltransferase of *Saccharomyces cerevisiae* is specific for Cys-Xaa-Xaa-Leu motif proteins and requires the *CDC43* gene product but not the *DPR1* gene product. *Proc. Natl. Acad. Sci. USA* 88: 4448–4452.

Fujiyama, A. and Tamanoi, F. (1990) RAS2 Protein of *Saccharomyces cerevisiae* undergoes removal of methionine at N terminus and removal of three amino acids at C terminus. *J. Biol. Chem.* 265: 3362–3368.

Gibbs, J.B. and Marshall, M.S. (1989) The ras oncogene – an important regulatory element in lower eucaryotic organisms. *Microbiol. Rev.* 53: 171–185.

Gilman, A. (1984) G proteins and dual control of adenylyl cyclase. *Cell* 36: 577–579.

Goodman, L.E., Perou, C.M., Fujiyama, A. and Tamanoi, F. (1988) Structure and expression of yeast *DPR1*, a gene essential for the processing and intracellular localisation of ras proteins. *Yeast* 4: 271–281.

Goodman, L.E., Judd, S.R., Farnsworth, C.C., Powers, S., Gelb, M.H., Glomset, J.A. and Tamanoi, F. (1990) Mutants of *Saccharomyces cerevisiae* defective in the farnesylation of Ras proteins. *Proc. Natl. Acad. Sci. USA* 87: 9665–9669.

Hartley, A.D., Ward, M.P. and Garret, S. (1994) The Yak1 protein kinase of *Saccharomyces cerevisiae* moderates thermotolerance and inhibits growth by an Sch9 protein kinase-independent mechanism. *Genetics* 136: 465–474.

Hartwell, L.H. (1974) *Saccharomyces cerevisiae* cell cycle. *Bacteriol. Rev.* 38: 164–198.

He, B. Chen, P., Chen, S.-Y., Vancura, K., Michaelis, S. and Powers, S. (1991) *RAM2*, an essential gene of yeast, and *RAM1* encode the two polypeptide components of the farnesyltransferase that prenylates a-factor and Ras proteins. *Proc. Natl. Acad. Sci. USA* 88: 11373–11377.

Hrycyna, C.A., Sapperstein, S.K., Clarke, S. and Michaelis, S. (1991) The *Saccharomyces cerevisiae STE14* gene encodes a methyltransferase that mediates C-terminal methylation of a-factor and RAS protein. *EMBO J.* 10: 1699–1709.

Jones, S., Vignais, M.L. and Broach, J.R. (1991) The CDC25 protein of *Saccharomyces cerevisiae* promotes exchange of guanine nucleotides bound to Ras. *Mol. Cell. Biol.* 11: 2641–2646.

Kataoka, T., Powers, S., McGill, C., Fasano, O., Strathern, J., Broach, J. and Wigler, M. (1984) Genetic analysis of yeast *RAS1* and *RAS2* genes. *Cell* 37: 437–445.

Kataoka, T., Powers, S., Cameron, S., Fasano, O., Goldfarb, M., Broach, J. and Wigler, M. (1985a) Functional homology of mammalian and yeast *RAS* genes. *Cell* 40: 19–26.

Kataoka, T., Broek, D. and Wigler, M. (1985b) DNA Sequence and characterization of the *S. cerevisiae* gene encoding adenylate cyclase. *Cell* 43: 493–505.

Kavounis, C., Verrotti, A.C., De Vendittis, E., Bozopoulos, A., Di Blasi, F., Zahn, R., Crechet, J.B., Parmeggiani, A., Tsernoglou, D. and Fasano, O. (1991) Role of glycine 82 as a pivot point during the transition from the inactive to the active form of the yeast Ras2 protein. *FEBS Lett.* 281: 235–239.

Kaziro, Y. (1978) The role of guanosine 5′-triphosphate in polypeptide chain elongation. *Biochim. Biophys. Acta* 505: 95–127.

Kim, J.H. and Powers, S. (1991) Overexpression of *RPI1*, a novel inhibitor of the Ras-cyclic AMP pathway, down-regulates normal but not mutationally activated Ras function. *Mol. Cell. Biol.* 11, 3894–3904.

Klein, C. and Struhl, K. (1994) Protein kinase A mediates growth-regulated expression of yeast ribosomal protein genes by modulating *RAP1* transcriptional activity. *Mol. Cell. Biol.* 14, 1920–1928.

Leberman, R. and Egner, U. (1984) Homologies in the primary structure of GTP-binding proteins. The nucleotide binding site of EF-Tu and p21. *EMBO J.* 3: 339–341.

Martin, G. A., Viskochill, D., Bollag, G., McCabe, P.C., Crosier, W.J., Haubruck, H., Conroy, L., Clark, R., O'Connell, P., Crawthon, R.M., Innis, M.A. and McCormick, F. (1990) The GAP-related domain of the neurofibromatosis type I gene product interacts with *ras* p21. *Cell* 63: 843–849.

Matsuura, A. and Anraku, Y. (1993) Characterization of the *MKS1* gene, a new negative regulator of the Ras-cyclic AMP pathway in *Saccharomyces cerevisiae*. *Mol. Gen. Genet.* 238: 6–16.

Mayer, M.L., Caplin, B.E. and Marshall, M.S. (1992) *CDC43* and *RAM2* encode the polypeptide subunits of a yeast type I protein geranylgeranyltransferase. *J. Biol. Chem.* 267: 20589–20593.

Mbonyi, K., Beullens, K., Detremerie, K., Geerts, L. and Thevelein, J.M. (1988) Requirement of one functional *RAS* gene and inability of an oncogenic *RAS* variant to mediate the glucose-induced cyclic AMP signal in the yeast *Saccharomyces cerevisiae*. *Mol. Cell. Biol.* 8: 3051–3057.

Miller, A.F., Papastavros, M.Z. and Redfield, A.G. (1992) NMR Studies on the conformational change in human N-p21[ras] produced by replacement of bound GDP with the GTP analog GTPgS. *Biochemistry* 31: 10208–10216.

Mirisola, M.G., Seidita, G., Verrotti, A.C., Di Blasi, F. and Fasano, O. (1994) Mutagenic alteration of the distal switch II region of RAS blocks CDC25-dependent signaling functions. *J. Biol. Chem.* 269: 15740–15748.

Mistou, M.Y., Jacquet, E., Poullet, P., Rensland, H., Gideon, P., Schlichting, I., Wittinghofer, A. and Parmeggiani, A. (1992) Mutations of Ha-*ras* p21 that define important regions for the molecular mechanism of the SDC25 C-domain, a guanine nucleotide dissociation stimulator. *EMBO J.* 11: 2391–2397.

Moodie, S. and Wolfman, A. (1994) The 3Rs of life: Ras, Raf and growth regulation. *Trends Genet.* 10: 44–48.

Mosteller, R.D., Han, J. and Broek, D. (1994) Identification of residues of the H-Ras protein critical for functional interaction with guanine nucleotide exchange factors. *Mol. Cell. Biol.* 4: 1104–1112.

Nikawa, J., Sass, P. and Wigler, M. (1987) Cloning and characterization of the low-affinity cyclic AMP phosphodiesterase gene of *Saccharomyces cerevisiae*. *Mol. Cell. Biol.* 7: 3629–3636.

Pai, E. (1991) p21 and other guanine-nucleotide-interacting proteins. *Curr. Biol.* 1: 941–945.

Powers, S., Michaelis, S., Broek, D., Santa Anna, A.S., Field, J., Herskowitz, I. and Wigler, M. (1986) *RAM*, a gene of yeast required for a functional modification of RAS proteins and for production of mating pheromone α-Factor. *Cell* 47: 413–422.

Quilliam, L.A., Kato, K.K., Rabun, K.M., Hisaka, M.M., Huff, S.Y., Campbell-Burk, S. and Der, D.J. (1994) Identification of residues critical for Ras(17N) growth-inhibitory phenotype and for Ras interaction with guanine nucleotide exchange factors. *Mol. Cell. Biol.* 14: 1113–1121.

Robinson, L.C. and Tatchell, K. (1998) *TSF1*: a suppressor of *cdc25* mutations in *Saccharomyces cerevisiae*. *Mol. Gen. Genet.* 230: 241–250.

Ruggieri, R., Tanaka, K., Nakafuku, M., Kaziro, Y., Toh-e, A. and Matsumoto, K. (1989) *MSI1*, a negative regulator of the RAS-cAMP pathway in *Saccharomyces cerevisiae*. *Proc. Natl. Acad. Sci. USA* 86: 8778–8782.

Ruggieri, R., Bender, A., Matsui, Y., Powers, S., Takai, Y., Pringle, J.R. and Matsumoto, K. (1992) *RSR1, a ras*-like gene homologous to *Krev-1 (smg21A/rap1A)*: role in the development of cell polarity and interactions with the Ras pathway in *Saccharomyces cerevisiae*. *Mol. Cell. Biol.* 12: 758–766.

Saas, P., Field, J., Nikawa, J. and Wigler, M. (1986) Cloning and characterization of the high affinity cAMP phosphodiesterase of *Saccharomyces cerevisiae*. *Proc. Natl. Acad. Sci. USA* 83: 9303–9307.

Schlichting, I., Almo, S.C., Rapp, G., Wilson, K., Petratos, K., Lentfer, A., Wittinghofer, A., Kabsh, W., Pai, E.F., Petsko, G.A. and Goody, R.S. (1990) Time-resolved X-ray crystallographic study of the conformational change in Ha-Ras p21 protein on GTP hydrolysis. *Nature* 345: 309–315.

Segal, M., Marbach, I., Engelberg, D., Simchen, G. and Levitki, A. (1992) Interaction between the *Saccharomyces cerevisiae CDC25* gene product and mammalian Ras. *J. Biol. Chem.* 267: 22747–22751.

Shirayama, M., Kawakami, K., Matsui, Y., Tanaka, K. and Toh-e A. (1993) *MSI3*, a multicopy suppressor of mutants hyperactivated in the *RAS*-cAMP pathway, encodes a novel HSP70 protein of *Saccharomyces cerevisiae*. *Mol. Gen. Genet.* 240: 323–332.

Sigal, I.S., Gibbs, J.B., D'Alonzo, J.S. and Scolnick E. (1986) Identification of effector residues and a neutralising epitope of Ha-*ras*-encoded p21. *Proc Natl. Acad. Sci. USA* 83: 4725–4729.

Stouten, P.F.W., Sander, C., Wittinghofer, A. and Valencia, A. (1993) How does the switch II region of G-domains work? *FEBS Lett.* 320: 1–6.

Suzuki, N., Tsujino, K., Minato, T., Nishida, Y., Okada, T. and Kataoka, T. (1993) Antibody mimiking the action of RAS proteins on yeast adenylyl cyclase: implication for RAS-effector interaction. *Mol. Cell. Biol.* 13: 769–774.

Tamanoi, F. (1988) Yeast *ras* genes. *Biochim. Biophys. Acta* 948: 1–16.

Tanaka, K., Matsumoto, K., and Toh-e, A (1989) *IRA1*, an inhibitory regulator of the *RAS*-cyclic AMP pathway in *Saccharomyces cerevisiae*. *Mol. Cell. Biol.* 9: 757–768.

Tanaka, K., Nakafuku, M., Satoh, T., Marshall, M.S., Gibbs, J.B., Matsumoto, K., Kaziro, Y., and Toh-e, A. (1990a) *S. cerevisiae* genes *IRA1* and *IRA2* encode proteins that may be functionally equivalent to mammalian *ras* GTPase-activating protein. *Cell* 60: 803–807.

Tanaka, K., Nakafuku, M., Tamanoi, F., Kaziro, Y., Matsumoto, K. and Toh-e, A. (1990b) *IRA2*, a second gene of *Saccharomyces cerevisiae* that encodes a protein with a domain homologous to mammalian *ras* GTPase-activating protein. *Mol. Cell. Biol.* 10: 4303–4313.

Tanaka, K., Wood, D.R., Lin, B.K., Khalil, M., Tamanoi, F. and Cannon, J.F. (1992) A dominant activating mutation in the effector region of RAS abolishes IRA2 sensitivity. *Mol. Cell. Biol.* 12: 631–637.

Thevelein, J.M. (1991) Fermentable sugars and intracellular acidification as specific activators of the RAS-adenylate cyclase signalling pathway in yeast: the relationship to nutrient-induced cell cycle control. *Mol. Microbiol.* 5: 1301–1307.

Thompson-Jaeger, S., François, J., Gaughran, P. and Tatchell, K. (1991) Deletion of *SNF1* affects the nutrient response of yeast and resembles mutations which activate the adenylate cyclase pathway. *Genetics* 129: 697–706.

Toda, T., Uno, I., Ishikawa, T., Powers, S., Kataoka, T., Broek, D., Cameron, S., Broach, J., Matsumoto, K. and Wigler, M. (1985) In yeast, RAS proteins are controlling elements of adenylate cyclase. *Cell* 40: 27–36.

Toda, T., Cameron, S., Sass, P., Zoller, M. and Wigler, M. (1987a) Three different genes in *S. cerevisiae* encode the catalytic subunits of the cAMP-dependent protein kinase. *Cell* 50: 277–287.

Toda, T., Cameron, S., Sass, P., Zoller, M., Scott, J.D., McMullen, B., Hurwitz, M., Krebs, E.G. and Wigler, M. (1987b) Cloning and characterization of *BCY1*, a locus encoding a regulatory subunit of the cAMP-dependent protein kinase. *Mol. Cell. Biol.* 7: 1371–1377.

Toda, T., Cameron, S., Sass, P. and Wigler, M. (1988) *SCH9*, a gene of *Saccharomyces cerevisiae* that encodes a protein distinct from, but functionally and structurally related to, cAMP-dependent protein kinase catalytic subunits. *Genes Develop.* 2: 517–527.

Uno, I., Mitsuzawa, H., Tanaka, K., Oshima, T. and Ishikawa, T. (1987) Identification of the domain of *Saccharomyces cerevisiae* adenylate cyclase associated with the regulatory function of *RAS* products. *Mol. Gen. Genet.* 210: 187–294.

Verrotti, A.C., Crechet, J.B., Di Blasi, F., Seidita, G., Mirisola, M., Kavounis, C., Nastopoulos, V., Burderi, E., De Vendittis, E., Parmeggiani, A. and Fasano, O. (1992) Ras residues that are distant from the GDP binding site play a critical role in dissociation factor-stimulated release of GDP. *EMBO J.* 11: 2855–2862.

Wang, J., Suzuki, N., Nishida, Y. and Kataoka, T. (1993) Analysis of the function of the 70-Kilodalton cyclase-associated protein (CAP) by using mutants of yeast adenylyl cyclase defective ion CAP binding. *Mol. Cell. Biol.* 13: 4087–4097.

Willumsen, B.M., Vass, W.C., Velu, T.J., Papageorge, A.G., Schiller, J.T. and Lowry, D.R. (1991) The bovine papillomavirus E5 oncogene can cooperate with *ras*: Identification of p21 amino acids critical for transformation by *c-ras*[H] but not v-*ras*[H]. *Mol. Cell. Biol.* 11: 6026–6033.

Wilson, R.B., Brenner, A.A., White, T.B., Engler, M.J., Gaughran, J.P. and Tatchell, K. (1991) The *Saccharomyces cerevisiae SRK1* gene, a suppressor of *bcy1* and *ins1*, may be involved in protein phosphatase function. *Mol. Cell. Biol.* 11: 3369–3373.

Xu, G., Lin, B., Tanaka, K., Dunn, D., Wood, D., Gesteland, R., White, R., Weiss, R. and Tamanoi, F. (1990) The catalytic domain of the neurofibromatosis type 1 gene product stimulates *ras* GTPase and complements *ira* mutants of *S. cerevisiae*. *Cell* 63: 835–841.

Biochemistry of Cell Membranes
ed. by S. Papa & J. M. Tager

Protein import into mitochondria

K. Hannavy[1] and G. Schatz[2]

[1]*Present address: Current Biology Ltd., Middlesex House, 34–42 Cleveland Street, London W1P 6LB, UK;* [2]*Department of Biochemistry, Biozentrum of the University of Basel, Klingelbergstrasse 70, CH-4056 Basel, Switzerland*

Summary. The key to the biogenesis of mitochondria is the import and sorting of cytosolically synthesized proteins to their final sub-mitochondrial destination. This review summarizes recent progress in our understanding of these processes and formulates some of the many questions to be answered in the field.

Introduction

Mitochondria are subcellular organelles essential for virtually all eukaryotic cells. Hundreds of different proteins are located in the mitochondria, performing manifold functions. About a dozen of these proteins are encoded by the mitochondrial genome, but the vast majority are nuclearly-encoded and synthesized on cytosolic ribosomes. How are the newly-synthesized proteins directed to, and recognized by the mitochondria? How do mitochondria sort the incoming proteins to their final submitochondrial location? How are these processes energized?

Research over the past two decades has provided a general outline of the import process; it has defined the nature of the targeting and sorting signals of a variety of imported proteins, identified and partially characterized a number of cytosolic and mitochondrial proteins which guide precursors to their ultimate destinations, and has identified the energy requirements of protein import. Most of the studies to date have used either mammalian (especially rat), or fungal (*Neurospora crassa* or yeast) mitochondria. This review will provide a brief general outline of the import process, highlighting steps for which our understanding is particularly limited.

Protein import into mitochondria can be conveniently divided into the following steps. Precursors are synthesized on cytosolic ribosomes, maintained in an import-competent state by interaction with a cytosolic chaperone, bound to the mitochondrial surface via proteinaceous receptors, and delivered to import sites at which the protein import machineries of the outer and inner mitochondrial membranes are in close proximity (membrane contact sites). Subsequent passage to the mitochondrial matrix requires the sequential action of an electrical potential

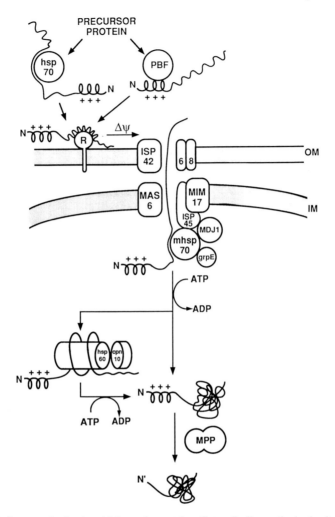

Figure 1. Import of mitochondrial matrix proteins. Cytosolically synthesized mitochondrial precursor proteins can associate with a variety of chaperones, such as hsp70 and PBF (presequence binding factor). The precursors bind receptor proteins (R) on the outer mitochondrial membrane (OM). Import of precursors into the matrix occurs via the closely apposed protein translocation channels of the outer and inner membranes (IM). Transit of the presequence across the inner membrane requires an electrical potential across the inner membrane, $\Delta\Psi$. Translocation of the remainder of the protein requires the ATP-dependent action of mhsp70. Cleavable N-terminal presequences are removed by MPPα/β (matrix processing peptidase). Folding and assembly steps may involve the chaperonins hsp60 and cpn10.

across the inner membrane (required for transit of the presequence), followed by hydrolysis of matrix ATP (believed to reflect the action of mhsp70 as a molecular motor, driving uptake of the remainder of the precursor into the matrix). Mitochondrial proteins not destined for the matrix usually follow part of this matrix pathway, but depart from it to reach the outer membrane, intermembrane space or inner membrane. Although most imported proteins studied so far conform to this general picture (see Fig. 1), exceptions are known (see below).

Precursors and signals

Cytosolically-synthesized mitochondrial proteins contain targeting information within their amino acid sequence. Most are synthesized as precursors comprising an N-terminal presequence which contains mitochondrial targeting and sorting information. These presequences are normally proteolytically removed upon import, although exceptions are known, for instance yeast and rat cpn10 (Rospert et al., 1993; Hartman et al., 1992) or mammalian rhodanese (Miller et al., 1991). Some mitochondrial proteins contain internal, non-cleaved targeting signals.

The primary structures of dozens of presequences are known. Proteins whose final destination is the mitochondrial matrix or inner membrane typically possess an N-terminal presequence of 20–30 amino acids. The primary structures of such matrix targeting sequences are quite divergent, but share a preponderance of positively charged and hydroxylated amino acids, a lack of acidic amino acids, and a predicted propensity to form amphipathic α-helices (Roise et al., 1986; von Heijne, 1986; Roise and Schatz, 1988). Studies with model peptides indicate that the α-helix forms on contact with membranes (Roise et al., 1986; Endo et al., 1989; Roise, 1993). It has been shown by experiments with heterologous fusion proteins that a matrix targeting sequence is both necessary and sufficient for targeting a protein to the mitochondrial matrix (Hurt et al., 1984).

Some proteins which reside in the inner membrane as components of multisubunit complexes are first targeted to the matrix by a matrix targeting sequence and are subsequently assembled with their partner subunits, thereby inserting into the inner membrane (for example, the Rieske Fe/S protein of the cytochrome bc_1 complex, subunit IV of cytochrome c oxidase and the $F_1\beta$ subunit of the F_1F_0 ATPase). The enzyme responsible for removal of the matrix targeting sequence is the heterodimeric matrix processing peptidase (MPPα/β) (Böhni et al., 1980; McAda and Douglas, 1982; Yang et al., 1988; Hawlitschek et al., 1988; Ou et al., 1989; Geli et al., 1990; Yang et al., 1991; Arretz et al., 1994). Cleavage of the matrix targeting sequence is not required for import (Reid et al., 1982; Zwizinski et al., 1984). Some precursors possess more complex presequences which contain an octapeptide C-terminal to the matrix targeting sequence (Kalousek et al., 1992). This octapeptide is removed by the mitochondrial intermediate protease (MIP) (Kalousek et al., 1992). The functional role played by the octapeptide is unclear.

Members of a group of proteins whose final destination is the mito-chondrial intermembrane space (IMS) possess bipartite presequences. The N-terminal portion resembles a typical matrix targeting sequence (and indeed can function as such; van Loon et al., 1987) and is removed by MPPα/β. The C-terminal region directs the attached protein to the IMS. Recent studies have defined two functionally important sub-do-mains within the sorting sequence; an N-terminal positively charged region followed by a stretch of predominantly non-polar residues (Jensen et al., 1992; Beasley et al., 1993; Schwarz et al., 1993). The removal of the sorting sequence is catalyzed by the heterodimeric inner membrane protease IMP1/2 protease (Schneider et al., 1991; Behrens et al., 1991; Nunnari et al., 1993; Schneider et al., 1994).

The best-characterized targeting sequence for an outer membrane protein is that of Mas70p (Hase et al., 1984; Nakai et al., 1989; Shore et al., 1992). The N-terminal 29 amino acids are both necessary and sufficient for targeting and anchoring an attached passenger protein to the outer membrane. This region includes a putative transmembrane α-helix of 19 amino acids. It also appears that this domain is involved in oligomerization of Mas70p (Millar and Shore, 1993). The nature of the sequences flanking the hydrophobic region appear to determine the membrane topology of the protein (Li and Shore, 1992a), a situation also found in the endoplasmic reticulum and *E. coli* cytoplasmic membrane (von Heijne, 1992). Interestingly, the signal-anchor sequence directs a fused passenger protein to be integrated into the inner membrane when the fusion protein is presented to mitoplasts (Li and Shore, 1992b). Yeast mitochondrial NADH-cytochrome b_5 reductase, another outer mem-brane protein, possesses a similar N-terminal signal-anchor sequence (Hahne et al., 1995). Other outer membrane proteins utilize different localization signals. The mammalian mitochondrial outer membrane proteins monamine oxidase B (Mitoma and Ito, 1992) and Bcl-2p (Nguyen et al., 1993; Lithgow et al., 1994b), as well as MOM22 of *N. crassa* (Kiebler et al., 1993) and Mas22p of yeast (T. Lithgow et al., in preparation) appear to be targeted and tethered by a C-terminal region, again by formation of a single transmembrane region. The targeting and integration signals of proteins such as porin and ISP42/MOM38 are unknown. These proteins are devoid of potential transmembrane α-heli-cal regions, but may form β-barrel type structures typical of the porins of bacterial outer membranes (Walian and Jap, 1990; Weiss et al., 1991).

Some proteins such as cytochrome *c* do not possess typical mitochon-drial targeting signals. Indeed, cytochrome *c* appears to localize to the IMS by a unique pathway (Stuart and Neupert, 1990). Yet other proteins, such as the ATP/ADP carrier (AAC), an integral membrane protein of the inner membrane, possess internal targeting signals. The region comprising amino acids 71–111 has been shown to be important in targeting AAC to the inner membrane (Smagula and Douglas, 1988).

It is presumed that in most cases the role of the presequence is solely to target the mature portion of the protein to its correct destination. However, there is one recently noted instance of a cleaved presequence forming a component of a multisubunit complex; the matrix targeting sequence of the bovine Rieske Fe/S protein is identical to subunit 9 of the cytochrome bc_1 complex (Brandt et al., 1993).

From the ribosome to the mitochondrial surface

Studies of a wide variety of protein translocating systems have resulted in the principle of "translocation competence". This means that in order to pass across membranes proteins must be largely devoid of a stable tertiary structure (Eilers et al., 1986; Randall and Hardy, 1986; Vestweber and Schatz, 1988a; Rassow et al., 1990). A number of possible ways of avoiding, or reversing, stable folding can be envisaged. Import could be cotranslational, minimizing the likelihood of extensive folding prior to import. Alternatively, cytosolic chaperones could associate with the precursor before it has adopted its native conformation, maintaining it in a "loosely-folded" conformation. Such an interaction would also prevent aggregation with other precursors, cytosolic proteins or the membrane surfaces of other organelles. Import of the precursor could occur upon release from the chaperone, which may require the input of energy. Another possibility would be that a folded precursor is actively unfolded, either prior to or during translocation. Finally, some precursors may fold at a slow rate relative to the rate of synthesis and import. Each of the suggested routes from ribosome to mitochondrial surface makes specific predictions about the energy requirements for this import step. Recent detailed *in vitro* studies of the energy requirements at different stages of import of a variety of proteins, coupled with the characterization of cytosolic proetins involved in import, indicate that several of the possible pathways described above may be utilized.

There is some uncertainty as to the pathway followed by cytosolically-synthesized proteins from the ribosome to the mitochondrial surface (for a recent review see Verner, 1993). That post-translational import can occur *in vitro* is well-established. However, there is indirect evidence suggesting that *in vivo* co-translational import may occur as well (Reid and Schatz, 1982; Schatz and Butow, 1983; Fujiki and Verner, 1991; Fujiki and Verner, 1993; Verner, 1993).

A variety of cytosolic proteins appears to be involved at this stage of import. It appears that some members (Ssb1p and Ssb2p in yeast) of the hsp70 family of chaperones associate transiently with all nascent polypeptides emerging from the ribosome (Nelson et al., 1992). Depletion of three (Ssa1p, 2p and 4p) out of six of the yeast cytosolic hsp70s leads to the *in vivo* accumulation of precursor proteins destined for mitochondria and the endoplasmic reticulum (Deshaies et al., 1988). Furthermore, addition of one of these proteins, Ssa1p, has a stimulatory effect on the *in vitro* import of the $F_1\beta$ precursor into yeast mitochondria (Murakami et al., 1988). Interestingly, recent studies on rat liver mitochondria have shown that a cytosolic hsp70 is firmly associated with the mitochondrial outer membrane (Lithgow et al., 1993).

Two independent studies indicated a role in mitochondrial import for Ydj1p/Mas5p, a yeast cytosolic protein homologous to *E. coli* DnaJ which interacts with the bacterial hsp70 DnaK (Atencio and Yaffe, 1992; Caplan et al., 1992; Caplan et al., 1993).

In contrast to Ydj1p/Mas5p and the cytosolic hsp70s, which are involved in the translocation of proteins across several organellar membranes, two proteins have recently been identified which appear to interact exclusively with mitochondrial precursors. These proteins were identified from rat liver lysates on the basis of their specific interaction with mitochondrial presequences. Presequence binding factor (PBF) is a 50 kDa homooligomer which appears to act by maintaining the unfolded state of a precursor and/or by preventing precursor aggregation (Murakami and Mori, 1990; Murakami et al., 1992). Mitochondrial import stimulating factor (MSF) is a heterodimer composed of 30 and 32 kDa subunits; it unfolds mitochondrial precursor proteins in an ATP-dependent fashion (Hachiya et al., 1993).

Thus, we are beginning to discover the various ways through which cytosolic factors ensure that precursors maintain or adopt a conformation compatible with translocation. Nonetheless, a number of key questions remains to be answered: What determines the interaction of a precursor with a particular chaperone? Do precursors interact with more than one chaperone during their journey from the ribosome to the mitochondrial surface? Is association with a chaperone important for recognition of a protein by the mitochondrial import machinery?

Arrival at the mitochondrial surface

For most precursors, binding to the mitochondrial surface requires the presence of trypsin-sensitive outer membrane proteins (Riezman et al., 1983; Zwizinski et al., 1984; for a recent review see Lithgow and Schatz, 1994). Subsequent translocation of bound precursor appears to occur through a proteinaceous channel spanning the outer membrane.

Several "receptor" proteins have been identified. In yeast, Mas70p and Mas20p were identified by antibodies against them which inhibited import (Hines et al., 1990; Ramage et al., 1993). Both proteins appear to be anchored in the outer membrane via their N-terminal regions with the bulk of the protein being exposed to the cytosol. Homologous proteins have been identified in *N. crassa* (Söllner et al., 1989; Söllner et al., 1990). It appears that different precursors differ in their reliance upon a particular receptor (Hines et al., 1990; Steger et al., 1990). The reason for this precursor-receptor preference is unclear. However, import studies with a variety of different proteins, using either strains disrupted in one or more of the receptors genes or anti-receptor antibodies, show that the receptors can largely substitute for one another (Hines et al., 1990; Ramage et al., 1993; Hines and Schatz, 1993; Moczko et al., 1994). Elimination of expression of either MAS70 or MAS20 is not lethal, but results in conditional phenotypes (Hines et al., 1990; Ramage et al., 1993; Moczko et al., 1994). Disruption of both receptors is lethal, at least in yeast (Ramage et al., 1993). However, this synthetic lethality can be suppressed by a dominant nuclear mutation (Lithgow et al., 1994a). This suppressor gene may encode yet another receptor protein. It is not clear whether the different receptors represent alternative means for binding precursors to the outer membrane or whether they are components acting on a single pathway.

The corresponding *N. crassa* receptors MOM72 and MOM19 were crosslinked to a precursor protein (AAC) arrested at an early stage during import; arrest at a slightly later step resulted in cross-linking to four other proteins; MOM7, 8, 30, and 38 (Söllner et al., 1992). It is argued that MOM7, 8, 30, and 38 form a translocation channel across the outer membrane. The yeast homologue of MOM38, ISP42, had been previously identified by crosslinking it to a stuck precursor and by demonstrating that depletion of the protein *in vivo* causes an inhibition of protein import into mitochondria (Vestweber et al., 1989; Baker et al., 1990). The yeast integral outer membrane protein ISP6 has been shown, genetically and biochemically, to interact with ISP42. Disruption of the ISP6 gene has no visible phenotype, but is synthetically lethal in combination with a temperature sensitive (ts) allele of ISP42. Furthermore, ISP6 co-immuno-precipitates with ISP42 (Kassenbrock et al., 1993). The relationship of ISP6 to MOM7 and 8 is not known. The same group has identified a large, inner membrane anchored protein as being another putative partner protein of ISP42 (M. Douglas and N. Martin, personal communication). Curiously, this protein contains an RNaseP domain.

The *N. crassa* proteins MOM7, 8, 30, 38, 19, and 72, together with an additional component, MOM22, were co-immunoprecipitated using antibodies against MOM19 (Kiebler et al., 1990). This complex is thought to represent the outer membrane import machinery. The same approach identified a similar complex in yeast (Moczko et al., 1992). Recent work in *N. crassa* has led to the suggestion that MOM22 mediates precursor transfer from the receptors to the import channel in the outer membrane (Kiebler et al., 1993). Furthermore, co-ordinate expression of MOM19 and MOM22 was demonstrated, indicating a physical interaction between the two proteins (Harkness et al., 1994).

A number of other proteins have been suggested to play a role at the outer membrane stage of mitochondrial import.

Recent genetic studies to identify proteins involved in intramitochondrial sorting, utilizing a fusion protein between the N-terminal 61 amino acids of Mas70p and the mature region of the IMS protein cytochrome c_1, led to the identification of a 40 kDa outer membrane protein, Msp1p (Nakai et al., 1993). Msp1p belongs to a family of proteins sharing a putative ATPase domain. It is suggested that Msp1p functions in the sorting of proteins to the outer membrane. Further characterization of Msp1p is required to confirm this role.

Earlier reports suggested that a 32 kDa mitochondrial membrane protein was able to recognize precursors and that it acted as a mitochondrial protein import receptor (Pain et al., 1990; Murakami et al., 1990). However, p32 has now been identified as the phosphate carrier of the mitochondrial inner membrane, casting doubts upon its proposed receptor role in protein import (Dietmeier et al., 1993).

A series of questions concerning the outer membrane stage of import remains to be answered. In what form do the receptors recognize precursors? How many receptors are there? To what degree do the components of the outer membrane import machinery function as complexes with each other? Is the composition of this complex static or is it dynamic?

An exciting recent development has been the development of an *in vitro* outer membrane import system (Mayer et al., 1993). Sealed right-side-out outer membrane vesicles were shown to integrate several outer membrane proteins, such as MOM19 and porin. Furthermore, the IMS protein cytochrome c heme lyase (CCHL) was fully imported in a receptor-dependent fashion. The energy source for translocation is unclear, but future experiments with this system should help to provide an answer. Interestingly, proteins normally routed via the inner membrane were bound to, but not imported by the vesicles. This result has been interpreted to mean that import of such proteins requires the coordinated action of the outer and inner membrane translocation machineries.

Several proteins have been shown to follow unusual import pathways. Cytochrome c is a heme protein of the IMS and is synthesized as an apoprotein in the cytosol. The signals directing it to the IMS are located in different parts of the apoprotein (Nye and Scarpulla, 1990a,b). Import is independent of the general outer membrane translocation machinery (Stuart and Neupert, 1990), but may involve a recently identified mitochondrial protein, Cyc2p (Dumont et al., 1993). Disruption of the *CYC2* gene results in a drastic reduction of the amount of both isoforms of cytochrome c, but has no effect on the import of other apocytochromes. It is thought that the import reaction is made irreversible by the covalent attachment of heme by CCHL, a peripheral membrane protein attached to the IMS side of the inner membrane (Nicholson et al., 1988).

CCHL itself has an interesting import pathway. Like most imported proteins studied so far, it utilizes the outer membrane translocation machinery, but not the inner membrane translocation machinery (Lill et al., 1992). Studies of the import of a fusion protein comprising the matrix targeting sequence of the β-subunit of F_1F_0 ATPase and CCHL showed that this protein was imported to the IMS in the absence of $\Delta\Psi$ across the inner membrane. Restoration of $\Delta\Psi$ resulted in import of the IMS-located fusion protein into the matrix (Segui-Real et al., 1993).

Import of the inner membrane protein cytochrome c oxidase subunit Va (Va) is unusual in several respects. It does not appear to use either Mas70p or Mas20p, as import is unaffected by "protease shaving" of mitochondria (Miller and Cumsky, 1991). Import depends upon $\Delta\Psi$ across the inner membrane, but appears to be independent of mhsp70 (Miller and Cumsky, 1993). Translocation is arrested by a region including a hydrophobic stretch of about 20 amino acids in the C-terminal third of the protein. Deletion of this region mislocalizes the protein to the matrix (Glaser et al., 1990). Therefore, the C-terminal inner membrane localization signal appears to function as a stop-transfer sequence for the inner membrane.

Transit across the inner membrane

Mitochondrial import is thought to occur largely at sites where the outer membrane and inner membrane are in close proximity (Kellems et al., 1975; Schleyer and Neupert, 1985; Schwaiger et al., 1987; Vestweber and Schatz, 1988b; Pon et al., 1989). Such "contact sites" have been visualized by electron microscopy (Hackenbrock, 1968; Kellems et al., 1975), but are poorly characterized biochemically. Initially it was suggested that the two membranes were spanned by a fixed channel (Schwaiger et al., 1987), but more recent data favor a model in which import into the matrix involves two separate, but closely apposed outer membrane and inner membrane translocation machineries (Glick et al., 1991; Pfanner et al., 1992). The inner membrane is itself translocation competent and possesses a translocation machinery capable of operating independently of that of the outer membrane. Mitoplasts (mitochondria in which the outer membrane has been disrupted), or inner membrane vesicles can efficiently import precursors. Unlike import into intact mitochondria, import into mitoplasts or inner membrane vesicles is unaffected by antibodies raised against outer membrane proteins, but is blocked by antisera to the inner membrane (Ohba and Schatz, 1987; Hwang et al., 1989). Furthermore, by judicious manipulation of the inner membrane $\Delta\Psi$ and ATP levels in the matrix, it has been possible to dissect the import of matrix targeted proteins into transit across the outer membrane followed by a separate translocation step across the inner membrane (Hwang et al., 1991; Jascur et al., 1992; Rassow and Pfanner, 1991; Segui-Real et al., 1993). The interaction between the two translocation machineries is suggested to be transient (Glick et al., 1991; Pfanner et al., 1992). The dynamic nature of the import machinery appears to be essential for the sorting of some proteins to the IMS (Glick et al., 1991, 1992a; Lill et al., 1992; Mayer et al., 1993; Segui-Real et al., 1993).

Mitochondrial precursors jammed in either the outer or the inner membranes are extractable by alkali or urea, indicating that they are in proteinaceous environments (Pfanner et al., 1987; Glick et al., 1992b; Söllner et al., 1992). These results have been interpreted to mean that translocating proteins cross a membrane through a protein channel. How these channels can accomodate proteins, and even branched polypeptide chains and protein-double-stranded DNA conjugates (Vestweber and Schatz, 1988c; Vestweber and Schatz, 1989) and yet remain proton-tight, is unknown.

As mentioned before, proteins destined for the matrix generally possess an N-terminally located matrix targeting sequence. The passage of this sequence across the inner membrane is dependent upon the electrical component of the $\Delta\Psi$ across the inner membrane (Martin et al., 1991). Transfer of the remainder of the protein into the matrix is dependent on matrix ATP (Hwang et al., 1991; reviewed in Rospert, 1994). It is thought that this ATP requirement reflects the role of the matrix protein mhsp70 as an "import motor" (Scherer et al., 1990; Neupert et al., 1990; Manning-Krieg et al., 1991; Schatz, 1993). mhsp70 has been cross-linked to precursors in transit across the inner membrane (Scherer et al., 1990; Ostermann et al., 1990; Manning-Krieg et al., 1991), and inactivation of mhsp70 causes matrix-targeted precursors to get stuck across both mitochondrial membranes (Kang et al., 1990). Recent studies using different ts versions of mhsp70 show that binding to mhsp70 is necessary for precursor translocation. Furthermore, binding to mhsp70 is ATP-dependent and requires that mhsp70 have a functional ATPase domain (Gambill et al., 1993; Voos et al., 1993). E. coli hsp70 (dnaK) acts in concert with at least two partner proteins, grpE and dnaJ (Georgopolous and Welch, 1993). Homologs of these proteins, GrpEp and Mdj1p, have very recently been identified in yeast mitochondria (Bolliger et al., 1994; Ikeda et al., 1994; Rowley et al., 1994). Mdj1p does not appear to be required for import, is not essential for viability, but is involved in the refolding of imported precursors (Rowley et al., 1994). GrpEp is essential for viability of yeast and physically interacts with mhsp70 (Bolliger et al., 1994). However, it remains to be seen whether it plays a role in import.

How mhsp70 acts to drive import is not known. In a variation of the "Brownian ratchet" model of protein translocation across biological membranes (Simon et al., 1992), it is envisaged that a precursor protein in the translocation channel oscillates back and forth across the inner membrane. Successive binding of mhsp70 molecules to newly emerged stretches of the precursor on the matrix side of the membrane prevents retrograde movement, eventually resulting in the precursor residing entirely within the matrix (Neupert et al., 1990). An extension of this model is proposed to explain the unfolding of proteins during translocation (Stuart et al., 1994a). However, it remains to be seen whether this model would be sufficient to explain unfolding of tightly-folded regions of precursor proteins such as the heme-binding domain of cytochrome b_2 (Glick et al., 1993; Gambill et al., 1993).

Only recently have components of the inner membrane translocation machinery been identified. A combination of biochemical and genetic approaches have identified mhsp70 (Manning-Krieg et al., 1991; Gambill et al., 1993; Voos et al., 1993), ISP45/Mim44p (Maarse et al., 1992; Scherer et al., 1992; Horst et al., 1993; Blom et al., 1993), Mas6p/Mim23p (Emtage and Jensen, 1993; Dekker et al., 1993) and Mim17p (Dekker et al., 1993). Mas6p and Mim17p are both integral inner membrane proteins. Whether ISP45 is a peripheral or integral protein of the inner membrane is still a moot point (Blom et al., 1993). Interestingly, Mas6p and Mim17p share a region of 90 amino acids with 75% of homology (Dekker et al., 1993). ISP45 was identified by antibodies which inhibited import into inner membrane vesicles, by cross-linking it to a stuck precursor (Scherer et al., 1992; Horst et al., 1993) and by a genetic selection for import mutants. This selection also yielded the genes for Mim17p and Mim23p (Maarse et al., 1992). Mim23p (Dekker et al., 1993) is identical to Mas6p, which had been identified from a yeast ts mutant which accumulates mitochondrial precursor proteins at the non-permissive temperature (Emtage and Jensen, 1993). Inhibition of import caused by defective ISP45 or Mim17p, but not Mas6p, can be suppressed by overexpression of mhsp70, suggesting a physical interaction between mhsp70 and these two proteins (M. Meijer, personal communication).

A host of questions remains to be addressed. What is the precise role of each of these proteins in import? In what order do they act? Are other proteins required to form an inner membrane translocation machinery? What is the role of the inner membrane translocation machinery in the sorting of proteins to the IMS and the inner membrane? Is this machinery involved in the integration of mitochondrially encoded proteins, such as subunit II of cytochrome oxidase, into the inner membrane? Are additional components required for these processes?

A particularly ticklish issue has been the pathway by which some proteins reach the IMS. These proteins are synthesized as precursors with a bipartite presequence; the N-terminal portion of the precursor is a matrix targeting sequence and the C-terminal part is a IMS sorting signal (van Loon et al., 1986; van Loon and Schatz, 1987; van Loon et al., 1987). The stop transfer model envisages this IMS sorting sequence to be recognized by the inner membrane translocation machinery as a signal to arrest further translocation into the matrix. The translocation machineries of the inner and outer membranes then separate with the result that the rest of the protein is pulled into the IMS. During this separation the precursor remains anchored to the inner membrane, but subsequent cleavage of the stop transfer sequence releases the mature protein into the IMS (van Loon et al., 1987; Glick et al., 1991; Glick et al., 1992b). The alternative model, known as conservative sorting, proposes that the entire precursor first passes into the matrix and is then reexported across the inner membane into the IMS (Hartl et al., 1987; Hartl and Neupert, 1990; Koll et al., 1992; Segui-Real et al., 1992).

The arguments for and against each model have been presented in a recent review (Glick et al., 1992a). In brief, all of the available evidence can be readily explained by the stop transfer model, but not by the conservative sorting model. Import of proteins into the matrix always requires matrix ATP, but import and sorting of cytochrome c_1 to the IMS does not. The same is true for cytochrome b_2 if tight folding of the heme-binding domain outside the mitochondria is prevented (Wachter et al., 1992; Wachter et al., 1994; Gambill et al., 1993; Stuart et al., 1994b). Furthermore, data taken as evidence for a matrix location of intermediate forms of cytochrome b_2 must now be reinterpreted (Glick et al., 1992; Rospert et al., 1994). In contrast to earlier reports (Hartl et al., 1987; Cheng et al., 1989; Koll et al., 1992), there is also no evidence for a role of the matrix chaperonin hsp60 in sorting to the IMS (Glick et al., 1992b; Rospert et al., 1994; Hallberg et al., 1993). Finally, the reexport machinery predicted to exist by the conservative sorting model has yet to be identified.

The Rieske Fe/S protein is imported into the matrix and only then incorporated into the inner membrane as a subunit of the cytochrome bc_1 complex where it is partially exposed to the IMS (Hartl et al., 1986). It has been argued that import of the Rieske Fe/S protein represents an example of conservative sorting; however, the targeting of this protein can equally be viewed as being composed of a matrix import step followed by a process of assembly together with the other components of the cytochrome bc_1 complex.

Conclusions

The import pathways of many different mitochondrial proteins have been defined, the corresponding targeting signals directing a wide variety of proteins to the mitochondrion and its component subcompartments have been thoroughly analyzed, the energetics of the import process are becoming reasonably well understood, and many components of the mitochondrial import machinery are known. The central biological importance of mitochondrial import is illustrated by the finding that most components of the mitochondrial protein import machinery identified so far are essential for life (Baker and Schatz, 1991). The essential nature of other components is unmasked when redundant proteins are comprised (Ramage et al., 1993; Kassenbrock et al., 1993).

Although much remains to be learned, general themes are emerging. Recently it has become apparent that the energy-requiring steps of protein translocation across mitochondrial (and other) membranes are coupled to the action of soluble chaperone proteins. These are the import components for which detailed three-dimensional structural information is most likely to be forthcoming. In fact, a high-resolution crystal structure of the ATP-binding domain of a cytosolic hsp70 has already been published (Flaherty et al. 1991). In contrast, the transmembrane components of translocation machineries appear to play a more passive role, providing a gated channel through which polypeptides pass, pulled (or pushed) by chaperones on one side or other of the membrane. How these channels are gated is unknown.

Doubtless, additional components of the import pathway will be discovered. We know little about how the individual components of the import machinery function and how they interact with each other. Although mitochondrial targeting signals have been identified, we have limited understanding as to how such signals are recognized by the import machinery. We are thus still far from a molecular understanding of import. A paramount goal will be the *in vitro* reconstitution of parts of the process, and ultimately of entire import and sorting pathways. Such studies will link up with work on the folding and assembly of imported proteins to provide a detailed picture of mitochondrial protein import.

Acknowledgements
We thank the members of our laboratory for helpful suggestions and discussions. Special thanks to Margrit Jäggi and Trevor Lithgow for preparing Figure 1 (ammended from Glick and Schatz, 1991). We thank Michael Douglas and Michel Meijer for sharing unpublished results. Work in our laboratory is supported by grants from the Swiss National Science Foundation (3-26189.89), the Human Capital and Mobility Program of the European Community, and the Human Frontier Science Program Organisation. KH is the recipient of a long-term, post-doctoral fellowship from the European Molecular Biology Organisation.

References

Arretz, M., Schneider, H., Guiard, B., Brunner, M. and Neupert, W. (1994) Characterization of the mitochondrial processing peptidase of *Neurospora crassa. J. Biol. Chem.* 269: 4959–4967.

Atencio, D.P. and Yaffe, M.P. (1992) *MAS5*, a yeast homolog of DnaJ involved in mitochondrial protein import. *Mol. Cell. Biol.* 12: 283–291.

Baker, K.P., Schaniel, A., Vestweber, D. and Schatz, G. (1990) A yeast mitochondrial outer membrane protein essential for protein import and cell viability. *Nature* 348: 606–609.

Baker, K.P. and Schatz, G. (1991) Mitochondrial proteins essential for viability mediate protein import into yeast mitochondria. *Nature* 349: 205–208.

Beasley, E.M., Müller, S. and Schatz, G. (1993) The signal that sorts yeast cytochrome b_2 to the mitochondrial intermembrane space contains three distinct functional regions. *EMBO J.* 12: 2302–2311.

Behrens, M., Michaelis, G. and Pratje, E. (1991) Mitochondrial inner membrane protease I of *Saccharomyces cerevisiae* shows sequence homology to the *Escherichia coli* leader peptidase. *Mol. Gen. Genet.* 228: 167–176.

Blom, J., Kübrich, M., Rassow, J., Voos, W., Dekker, P.J.T., Maarse, A.C., Meijer, M. and Pfanner, N. (1993) The essential yeast protein MIM44 (encoded by *MPI1*) is involved in an early step of preprotein translocation across the mitochondrial inner membrane. *Mol. Cell. Biol.* 13: 7364–7371.

Böhni, P.C., Gasser, S., Leaver, C. and Schatz, G. (1980) A matrix-localized mitochondrial protease processing cytoplasmically-made precursors to mitochondrial proteins. *In:* A.M. Kroon and C. Saccone (eds): *The Organization and Expression of the Mitochondrial Genome.* Elsevier/North Holland, Amsterdam, pp. 423–433.

Bolliger, L., Deloche, O., Glick, B.S., Georgopoulos, C., Jenö, P., Krondiou, N., Horst, M., Morishima, N. and Schatz, G. (1994) A mitochondrial homolog of bacterial GrpE interacts with mitochondrial hsp70 and is essential for viability. *EMBO J.* 13: 1998–2006.

Brandt, U., Yu, L., C.-A. and Trumpower, B.L. (1993) The mitochondrial targeting presequence of the Rieske iron-sulfur protein is processed in a single step after insertion into the cytochrome bc_1 complex in mammals and retained as a subunit in the complex. *J. Biol. Chem.* 268: 8387–8390.

Caplan, A.J., Cyr, D.M. and Douglas, M.G. (1992) YDJ1p facilitates polypeptide translocation across different intracellular membranes by a conserved mechanism. *Cell* 71: 1143–1155.

Caplan, A.J., Cyr, D.M. and Douglas, M.G. (1993) Eukaryotic homologues of *Escherichia coli* dnaJ: a diverse protein family that functions with hsp70 stress proteins. *Mol. Biol. Cell* 4: 555–563.

Cheng, M.Y., Hartl, F.-U., Martin, J., Pollock, R.A., Kalousek, F. and Neupert, W. (1989) Mitochondrial heat-shock protein hsp60 is essential for assembly of proteins imported into yeast mitochondria. *Nature* 337: 620–625.

Dekker, P.J.T., Keil, P., Rassow, J., Maarse, A.C., Pfanner, N. and Meijer, M. (1993) Identification of MIM23, a putative component of the protein import machinery of the mitochondrial inner membrane. *FEBS Lett.* 330: 66–70.

Deshaies, R.J., Koch, B.D., Werner-Washburne, M., Craig, E.A. and Schekman, R. (1988) A subfamily of stress proteins facilitates translocation of secretory and mitochondrial precursor polypeptides. *Nature* 332: 800–805.

Dietmeier, K., Zara, V., Palmisano, A., Palmieri, F., Voos, W., Schlossmann, J., Moczko, M., Kispal, G. and Pfanner, N. (1993) Targeting and translocation of the phosphate carrier/p32 to the inner membrane of yeast mitochondria. *J. Biol. Chem.* 258: 25958–25964.

Dumont, M.E., Schlichter, J.B., Cardillo, T.S., Hayes, M.K., Bethlendy, G. and Sherman, F. (1993) *CYC2* encodes a factor involved in mitochondrial import of yeast cytochrome *c*. *Mol. Cell. Biol.* 13: 6442–6451.

Eilers, M. and Schatz, G. (1986) Binding of a specific ligand inhibits import of a purified precursor protein into mitochondria. *Nature* 322: 228–232.

Emtage, J.L.T. and Jensen, R.E. (1993) *MAS6* encodes an essential inner membrane component of the yeast mitochondrial protein import pathway. *J. Cell. Biol.* 122: 1003–1012.

Endo, T., Shimada, L., Roise, D. and Inagaki, F. (1989) N-terminal half of a mitochondrial presequence peptide takes a helical conformation when bound to dodecyl-phosphocholine micelles: a proton nuclear magnetic resonance study. *J. Biochem.* 106: 396–400.

Flaherty, K.M., McKay, D.B., Kabsch, W. and Holmes, K.C. (1991) Similarity of the 3-dimensional structures of actin and the ATPase fragment of a 70-kDa heat shock cognate protein. *proc. Natl. Acad. Sci. USA* 88: 5041–5045.

Fujiki, M. and Verner, K. (1991) Coupling of protein synthesis and mitochondrial import in a homologous yeast *in vitro* system. *J. Biol. Chem.* 266: 6841–6847.

Fujiki, M. and Verner, K. (1993) Coupling of cytosolic protein synthesis and mitochondrial protein import in yeast: evidence for cotranslational import *in vivo*. *J. Biol. Chem.* 268: 1914–1920.

Gambill, B.D., Voss, W., Kang, P.J., Miao, B., Langer, T., Craig, E.A. and Pfanner, N. (1993) A dual role for mitochondrial heat shock protein 70 in membrane translocation of preproteins. *J. Cell. Biol.* 123: 109–117.

Geli, V., Yang, M., Suda, K., Lustig, A. and Schatz, G. (1990) The *MAS*-encoded processing protease of yeast mitochondria – over production and characterization of its two nonidentical subunits. *J. Biol. Chem.* 265: 19216–19222.

Georgopolous, C. and Welch, W. (1993) Role of the major heat shock proteins as molecular chaperones. *Annu. Rev. Cell. Biol.* 9: 601–635.

Glaser, S.M., Miller, B.R. and Cumsky, M.G. (1990) Removal of a hydrophobic domain within the mature portion of a mitochondrial inner membrane protein causes its mislocalization to the matrix. *Mol. Cell. Biol* 10: 1873–1881.

Glick, B. and Schatz, G. (1991) Import of proteins into mitochondria. *Annu. Rev. Genet.* 25: 21–44.

Glick, B., Wachter, C. and Schatz, G. (1991) Protein import into mitochondria: two systems acting in tandem. *Trends Cell Biol.* 1: 99–103.

Glick, B.S., Beasley, E.M. and Schatz, G. (1992a) Protein sorting in mitochondria. *Trends Biochem. Sci.* 17: 453–459.

Glick, B.S., Brandt, A., Cunningham, K., Muller, S., Hallberg, R.L. and Schatz, G. (1992b) Cytochromes c_1 and b_2 are sorted to the intermembrane space of yeast mitochondria by a stop-transfer mechanism. *Cell* 69: 809–822.

Glick, B.S., Wachter, C., Reid, G.A. and Schatz, G. (1993) Import of cytochrome b_2 to the mitochondrial intermembrane space: The tightly folded heme-binding domain makes import dependent upon matrix ATP. *Protein Sci.* 2: 1901–1917.

Hachiya, N., Alam, R., Sakasegawa, Y., Sakaguchi, M., Mihara, K. and Omura, T. (1993) A mitochondrial import factor purified from rat liver cytosol is an ATP-dependent conformational modulator for precursor proteins. *EMBO J.* 12: 1579–1586.

Hackenbrock, C.R. (1968) Chemical and physical fixation of isolated mitochondria in low-energy and high-energy states. *Proc. Natl. Acad. Sci. USA* 61: 598–605.

Hallberg, E.M., Shu, Y. and Hallberg, R.L. (1993) Loss of mitochondrial hsp60 function: nonequivalent effects on matrix-targeted and intermembrane-targeted proteins. *Mol. Cell. Biol.* 13: 3050–3057.

Harkness, T.A.A., Nargang, F.E., van der Klei, I., Neupert, W. and Lill, R. (1994) A crucial role of the mitochondrial protein import receptor MOM19 for the biogenesis of mitochondria. *J. Cell. Biol.* 124: 637–648.

Hartl, F.-U., Ostermann, J., Guiard, B. and Neupert, W. (1987) Successive translocation into and out of the mitochondrial matrix: targeting of proteins to the intermembrane space by a bipartite signal peptide. *Cell* 51: 1027–1037.

Hartl, F.-U., Schmidt, B., Wachter, E., Weiss, H. and Neupert, W. (1986) Transport into mitochondria and intramitochondrial sorting of the Fe/S protein of ubiquinol-cytochrome *c* reductase. *Cell* 47: 939–951.

Hartl, F.U. and Neupert, W. (1990) Protein sorting to mitochondria – evolutionary conservation of folding and assembly. *Science* 247: 930–938.

Hartman, D.J., Hoogenraad, N.J., Condron, R. and Høj, P.B. (1992) Identification of a mammalian 10-kDa heat shock protein, a mitochondrial chaperonin 10 homologue essential for assisted folding of trimeric ornithine transcarbamoylase *in vitro*. *Proc. Natl. Acad. Sci. USA* 89: 3394–3398.

Hase, T., Mueller, U., Riezman, H. and Schatz, G. (1984) A 70-kd protein of the yeast mitochondrial outer membrane is targeted and anchored via its extreme amino terminus. *EMBO J.* 3: 3157–3164.

Hawlitschek, G., Schneider, H., Schmidt, B., Tropschug, M., Hartl, F.-U. and Neupert, W. (1988) Mitochondrial protein import: identification of processing peptidase and of PEP, a processing enhancing protein. *Cell* 53: 795–806.

Hines, V., Brandt, A., Griffiths, G., Horstmann, H., Brütsch, H. and Schatz, G. (1990) Protein import into yeast mitochondria is accelerated by the outer membrane protein Mas70. *EMBO J.* 9: 3191–3200.

Hines, V. and Schatz, G. (1993) Precursor binding to yeast mitochondria: a general role for the outer membrane protein Mas70p. *J. Biol. Chem.* 268: 449–454.

Horst, M., Jenö, P., Kronidou, N.G., Bolliger, L., Oppliger, W., Scherer, P., Manning-Krieg, U., Jascur, T. and Schatz, G. (1993) Protein import into yeast mitochondria: the inner membrane import site protein ISP45 is the *MPI1* gene product. *EMBO J.* 12: 3035–3041.

Hurt, E.C., Pesold-Hurt, B. and Schatz, G. (1984) The cleavable prepiece of an imported mitochondrial protein is sufficient to direct cytosolic dihydrofolate reductase into the mitochondrial matrix. *FEBS Lett.* 178: 306–310.

Hwang, S., Jascur, T., Vestweber, D., Pon, L. and Schatz, G. (1989) Disrupted yeast mitochondria can import precursor proteins directly through their inner membrane. *J. Cell. Biol.* 109: 487–493.

Hwang, S.T., Wachter, C. and Schatz, G. (1991) Protein import into the yeast mitochondrial matrix – a new translocation intermediate between the two mitochondrial membranes. *J. Biol. Chem.* 266: 21083–21089.

Ikeda, E., Yoshida, S., Mitsuzawa, H., Uno, I. and Toh-e, A. (1994) *YGE1* is a yeast homologue of *Escherichia coli grpE* and is required for maintenance of mitochondrial functions. *FEBS Lett.* 339: 265–268.

Jascur, T., Goldenberg, D.P., Vestweber, D. and Schatz, G. (1992) Sequential translocation of an artificial precursor protein across the two mitochondrial membranes. *J. Biol. Chem.* 267: 13636–13641.

Jensen, R.E., Schmidt, S. and Mark, R.J. (1992) Mutations in a 19-amino-acid hydrophobic region of the yeast cytochrome c_1 presequence prevent sorting to the mitochondrial intermembrane space. *Mol. Cell. Biol.* 12: 4677–4686.

Kalousek, F., Isaya, G. and Rosenberg, L.E. (1992) Rat liver mitochondrial intermediate peptidase (MIP): purification and initial characterization. *EMBO J.* 11: 2803–2809.

Kang, P.J., Ostermann, J., Shilling, J., Neupert, W., Craig, E.A. and Pfanner, N. (1990) Requirement for hsp70 in the mitochondrial matrix for translocation and folding of precursor proteins. *Nature* 348: 137–143.

Kassenbrock, C.K., Cao, W. and Douglas, M.G. (1993) Genetic and biochemical characterization of ISP6; a small mitochondrial outer membrane protein associated with the protein translocation complex. *EMBO J.* 12: 3023–3034.

Kellems, R.E., Allison, V.F. and Butow, R.A. (1975) Cytoplasmic type 80S ribosomes associated with yeast mitochondria IV: attachment of ribosomes to the outer membrane of isolated mitochondria. *J. Cell. Biol.* 65: 1–14.

Kiebler, M., Keil, P., Schneider, H., van der Klei, I.J., Pfanner, N. and Neupert, W. (1993) The mitochondrial receptor complex: a central role of MOM22 in mediating preprotein transfer from receptors to the general insertion pore. *Cell* 74: 483–492.

Kiebler, M., Pfaller, R., Söllner, T., Griffiths, G., Horstmann, H., Pfanner, N. and Neupert, W. (1990) Identification of a mitochondrial receptor complex required for recognition and membrane insertion of precursor proteins. *Nature* 348: 610–616.

Koll, H., Guiard, B., Rassow, J., Ostermann, J., Horwich, A.L., Neupert, W. and Hartl, F.-U. (1992) Antifolding activity of hsp60 couples protein import into the mitochondrial matrix with export to the intermembrane space. *Cell* 68: 1163–1175.

Li, J.-M. and Shore, G.C. (1992a) Reversal of the orientation of an integral protein of the mitochondrial outer membrane. *Science* 256: 1815–1817.

Li, J.-M. and Shore, G.C. (1992b) Protein sorting between mitochondrial outer and inner membranes. Insertion of an outer membrane protein into the inner membrane. *Biochim. Biophys. Acta* 1106: 233–241.

Lill, R., Stuart, R.A., Drygas, M.E., Nargang, F.E. and Neupert, W. (1992) Import of cytochrome *c* heme lyase into mitochondria – a novel pathway into the intermembrane space. *EMBO J.* 11: 449–456.

Lithgow, T., Glick, B.S. and Schatz, G. (1994) The protein import receptor of mitochondria. *Trends Biochem. Sci.* 20: 98–101.

Lithgow, T., Junne, T., Wachter, C. and Schatz, G. (1994a) Yeast mitochondria lacking the two import receptors Mas20p and Mas70p can efficiently and specifically import precursor proteins. *J. Biol. Chem.* 269: 15325–15330.

Lithgow, T., Ryan M., Anderson, R.L., Høj, P.B. and Hoogenraad, N.J. (1993) A constitutive form of heat-shock protein 70 is located in the outer membranes of mitochondria from rat liver. *FEBS Lett.* 332: 277–281.

Lithgow, T., van Driel, R., Bertram, J.F. and Strasser, A. (1994b) The protein product of the oncogene *bcl-2* is a component of the nuclear envelope, the endoplasmic reticulum and the outer mitochondrial membrane. *Cell Growth & Diff.* 5: 411–417.

Maarse, A.C., Blom, J., Grivell, L.A. and Meijer, M. (1992) *MPI1*, an essential gene encoding a mitochondrial membrane protein, is possibly involved in protein import into yeast mitochondria. *EMBO J.* 11: 3619–3628.

Manning-Krieg, U.C., Scherer, P.E. and Schatz, G. (1991) Sequential action of mitochondrial chaperones in protein import into the matrix. *EMBO J.* 10: 3273–3280.

Martin, J., Mahlke, K. and Pfanner, N. (1991) Role of an energized inner membrane in mitochondrial protein import. *J. Biol. Chem.* 266: 18051–18057.

Mayer, A., Lill, R. and Neupert, W. (1993) Translocation and insertion of precursor proteins into isolated outer membranes of mitochondria. *J. Cell. Biol.* 121: 1233–1243.

McAda, P. and Douglas, M.G. (1982) A neutral metallo-endoprotease involved in the processing of an F_1-ATPase subunit precursor in mitochondria. *J. Biol. Chem.* 257: 3177–3182.

Millar, D.G. and Shore, G.C. (1993) The signal anchor sequence of mitochondrial Mas70p contains an oligomerization domain. *J. Biol. Chem.* 268: 18403–18406.

Miller, B.R. and Cumsky, M.G. (1991) An unusual mitochondrial import pathway for the precursor to yeast cytochrome *c* oxidase subunit Va. *J. Cell. Biol.* 112: 833–841.

Miller, B.R. and Cumsky, M.G. (1993) Intramitochondrial sorting of the precursor to yeast cytochrome *c* oxidase subunit Va. *J. Cell. Biol.* 121: 1021–1029.

Miller, D.M., Delgado, R., Chirgwin, J.M., Hardies, S.C. and Horowitz, P.M. (1991) Expression of cloned bovine adrenal rhodanese. *J. Biol. Chem.* 266: 4686–4691.

Mitoma, J. and Ito, A. (1992) Mitochondrial targeting signal of rat liver monoamine oxidase B is located at its carboxy terminus. *J. Biochem.* 111: 20–24.

Moczko, M., Dietmeier, K., Söllner, T., Segui, B., Steger, H.F., Neupert, W. and Pfanner, N. (1992) Identification of the mitochondrial receptor complex in *Saccharomyces cerevisiae*. *FEBS Lett.* 310: 265–268.

Moczko, M., Ehmann, B., Gärtner, F., Hönlinger, A., Schäfer, E. and Pfanner, N. (1994) Deletion of the receptor MOM19 strongly impairs import of cleavable preproteins into *Saccharomyces cerevisiae* mitochondria. *J. Biol. Chem.* 269: 9045–9051.

Murakami, H., Blobel, G. and Pain, D. (1990) Isolation and characterization of the gene for a yeast mitochondrial import receptor. *Nature* 347: 488–491.

Murakami, H., Pain, D. and Blobel, G. (1988) 70-kD heat shock-related protein is one of at least two distinct cytosolic factors stimulating protein import into mitochondria. *J. Cell. Biol.* 107: 2051–2057.

Murakami, K. and Mori, M. (1990) Purified presequence binding factor (PBF) forms an import-competent complex with a purified mitochondrial precursor protein. *EMBO J.* 9: 3201–3208.

Murakami, K., Tannase, S., Morino, Y. and Mori, M. (1992) Presequence binding factor-dependent and -independent import of proteins into mitochondria. *J. Biol. Chem.* 267: 13119–13122.

Nakai, M., Endo, T., Hase, T. and Matsubara, H. (1993) Isolation and characterization of the yeast *MSP1* gene which belongs to a novel family of putative ATPases. *J. Biol. Chem.* 268: 24262–24269.

Nakai, M., Hase, T. and Matsubara, H. (1989) Precise determination of the mitochondrial import signal contained in a 70 kDa protein of yeast mitochondrial outer membrane. *J. Biochem.* 105: 513–519.

Nelson, R.J., Zieglehoffer, T., Nicolet, C., Werner-Washburne, M. and Craig, E.A. (1992) The translocation machinery and 70 kDa heat shock protein cooperate in protein synthesis. *Cell* 71: 97–105.

Neupert, W., Hartl, F.-U., Craig, E.A. and Pfanner, N. (1990) How do polypeptides cross the mitochondrial membranes? *Cell* 63: 447–450.

Nguyen, M., Millar, D.G., Yong, W., Korsmeyer, S.J. and Shore, G.C. (1993) Targeting of Bcl-2 to the mitochondrial outer membrane by a COOH-terminal signal anchor sequence. *J. Biol. Chem.* 268: 26265–265268.

Nicholson, D.W., Hergersberg, C. and Neupert, W. (1988) Role of cytochrome *c* heme lyase in the import of cytochrome *c* into mitochondria. *J. Biol. Chem.* 263: 19034–19042.

Nunnari, J., Fox, T.D. and Walter, P. (1993) A mitochondrial protease with two catalytic subunits of nonoverlapping specificities. *Science* 262: 1997–2004.

Nye, S.H. and Scarpulla, R.C. (1990a) *In vivo* expression and mitochondrial targeting of yeast apoiso-1-cytochrome *c* fusion proteins. *Mol. Cell. Biol.* 10: 5753–5762.

Nye, S.H. and Scarpulla, R.C. (1990b) Mitochondrial targeting of yeast apoiso-1-cytochrome *c* is mediated through functionally independent structural domains. *Mol. Cell. Biol.* 10: 5763–5771.

Ohba, M. and Schatz, G. (1987) Disruption of the outer membrane restores protein import to trypsin treated yeast mitochondria. *EMBO J.* 6: 2117–2122.

Ostermann, J., Voos, W., Kang, P.J., Craig, E.A. and Neupert, W. (1990) Precursor proteins in transit through mitochondrial contact sites interact with hsp70 in the matrix. *FEBS Lett.* 277: 281–284.

Ou, W.-J., Ito, A., Okazaki, H. and Omura, T. (1989) Purification and characterization of a processing protease from rat liver mitochondria. *EMBO J.* 8: 2605–2612.

Pain, D., Murakami, H. and Blobel, G. (1990) Identification of a receptor for protein import into mitochondria. *Nature* 347: 444–449.

Pfanner, N., Hartl, F.-U., Guiard, B and Neupert, W. (1987) Mitochondrial precursor proteins are imported through a hydrophilic membrane environment. *Eur. J. Biochem.* 169: 289–293.

Pfanner, N., Rassow, J., van der Klei, I.J. and Neupert, W. (1992) A dynamic model of the mitochondrial protein import machinery. *Cell* 68: 999–1002.

Pon, L., Moll, T., Vestweber, D., Marshallsay, B. and Schatz, G. (1989) Protein import into mitochondria: ATP-dependent protein translocation activity in a submitochondrial fraction enriched in membrane contact sites and specific proteins. *J. Cell. Biol.* 109: 2603–2616.

Ramage, L., Junne, T., Hahne, K., Lithgow, T. and Schatz, G. (1993) Functional cooperation of mitochondrial protein import receptors in yeast. *EMBO J.* 12: 4115–4123.

Randall, L.L. and Hardy, S.J.S. (1986) Correlation of competence for export with lack of tertiary structure of the mature species: a study *in vivo* of maltose-binding protein in *E. coli.* *Cell* 46: 921–928.

Rassow, J., Hartl, F.U., Guiard, B., Pfanner, N. and Neupert, W. (1990) Polypeptides traverse the mitochondrial envelope in an extended state. *FEBS Lett.* 275: 190–194.

Rassow, J. and Pfanner, N. (1991) Mitochondrial preproteins en route from the outer membrane to the inner membrane are exposed to the intermembrane space. *FEBS Lett.* 293: 85–88.

Reid, G.A. and Schatz, G. (1982) Import of proteins into mitochondria: extramitochondrial pools and post-translational import of mitochondrial protein precursors *in vivo. J. Biol. Chem.* 257: 13062–13067.

Riezman, H., Hay, R., Witte, C., Nelson, N. and Schatz, G. (1983) Yeast mitochondrial outer membrane specifically binds cytoplasmically-synthesized precursors of mitochondrial proteins. *EMBO J.* 2: 1113–1118.

Roise, D. (1993) The amphipathic helix in mitochondrial targeting sequences. *In:* R.M. Epand (ed): *The Amphipathic Helix.* CRC Press, Boca Raton, pp. 257–279.

Roise, D., Horvath, S.J., Tomich, J.M., Richards, J.H. and Schatz, G. (1986) A chemically synthesized mitochondrial presequence of an imported protein can form an amphiphilic helix and perturb natural and artificial phospholipid bilayers. *EMBO J.* 5: 1327–1334.

Roise, D. and Schatz, G. (1988) Mitochondrial presequences. *J. Biol. Chem.* 263: 4509–45011.

Rospert, S. (1994) Energies of mitochondrial protein import and intramitochondrial protein sorting. *Adv. Mol. Cell Biol; in press.*

Rospert, S., Junne, T., Glick, B.S. and Schatz, G. (1993) Cloning and disruption of the gene encoding yeast mitochondrial chaperonin 10, the homolog of *E. coli* groES. *FEBS Lett.* 335: 358–360.

Rospert, S., Müller, S., Schatz, G. and Glick, B.S. (1994) Fusion proteins containing the cytochrome b_2 presequence are sorted to the mitochondrial intermembrane space independently of hsp60. *J. Biol. Chem.* 269: 17279–17288.

Rowley, N., Prip-Buus, C., Westermann, B., Brown, C., Schwarz, E., Barrell, B. and Neupert, W. (1994) Mdj1p, a novel chaperone of the DnaJ family, is involved in mitochondrial biogenesis and protein folding. *Cell.* 77: 249–259.

Schatz, G. (1993) The protein import machinery of mitochondria. *Protein Sci.* 2: 141–146.

Schatz, G. and Butow, R.A. (1983) How are proteins imported into mitochondria? *Cell* 32: 316–318.

Scherer, P.E., Krieg, U.C., Hwang, S.T., Vestweber, D. and Schatz, G. (1990) A precursor protein partly translocated into yeast mitochondria is bound to a 70 kd mitochondrial stress protein. *EMBO J.* 9: 4315–4322.

Scherer, P.E., Manning-Krieg, U.C., Jenö, P., Schatz, G. and Horst, M. (1992) Identification of a 45-kDa protein at the protein import site of the yeast mitochondrial inner membrane. *Proc. Natl. Acad. Sci. USA* 89: 11930–11934.

Schleyer, M. and Neupert, W. (1985) Transport of proteins into mitochondria: translocational intermediates spanning contact sites between outer and inner membranes. *Cell* 43: 339–350.

Schneider, A., Behrens, M., Scherer, P., Pratje, E., Michaelis, G. and Schatz, G. (1991) Inner membrane protease I, an enzyme mediating intramitochondrial protein sorting in yeast. *EMBO J.* 10: 247–254.

Schneider, A., Oppliger, W. and Jenö, P. (1994) Purified inner membrane protease I of yeast mitochondria is a heterodimer. *J. Biol. Chem.* 269: 8635–8638.

Schwaiger, M., Herzog, V. and Neupert, W. (1987) Characterization of translocation contact sites involved in the import of mitochondrial proteins. *J. Cell. Biol.* 105: 235–246.

Schwarz, E., Seytter, T., Guiard, B. and Neupert, W. (1993) Targeting of cytochrome b_2 into the mitochondrial intermembrane space: specific recognition of the sorting signal. *EMBO J.* 12: 2295–2302.

Segui-Real, B., Kispal, G., Lill, R. and Neupert, W. (1993) Functional independence of the protein translocation machineries in mitochondrial outer and inner membranes: passage of preproteins through the intermembrane space. *EMBO J.* 12: 2211–2218.

Segui-Real, B., Stuart, R.A. and Neupert, W. (1992) Transport of proteins into the various subcompartments of mitochondria. *FEBS Lett.* 313: 2–7.

Shore, G.C., Millar, D.G. and Li, J.-M. (1992) Protein insertion into mitochondrial outer and inner membranes via the stop-transfer sorting pathway. *In:* W. Neupert and R. Lill (eds): *Membrane biogenesis and Protein Targeting.* Elsevier Science Publishers B.V., Amsterdam, pp. 253–262.

Simon, S.M., Peskin, S.P. and Oster, G.F. (1992) What drives the translocation of proteins? *Proc. Natl. Acad. Sci. USA* 89: 3770–3774.

Smagula, C. and Douglas, M.G. (1988) Mitochondrial import of the ADP/ATP carrier protein in *S. cerevisiae.* Sequences required for receptor binding and membrane translocation. *J. Biol. Chem.* 263: 6783–6790.

Söllner, T., Griffiths, G., Pfaller, R., Pfanner, N. and Neupert, W. (1989) MOM19, an import receptor for mitochondrial precursor proteins. *Cell* 59: 1061–1070.

Söllner, T., Pfaller, R., Griffiths, G., Pfanner, N. and Neupert, W. (1990) A mitochondrial import receptor for the ADP/ATP carrier. *Cell* 62: 107–115.

Söllner, T., Rassow, J., Wiedmann, M., Schlossmann, J., Keil, P., Neupert, W. and Pfanner, N. (1992) Mapping of the protein import machinery in the mitochondrial outer membrane by crosslinking of translocation intermediates. *Nature* 355: 84–87.

Steger, H.F., Söllner, T., Kiebler, M., Dietmeier, K.A., Pfaller, R., Trülzsch, K.S., Tropschug, M., Neupert, W. and Pfanner, N. (1990) Import of ADP/ATP carrier into mitochondria: two receptors act in parallel. *J. Cell. Biol.* 111: 2353–2363.

Stuart, R.A., Cyr, D.M., Craig, E.A. and Neupert, W. (1994a) Mitochondrial molecular chaperones: their role in protein translocation. *Trends Biochem. Sci.* 19: 87–92.

Stuart, R.A., Gruhler, A., van der Klei, I., Guiard, B., Koll, H. and Neupert, W. (1994b) The requirement of matrix ATP for the import of precursor proteins into the mitochondrial matrix and intermembrane space. *Eur. J. Biochem.* 220: 9–18.

Stuart, R.A. and Neupert, W. (1990) Apocytochrome *c* – an exceptional mitochondrial precursor protein usig an exceptional import pathway. *Biochimie* 72: 115–121.

van Loon, A.P.G.M., Brändli, A.W., Pesold-Hurt, B., Blank, D. and Schatz, G. (1987) Transport of proteins to the mitochondrial intermembrane space: the 'matrix-targeting' and the 'sorting' domains in the cytochrome c_1 presequence. *EMBO J.* 6: 2433–2439.

van Loon, A.P.G.M., Brändli, A.W. and Schatz, G. (1986) The presequences of two imported mitochondrial proteins contain information for intracellular and intramitochondrial sorting. *Cell* 44: 801–812.

van Loon, A.P.G.M. and Schatz, G. (1987) Transport of proteins to the mitochondrial intermembrane space: the 'sorting' domain of the cytochrome c_1 presequence is a stop-transfer sequence specific for the mitochondrial inner membrane. *EMBO J.* 6: 2441–2448.

Verner, K. (1993) Co-translational import into mitochondria: an alternative view. *Trends Biochem. Sci.* 18: 366–371.

Vestweber, D., Brunner, J., Baker, A. and Schatz, G. (1989) A 42K outer-membrane protein is a component of the yeast mitochondrial protein import site. *Nature* 341: 205–209.

Vestweber, D. and Schatz, G. (1988a) Point mutations destabilizing a precursor protein enhance its post-translational import into mitochondria. *EMBO J.* 7: 1147–1151.

Vestweber, D. and Schatz, G. (1988b) A chimeric mitochondrial precursor protein with internal disulfide bridges blocks import of authentic precursors into mitochondria and allows quantitation of import sites. *J. Biol. Chem.* 107: 2037–2043.

Vestweber, D. and Schatz, G. (1988c) Mitochondria can import artificial precursor proteins containing a branched polypeptide chain or a carboxy-terminal stilbene disulfonate. *J. Cell. Biol.* 107: 2045–2049.

Vestweber, D. and Schatz, G. (1989) DNA-protein conjugates can enter mitochondria via the protein import pathway. *Nature* 338: 170–172.

von Heijne, G. (1986) Mitochondrial targeting sequences may form amphiphilic helices. *EMBO J.* 5: 1335–1342.

von Heijne, G. (1992) Membrane protein structure prediction: hydrophobicity analysis and the positive-inside rule. *J. Mol. Biol.* 225: 487–494.

Voos, W., Gambill, B.D., Guiard, B., Pfanner, N. and Craig, E.A. (1993) Presequence and mature part of preproteins strongly influence the dependence of mitochondrial protein import on heat shock protein 70 in the matrix. *J. Cell. Biol.* 123: 119–126.

Wachter, C., Schatz, G. and Glick, B. (1992) Role of ATP in the intramitochondrial sorting of cytochrome c_1 and the adenine nucleotide translocator. *EMBO J.* 11: 4787–4794.

Wachter, C., Schatz, G. and Glick, B. (1994) Protein import into mitochondria: the requirement for external ATP is precursor-specific whereas intramitochondrial ATP is universally needed for translocation into the matrix. *Mol. Biol. Cell* 5: 465–474.

Walian, P.J. and Jap, B.K. (1990) Three-dimensional electron diffraction of PhoE porin to 2.8 Å resolution. *J. Mol. Biol.* 215: 429–438.

Weiss, M.S., Kreusch, A., Schiltz, E., Nestel, U., Welte, W., Weckesser, J. and Schulz, G.E. (1991) The structure of porin from *Rhodobacter capsulatus* at 1.8 Å resolution. *FEBS Lett.* 280: 379–382.

Yang, M., Geli, V., Oppliger, W., Suda, K., James, P. and Schatz, G. (1991) The *MAS*-encoded processing protease of yeast mitochondria – interaction of the purified enzyme with signal peptides and a purified precursor protein. *J. Biol. Chem.* 266: 6416–6423.

Yang, M., Jensen, R.E., Yaffe, M.P., Oppliger, W. and Schatz, G. (1988) Import of proteins into yeast mitochondria: the purified matrix processing protease contains two subunits which are encoded by the nuclear *MAS1* and *MAS2* genes. *EMBO J.* 7: 3857–3862.

Zwizinski, C., Schleyer, M. and Neupert, W. (1984) Proteinaceous receptors for the import of mitochondrial precursor proteins. *J. Biol. Chem.* 259: 7850–7856.

Biochemistry of Cell Membranes
ed. by S. Papa & J. M. Tager
© 1995 Birkhäuser Verlag Basel/Switzerland

Translocation of bacterial protein toxins across membranes

C. Montecucco, G. Schiavo, E. Papini, O. Rossetto, M. De Bernard,
F. Tonello, G.N. Moll and P.E. Washbourne

*Centro CNR Biomembrane and Dipartimento di Scienze Biomediche, Università di Padova,
Via Trieste, 75, I-35121 Padova, Italy*

Summary. Many bacterial protein toxins act inside cells by modifying a variety of cytosolic
targets. To intoxicate cells, these toxins perform a four-step process which consists of: (1)
binding, (2) internalization, (3) membrane translocation, and (4) target modification. All of
them form ion channels across planar lipid bilayers and plasma membrane of cells. A relation
between ion channel and membrane translocation may be inferred and two different models
have been put forward to account for these phenomena. The two models are discussed on the
basis of the available experimental evidence and in terms of the main points of difference to
be tested in future investigations.

Introduction

Many bacteria which are pathogenic to humans and other animals
produce protein toxins that are important or essential virulent factors in
pathogenesis (Alouf and Freer, 1991; Menestrina et al., 1994). These
protein toxins can be divided into two groups: (a) toxins acting on
protein or lipid components of the plasma membrane; (b) toxins acting
inside cells by modifying a specific cytosolic target. Some of the toxins
belonging to this second group, such as diphtheria toxin (DT), exotoxin
A of *Pseudomonas aeruginosa* (ETA), anthrax toxic complex of *Bacillus
anthracis* (protective antigen (PA), edema factor (EF), and lethal
factor (LF)), tetanus neurotoxin (TeNT), and botulinum neurotoxins
(BoNT, seven different serotypes termed BoNT/A, /B, /C, /D, /E, /F,
and /G), are the sole factor responsible for the corresponding disease.
Hence, understanding the mechanism of cell intoxication leads to
knowledge of the molecular pathogenesis of the disease as well as to the
discovery of new aspects of cell physiology. Toxins acting inside cells
are composed of two protomers generally held together by a single
disulfide bond: protomer A has a catalytic activity which is expressed
inside the cell, while protomer B is responsible for cell binding and
translocation into the cell (Menestrina et al., 1994) (Fig. 1). The
integrity of this structural organization is required for cell intoxication.
The reduced toxin or the isolated protomers are non toxic when added

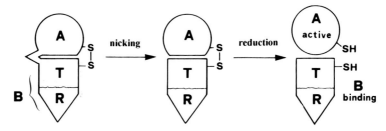

Figure 1. Schematic structure of bacterial protein toxins with intracellular targets. These toxins consist of an active protomer A, endowed with catalytic activity, linked via a disulfide bond to a B protomer, responsible for cell binding and penetration. The toxin is synthesized as a single polypeptide chain and cleaved at a single site by proteases. Reduction is required to free the enzymic activity of A. A group of toxins including diphtheria toxin, exotoxin A, tetanus and botulinum neurotoxins and the anthrax toxic complex is organized in three domains: A is the enzymic part, R is responsible for cell binding, and T is involved in the membrane translocation of A.

to a cell and toxicity can be recovered by reforming the disulfide link. Recent crystallographic studies have revealed the structure of several of these toxins (Allured et all., 1986; Sixma et al., 1991, 1993; Choe et al.,1992; Stein et al., 1992, 1994). These toxins show two types of structural organizations. Toxins acting on body surfaces which are rapidly and efficiently cleared by appropriate mechanisms, such as the respiratory and gastrointestinal tracts, have a pentameric B oligomer with multiple binding sites for oligosaccharides or glycolipids or glycoproteins (Menestrina et al., 1994). In contrast, bacterial protein toxins released in the tissue fluids, which include DT, ETA, TeNT, BoNT and the anthrax complex, are composed of three domains (Allured et al., 1986; Robinson et al., 1988; Leppla, 1991; Choe et al., 1992; Montecucco and Schiavo; 1993, 1994; Menestrina et al., 1994). The B promoter comprises two domains: R, responsible for binding to a cell surface receptor and T, involved in membrane translocation; domain A is the enzymic moiety active in the cytosol.

The process of cell intoxication can be dissected into four different steps, as depicted in Figure 2: (1) binding, (2) internalization, (3) membrane translocation, and (4) target modification. Here, we will discuss these passages with emphasis on the membrane translocation step.

Overview

Cell binding

The cell surface receptors of DT, ETA and BoNT/B have been recently identified. DT binds to a heparin-binding EGF-like growth factor

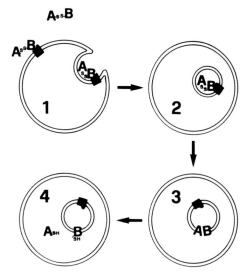

Figure 2. Schematic picture of the four steps of the mechanism of cell intoxication with A-B type bacterial protein toxins. This process can be sub-divided into four different steps with location of the toxin in the cell. (1) Cell binding occurs via interaction of protomer B with a lipid and/or protein component of the cell surface; (2) binding is followed by internalization via endocytosis and intracellular routing; (3) to reach the cytosol, the catalytic A subunit of the toxin has to cross the vesicle membrane, a process assisted by the T domain; (4) the enzymatic A subunit performs its catalytic activity toward a specific cytosolic target with ensuing cell intoxication.

precursor (Naglich et al., 1992). ETA binds the α_2-macroglobulin receptor (Kounnas et al., 1992), while BoNT/B binds to synaptotagmin (Nishiki et al., 1994). The receptors of DT and ETA are integral proteins of the plasma membrane that internalize as toxin-receptor complexes inside coated vesicles, which eventually merge into early endosomes. In contrast synaptotagmin is a membrane protein of the synaptic vesicles with cytosolic, transmembrane and lumenal segments (Perin et al., 1991; Perin, 1994). The lumenal segment is exposed to the cell surface only upon synaptic vesicle fusion with the plasma membrane (Matteoli et al., 1992). Hence, BoNT/B can bind to the cell only during the transient period between vesicle fusion and reuptake. It is likely that all BoNTs bind to the lumenal domain of synaptic vesicle membrane proteins. Nishiki et al. (1994) have also shown that BoNT/B binding to synaptotagmin is dependent on polysialogangliosides. The receptors of the other clostridial neurotoxins is not known, but there is evidence that a protein receptor is also involved in the binding of TeNT (Yavin and Nathan, 1986; Schiavo et al., 1991b). The interaction of BoNT and TeNT with polysialogangliosides has been extensively studied after the proposal of van Heyningen (1968) that they are the cellular receptors of

TeNT as they are of cholera toxin (CLT) (Mellanby and Green, 1981). On the basis of available data and of some theoretical considerations, it was proposed that these neurotoxins bind to a "double lipid-and-protein receptor" composed of gangliosides and a protein (Montecucco, 1986). Nishiki et al. (1994) have provided the first experimental evidence in favor of the existence of such a receptor. There is evidence that this situation may hold true also for DT (Olsnes et al., 1985; Papini et al., 1987a). A consequence of such a mode of binding is that the T domain of the toxin is placed in direct contact with the lipid bilayer surface, ready to perform the next step in the intoxication sequence.

Internalization

After binding, the toxin-receptor complex is internalized by the cell inside vesicles (Fig. 2). This stage is very relevant in the therapy with anti-toxin specific antibodies, because, after internalization, the toxin is no longer neutralized. The different temperature dependence (binding can occur in the cold and internalization only above 10°C), and accessibility to external ligands differentiates binding from internalization.

Endocytosis may take place via coated vesicles, as in the case for DT, or via non coated vesicles, as found for CLT and TeNT (Moya et al., 1985; Montesano et al., 1982; Sandvig and Olsnes, 1991; London, 1992). Only partial information is available on the intracellular routing of these toxins. Morphological studies indicate that some toxins undergo a retroaxonal transport to the TGN, Golgi, and ER (Sandvig et al., 1992). The presence in ETA of a sequence similar to the ER-retention sequence may be involved in the recycling of this toxin inside ER (Pelham, 1991; Seetharam et al., 1991).

DT is the most extensively studied toxin. After a few minutes upon warming of a cell, DT bound to the cell surface is internalized inside early endosomes, where about one-third of DT adopts a state competent to membrane translocation and release of A in the cytosol (Papini et al., 1993). The remaining two-thirds are unable to perform such a step (discussed in the next section) and are degraded inside late endosomes and lysosomes in a process blocked by inhibitors of lumenal acidification such as monensin or bafilomycin A1 (Uchida et al., 1990; Papini et al., 1993). From intoxicated cells, Beaumelle et al. (1992) have isolated endosomal fractions containing DT and have shown that the A subunit can be induced to translocate by lowering the intralumenal pH, thus providing a clear-cut demonstration that internalization is distinct from translocation. There is indirect evidence that other three-domain toxins such as TeNT and BoNT and the anthrax edema (EF) and lethal (LF) factors enter the cell cytosol via an acidic intracellular compartments since cell intoxication is prevented by agents that quench intracellular proton gradients (Simpson, 1982, 1989; Friedlander, 1986; Leppla, 1991; Neale and Williamson, 1994).

Membrane translocation

The ion channel of toxins in planar lipid bilayers
These toxins at low pH form voltage-dependent ion channels across planar lipid bilayers (Donovan et al., 1981, 1982; Kagan et al., 1981; Deelers et al., 1993; Hoch et al., 1985; Donovan and Middlebrook, 1986; Gambale and Montal, 1988; Blaustein et al., 1989) with the following main properties:

1. These channels have low conductance at physiological ionic strength.
2. These toxins induce an increase of membrane conductance only at low pH.
3. The channel is cation selective at physiological pH values.
4. The conductance of planar lipid bilayers containing acidic phospho-lipids in the presence of DT at low pH is higher than those made only of zwitterionic lipids.
5. The rate of conductance increase is strongly dependent on the existence of a positive potential.

Similar conclusions were reached by analysis of the change of permeability induced by toxins on potassium-loaded liposomes (Boquet and Duflot, 1982; Shiver and Donovan, 1987; Shone et al., 1987). These studies provided the first indication that these toxins undergo a pH-driven structural change to a lipid-interacting acid form and gave the first experimental support for the *tunnel model* for the translocation of bacterial protein toxins across membranes proposed by Boquet et al. (1976) on the basis of experiments of detergent binding to DT. The main features of the tunnel model are schematically summarized in Figure 3. Fragment B was suggested to undergo a conformational change from a neutral form to an acidic form, characterized by the presence of hydrophobic surfaces which mediate detergent and lipid binding. The acid form of fragment B is thus able to insert into lipid bilayers forming a transmembrane hydrophilic pore large enough to allow the passage of the unfolded A chain, which shields A from the intersection with lipids. Ions are supposed to flow through the same channel used by the unfolded A chain: the formation of the channel is a prerequisite for the membrane translocation of A. On the basis of this model, one may expect that a single pore incorporated in a planar lipid bilayer is able to mediate the passage of more than one A chain. In the tunnel model the A and B subunits of DT are seen as two independent domains with respect to the conformational change triggered by pH: protomer B is suggested to become hydrophobic and penetrate the lipid bilayer, while A unfolds and inserts into the channel.

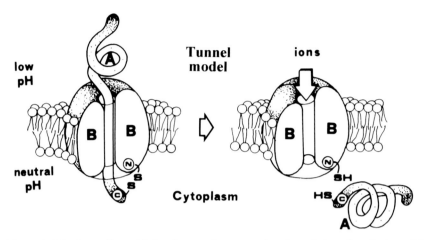

Figure 3. The "tunnel" model for the membrane translocation of three-domain bacterial toxins. At acid pH, protomer B becomes hydrophobic, inserts in the lipid bilayer and forms a hydrophilic pore or tunnel. Domain A unfolds and crosses the membrane inside the tunnel protected from contact with lipids (left panel). After the passage of the A chain, the protomer B tunnel is an ion channel (right panel).

A controversial aspect of toxin channels is their size. Though there is no direct relation between the size of a channel and its conductance, values of the order of a few picoSiemens do not fit with the dimensions expected for a protein channel that has to accomodate a polypeptide chain with its lateral chains of different volume, charge and hydrophilicity. Also, the properties of the first protein-conducting channel characterized in planar lipid bilayers lend support to such a contention (Simon and Blobel, 1991, 1992). These authors have shown that the rough endoplasmic reticulum and the inner membrane of *Escherichia coli* contain a channel which is closed when plugged by the nascent polypeptide chain and it is opened by puromycin, a drug that induces the release of the peptidyl-puromycin complex. The channel shows a large conductance (220 picoSiemens) with no ion selectivity. A change in size of polarity of the applied voltage does not influence either the conductance or the gating of these channels. These properties are those that one would predict for a pore that has to accommodate amino acids with lateral chains differing in size, charge, and polarity.

DT, PA, and TeNT form channels also in the plasma membrane of cells (Sandvig and Olsnes, 1988; Papini et al., 1988; Milne and Collier, 1993; Beise et al., 1994) with the following main properties:

a. The channel has a conductance of about 40 pS.
b. Their pH dependence is almost superimposable to that of cellular intoxication.

Table 1. Comparison of the properties of toxin ion channels and protein conducting channels of endoplasmic reticulum and bacterial membrane

Property	Toxin channels	Protein-conducting channel
Conductance	5–43 pS	220 pS
Ion selectivity	X^+	non selective
Size of permeant ion	<glucosamine	>gluconate
pH dependence	4–6	n.d.
Voltage gating	+	−
Voltage dependence of conductance	+	−

For references see text.

c. The channel is specific for monovalent cations, does not leak out amino acids or phosphorylated metabolites, and does not allow the uptake of glucoasmine.
d. The membrane remains permeable to monovalent cations as long as a transmembrane pH gradient is maintained; if the extracellular medium is neutralized, the cell quickly recovers its normal K^+ and Na^+ content.

Table 1 compares the main properties of bacterial toxins channels in planar lipid bilayers and in the plasmalemma with those of the protein channels identified by Simon and Blobel (1991, 1992). This comparison highlights the differences among toxin channels and *bona fide* protein-conducting channels.

The low pH-induced toxin insertion in lipid bilayers
The structural changes of these toxins induced by pH in relation to model membranes have been studied with a variety of techniques and these studies show substantial agreement with a good correlation with results obtained with cells (Alving et al., 1980; Draper and Simon, 1980; Sandvig and Olsnes, 1980; Boquet and Duflot, 1982; Cabiaux et al., 1984, 1989, 1994; Blewitt et al., 1985; Montecucco et al., 1985, 1986, 1988, 1989; Brasseur et al., 1986; Moskaug et al., 1987, 1989, 1991; Papini et al., 1987b,c; Defrise-Quertain et al., 1989; Jiang et al., 1989, 1991; Menestrina et al., 1989; Schiavo et al., 1990, 1991a; Demel et al., 1991; London, 1992; de Paiva et al., 1993). The more relevant conclusions, derived from the above-quoted studies, can be summarized as follows:

a. At neutral pH, DT, ETA, TeNT, and BoNTs interact with the surface of negatively charged membranes. The possible relevance of this membrane surface interaction to cell surface binding has been mentioned above.

b. Acidification induces a true structural change from a "neutral" form to an "acidic" form. The neutral form is water-soluble, while the acidic form aggregates following the exposure of previously hidden hydrophobic regions. The hydrophobic acid form penetrates into the hydrophobic core of lipid micelles or bilayers with a strong preference for negatively charged lipids.

c. Both the A and B protomers are involved in the conformational transition and in the low pH-induced interaction of the toxin with the fatty acid portion of phospholipids: such interaction extends all along the hydrocarbon chains. In this process, α-helices of the inner part of the B chain of DT orientate parallel to the lipid acyl chains, while β-sheets of the carboxy terminal part of the B chain orient parallel to the membrane surface. While the low pH-induced lipid interaction was predicted by the tunnel model, the finding that also the catalytic moiety does interact with the hydrophobic lipid core of the membrane at low pH was unexpected.

d. The low pH-induced lipid interaction of chain A of DT is fully reversible, i.e., chain A appears to insert in the lipid bilayer at low pH and to re-escape when pH is brought back to neutrality.

e. The neutral and acid structures have different shapes with different apparent molecular areas.

f. The decrease of pH causes the exposure of the interchain disulfide bond, which becomes accessible to reduction by thioreductase.

g. The transition between the neutral and acid forms of DT in the presence of negatively charged liposomes occurs in a range of pH values present in endosomes.

The low pH lipid interaction of A and the small size of the toxin channels, cannot be accommodated in the tunnel model. Moreover, the properties of the protein-conducting channel of the endoplasmic reticulum and of *E. coli* are very different from those formed by the toxins. To account for the experimental findings, we have introduced a *cleft* model for the membrane translocation of bacterial protein toxins (Montecucco et al., 1991; 1992; 1994), which is schematized in Figure 4. Toxin cell binding is the result of the interactions of DT with both a protein receptor and with the headgroups of negatively charged lipids. Binding to the protein receptor takes place via the R domain in such a way that the T domain is interacting with the lipid bilayer surface. In DT there are three lysine residues at the bottom of the T domain that could be responsible for electrostatic interactions. Several pieces of evidence support such a double interaction (Alving et al., 1980; Olsnes et al., 1985; Montecucco, 1986; Papini et al., 1987a; Nishiki et al., 1994).

Receptor-bound DT enters endosomes, where acidification induces the exposure of hydrophobic regions, previously hidden in the protein interior. Such hydrophobic surfaces can be shielded from the contact

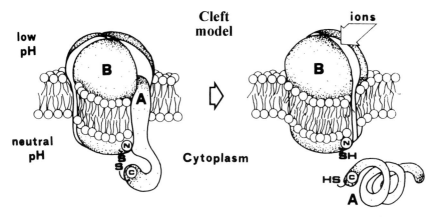

Figure 4. The "cleft" model for the membrane translocation of three-domain bacterial toxins. Inside the intracellular compartment with low lumenal pH, the toxin changes conformation. In a monomeric or multimeric form, protomer B forms a hydrophilic cleft that wraps the hydrophilic surfaces or protomer A during its membrane penetration, while the hydrophobic surfaces of chain A are exposed to the lipids (left panel). Toxin charges are neutralized during membrane translocation by counterions of phospholipid head groups or small anions. A matching of hydrophilic (protein-protein and protein-water) and hydrophobic (protein-protein and protein-lipid) interactions is required to render the process energetically feasible. The neutral cytosolic pH of the cytosol induces the refolding or protomer A to its neutral and catalytically active form (right panel). After domain A has left the membrane, the margins of the B cleft come together in order to minimize the energetically unfavorable interactions with lipids. This leaves a transmembrane alignment of hydrophilic residues that constitutes a transmembrane ion channel (right panel). Due to its reduced size, this channel can accommodate only small ions.

with water either by involvement in protein-protein contacts or by insertion into the hydrophobic core of the lipid bilayer. Parker and Pattus (1993) have suggested that the first part of the T domain that inserts in the lipid bilayer is a hydrophobic helical hairpin that is present in all membranes penetrating toxins. Around this part, the toxin organizes a hydrophilic cleft that drives the insertion of chain A with its hydrophobic segments exposed to lipids and its hydrophilic segments interacting with corresponding regions of the hydrophilic cleft. Such a matching of hydrophobic and hydrophilic protein-lipid and protein-protein interactions will minimize the energetic cost of the process. The open structure of the cleft can accommodate portions of the A chain of different bulkiness. Moreover, such a transmembrane structure is expected to be able to translocate across the membrane peptide segments added to the N-terminus of the A chain, but not to the C-terminus, as recently found by Stenmark et al. (1991). The cleft model proposes that A undergoes a conformational change in concert with B, which involves a rearrangement of its structure, but not a complete unfolding as required in the tunnel model.

It is possible that the membrane penetrating form of the toxin has a net positive charge and that some Lys, Arg or His residues are not in the position to form salt bridges with corresponding negatively charged Asp and Glu residues. Due to the high energetic cost of driving charged groups across the hydrophobic core of the lipid bilayer (particularly the positive charges of lysines and arginines), the process would be made less energetically unfavorable if these charges were neutralized and shielded via formation of ionic couples with negatively charged phospholipid head groups or small anions such as chloride, bromide or thiocyanate. This effect would help to explain the finding that anions are involved in the penetration of DT into cells and that the more hydrophobic membrane permeant anions are the more efficient (Moskaug et al., 1989). It is possible that the membrane translocating form of the toxin is a transient anion-phospholipid-multimeric toxin complex that assembles on the low pH lumen side of the endosomal membrane and lasts until it reaches the cytoplasmic face of the membrane, where the complex disassembles because its ion couples are released by water solvation. Such a proposal emphasizes the role of non-bilayer lipidic configurations (Nemeyanova, 1982). This proposal is in keeping with the well documented ability of these toxins to induce fusion of lipid vesicles at low pH, which is believed to occur via zones of DT-induced lipid destabilization (Cabiaux et al., 1984; Defrise-Quertain et al., 1989). Available data are insufficient to draw a picture of protomer B folding in the membrane for any of the known toxins. Efforts are concentrated on DT and the protein segments of DT embedded in the lipid bilayer at low pH are being identified (Cabiaux et al., 1994). The finding that chain A can escape from liposomes once the pH is returned to neutrality (Montecucco et al., 1985) suggests that, after reduction of the interchain disulfide bond, fragment A can re-acquire its neutral water-soluble form as soon as it faces the neutral cytoplasmic pH.

The cleft model suggests that, after chain A has reached the cytoplasm, the margins of the hydrophilic cleft embedded in the lipid bilayer come closer to minimize the amount of hydrophilic protein surface exposed to the hydrocarbon chain of lipids. This forms a transmembrane alignment of hydrophilic residues that constitutes a flat-shaped channel with protein walls and a flexible lipid side. Such a channel is expected to have a low conductance because of its reduced size and, at the same time, to be able to accommodate large charged objects provided that they have a hydrophobic portion that can interact with the hydrocarbon chains. Such features would account for the finding that the DT channel on cells has a low permeability to choline, glucosamine, amino acids, and phosphorylated intermediates (Sandvig and Olsnes, 1988; Alder et al., 1990). The driving force for membrane insertion and translocation is the transmembrane proton gradient present across the endosomal membrane (for DT) and the synaptic vesicle membrane (for BoNTs). However a contribution of membrane potential is suggested by cellular studies (Neville and Hudson, 1986; Hudson et al., 1988). Moreover, the voltage opening of DT channels at low pH with no transmembrane pH gradient (Donovan et al., 1981; Kagan et al., 1981) suggests that voltage is able to unplug the channel by inducing the release of the A chain. Also, the role of disulfide reduction in membrane translocation and in the refolding of protomer A in the cytoplasm remains to be investigated. Reduction is the rate-limiting step of the entire cell intoxication process (Schiavo et al., 1990; Papini et al., 1993; de Paiva et al., 1993).

The tunnel model and the cleft model differ in several respects. The main differences that can be tested experimentally are the low pH-driven conformational change of these toxins, the lipid interaction and the properties of the toxin ion channel.

The low pH-driven conformational change of these toxins. In the tunnel model the A and B protomers are supposed to change conformation independently, while in the cleft model this is a concerted phenomenon involving B and A at the same time. There is no experimental evidence for an unfolding of A at low pH, rather it appears that the catalytic fragment of DT adopts an acidic conformation that can be crystallized and characterized spectroscopically (Cabiaux et al., 1989; Kantardjieff et al., 1989; Jiang et al., 1991; London, 1992).

Lipid interaction. In the tunnel model fragment A crosses the membrane inside a hydrophilic pore with no contact with lipids. The model does not explain how bulky hydrophobic residues of the A chain can pass through a hydrophilic pore. In the cleft model the membrane translocation of fragment A occurs at the lipid protein boundary in a way compatible with the different properties of the lateral chains of the different amino acids.

Properties of the toxin ion channel. In the tunnel model the channel that mediates the translocation of A is also the one that carries ions across the membrane: the protein conducting channel and the ion channel are the same entity with the same dimensions. The formation of the tunnel is a prerequisite for the membrane translocation of fragment A. In the cleft model the ion channel is related to the structure that has allowed the translocation of A, but it is not the same entity; it does not have the same size and properties; the ion channel is a consequence of the membrane translocation of A. This difference is such that the tunnel model would be dismissed by finding a mutation or an agent that allows cell intoxication without channel formation.

Target modification

This fourth step is the final goal of the whole process. Table 2 lists the four enzymatic activities displayed inside cells by known toxins. The largest group is that of the ADP-ribosyltransferases: they bind cytosolic NAD$^+$ and transfer the ADP-ribose moiety to a variety of cytosolic toxin-specific targets. DT and ETA specifically modity elongation factor 2 and block protein synthesis with consequent cell death (Pappen-heimer, 1982; Collier, 1990). It has been estimated that a single molecule of A is sufficient to kill a cell (Yamaizumi et al., 1978). To the contrary, cells intoxicated by other ADP-ribosyltransferases such as CLT, LT, and PT do not die, but have an altered physiology due to a large increase in c-AMP level. The increase in cellular c-AMP follows the modification of specific G proteins involved in the control of adenylate cyclase activity (Rappuoli and Pizza, 1991). Cellular effects differ as a function of the type of cell intoxicated.

Table 2. Enzymatic activities and cytosolic targets of bacterial protein toxins with intracellular targets

Toxin	Activity	Target	Effect
DT	ADP-ribosyltransferase	EF-2	Blockade of protein synthesis and cell death
ETA	ADP-ribosyltransferase	EF-2	Blockade of protein synthesis and cell death
CLT	ADP-ribosyltransferase	Gαs	Increase c-AMP (alteration of permeability)
LT	ADP-ribosyltransferase	Gαs	Increase c-AMP (alteration of permeability)
PT	ADP-ribosyltransferase	Gαi, Gt	Increase c-AMP (various effects)
C2	ADP-ribosyltransferase	actin	Cell rounding and detachment
C3	ADP-ribosyltransferase	Rho	Cell rounding and detachment
STs	adenine-glycohydrolase	r-RNA 28S	Blockade or protein synthesis and cell death
EF	adenylcyclase	none	Increase c-AMP
BP-ADC	adenylcyclase	none	Increase c-AMP
TeNT	zinc-protease	VAMP	Blockade of exocytosis
BoNT/A	zinc-protease	SNAP-25	Blockade of exocytosis
BoNT/B	zinc-protease	VAMP	Blockade of exocytosis
BoNT/C	zinc-protease	syntaxin	Blockade of exocytosis
BoNT/D	zinc-protease	VAMP	Blockade of exocytosis
BoNT/E	zinc-protease	SNAP-25	Blockade of exocytosis
BoNT/F	zinc-protease	VAMP	Blockade of exocytosis
BoNT/G	zinc-protease	VAMP	Blockade of exocytosis

For abbreviations and references see text.

Other ADP-ribosyltransferases are produced by several *Clostridium spp.* and direct their action on actin polymerization. C-2 toxin specifically modifies G-actin, thus preventing formation of actin filaments (Aktories et al., 1986; Boquet and Gill, 1991; Aktories and Wegner, 1992), while the C-3 enzyme ADP-ribosylates Rho, a protein involved in the control of actin polymerization (Boquet and Gill, 1991; Aktories et al., 1992). As a result, all those phenomena depending on a functional contractile apparatus are altered and the cell may eventually die.

The edema factor of the anthrax toxin complex (EF) is a calmodulin-dependent adenylate cyclase that causes a rapid rise of c-AMP and cell rounding (Leppla, 1982). Also *Bordetella pertussis* produces an adenylate cyclase toxin (Bp-ADC) (Glaser et al., 1988). The increase of the cellular level c-AMP produced by these two toxins is only transient, because they are rapidly degraded by cellular proteases (Leppla, 1991).

Yet another kind of activity is displayed by the Shiga toxins (STs). Like several plant toxins named RIPs (Stirpe and Barbieri, 1986), STs remove a single adenine residue from the 28S ribosomal RNA (Endo et al., 1988). This impairs the function of the 60S ribosomal subunit and blocks protein synthesis (Brown et al., 1980). The final result is the same as that caused by DT and ETA: cell and tissue necrosis.

The most recent addition to the bacterial toxin enzymic activities is that of the clostridial neurotoxins responsible for tetanus and botulism. Their catalytic domain (termed L chain) is zinc-dependent proteinases that specifically attack protein components of the neuroexocytosis apparatus (Schiavo et al., 1992a,b,c; Montecucco and Schiavo, 1993, 1994; Blasi et al., 1993a,b; Schiavo et al., 1993a,b,c,d; Schiavo et al., 1994). TeNT and serotypes B, D, F, and G of BoNT cleave at single different peptide bonds VAMP, a membrane protein of the synaptic vesicle implicated in the docking and fusion of the vesicles with the presynaptic membrane. Serotypes A, C, and E attack SNAP-25 and syntaxin, two proteins of the presynaptic membrane. As a result, the intoxicated neurons remain alive, but the synapes loses functionality and degenerates (Jankovic and Brin, 1994). This result indicates that these three proteins play a fundamental role in exocytosis and, at the same time, points to their possible role in neuronal plasticity (Oyler et al., 1989; Bennett et al., 1992, 1993; Osen-Sand et al., 1993; Söllner et al., 1993a,b; Bennett and Scheller, 1994).

Conclusions

Bacterial protein toxins concentrate in one molecule a variety of functions that enables them to bind and penetrate cells and to alter their physiology. The study of their mechanism of action has already revealed important aspects of cell physiology and promises to provide further information useful for the treatment of diseases, for the development of new vaccines, and for the understanding of fundamental cellular processes.

Acknowledgements
Work in the authors' laboratory is supported by CNR, Telethon-Italia and MURST.

References

Aktories, K., Barmann, M., Ohishi, I., Tsuyama, S., Jacobs, K.H. and Habermann, E. (1986) Botulinum C2 toxin ADP-ribosylates actin. *Nature* 322: 390–392.

Aktories, K. and Wegner, A. (1992) Mechanism of the cytopathic action of actin-ADP-ribosylating toxins. *Microbiol. Rev.* 6: 2905–2908.

Aktories, K., Mohr, C. and Koch, G. (1992) *Clostridium botulinum* C3 ADP-ribosyltransferase. *Curr. Top. Microbiol. Immunol.* 175: 115–131.

Alder, G.M., Bashford, C.L. and Pasternak, C.A. (1990) Action of diphtheria toxin does not depend on the induction of large, stable pores across biological membranes. *J. Membr. Biol.* 113: 67–74.

Allured, V.C., Collier, R.J., Carroll, S.F. and McKay, D.B. (1986) Structure of exotoxin A from *Pseudomonas aeruginosa* at 3.0 Angstrom resolution. *Proc. Natl. Acad. Sci. USA* 83: 1320–1324.

Alouf, J.E. and Freer, J.H. (1991) *A Sourcebook of Bacterial Protein Toxing.* Academic Press, London.

Alving, C.R., Iglewski, B.H., Urban, K.A., Moss, J., Richards, R.L. and Sadoff, J.C. (1980) Binding of diphtheria toxin to phospholipids in liposomes. *Proc. Natl. Acad. Sci. USA* 77: 1986–1990.

Beaumelle, B., Bensammar, L. and Bienvenue, A. (1992) Selective translocation of the A chain of diphtheria toxin across the membrane of purified endosomes. *J. Biol. Chem.* 267: 11525–11531.

Beise, J., Hahnen, J., Andersenbeckh, B. and Dreyer, F. (1994) Pore formation by tetanus toxin, its chain and fragments in neuronal membranes and evaluation of the underlying motifs in the structure of the toxin molecule. *Naunyn-Schmiedbergs Arch. Pharmacol.* 349: 66–73.

Bennett, M.K., Calakos, N. and Scheller, R.H. (1992) Syntaxin: a synaptic protein implicated in docking of synaptic vesicles at presynaptic active zones. *Science* 257: 255–259.

Bennett, M.K., Garcia-Arras, J.E., Elferink, L.A., Peterson, K., Fleming, A.M., Hazuka, C.D. and Scheller, R.H. (1993) The syntaxin family of vesicular transport receptors. *Cell* 74: 863–873.

Bennett, M.K. and Scheller, R.H. (1994) Cellular and molecular biology of the presynaptic nerve terminal. *Annu. Rev. Neurosci.* 14: 93–122.

Blasi, J., Chapman, E.R., Link, E., Binz, T., Yamasaki, S., De Camilli, P., Sudhof, T., Niemann, H. and Jahn, R. (1993a) Botulinum neurotoxin A selectively cleaves the synaptic protein SNAP-25. *Nature* 365: 160–163.

Blasi, J., Chapman, E.R., Yamasaki, S., Binz, T., Niemann, H. and Jahn, R. (1993b) Botulinum neurotoxin C blocks neurotransmitter release by means of cleaving HPC-1/syntaxin. *EMBO J.* 12: 4821–4828.

Blaustein, R.O., Koehler, T.M., Collier, R.J. and Finkelstein, A. (1989) Anthrax toxin: channel-forming activity of protective antigen in planar phospholipid bilayers. *Proc. Natl. Acad. Sci. USA* 86: 2209–2213.

Blewitt, M.G., Chung, L.A. and London, E. (1985) Effect of pH on the conformation of diphtheria toxin and its implications for membrane penetration. *Biochemistry* 24: 5458–5464.

Boquet, P., Silverman, M.S., Pappenheimer, A.M. and Vernon, W.B. (1976) Binding of Triton X-100 to diphtheria toxin, crossreacting material 45 and their fragments. *Proc. Natl. Acad. Sci. USA* 73: 4449–4453.

Boquet, P. and Euflot, E. (1982) Tetanus toxin fragment forms channels in lipid vesicles at low pH. *Proc. Natl. Acad. Sci. USA* 79: 7614–7618.

Boquet, P. and Gill, D.M. (1991) Modulation of Cell Functions by ADP-Ribosylating Bacterial toxins. *In:* J.E. Alouf and J.H. Freer (eds): *A Sourcebook of Bacterial Protein Toxins.* Academic Press, London, pp. 23–44.

Brasseur, R., Cabiaux, V., Falmagne, P. and Ruysschaert, J.M. (1986) pH-dependent insertion of the diphtheria toxin B fragment peptide into the lipid membrane: a conformational analysis. *Biochem. Biophys. Res. Commun.* 136: 160–168.

Brown, J.E., Ussery, M.A., Leppla, S.H. and Rothman, S.W. (1980) Inhibition of protein synthesis by Shiga toxin. Activation of the toxin and inhibition of peptide elongation. *FEBS Lett.* 117: 84–88.

Cabiaux, V., Vandenbranden, M., Falmagne, P. and Ruysschaert, J.M. (1984) Diphtheria toxin induces fusion of small unilamellar vesicles at low pH. *Biochim. Biophys. Acta* 775: 31–36.

Cabiaus, V., Brasseur, R., Wattiez, R., Falmagne, P., Ruysschaert, J.M. and Goormaghtigh, E. (1989) Secondary structure of diphtheria toxin and its fragments interacting with acidic liposomes studied by polarized infrared spectroscopy. *J. Biol. Chem* 264: 4928–4938.

Cabiaux, V., Quertermont, P., Conrath, K., Brasseur, R., Capiau, C. and Ruysschaert, J.M. (1994) Topology of diphtheria toxin B fragment inserted in lipid vesicles. *Mol. Microbiol.* 11: 43–50.

Choe, S., Bennett, M.J., Fujii, G., Curmi, P.M.G., Kantardjieff, K.A., Collier, R.J. and Eisenberg, D. (1992) The crystal structure of diphtheria toxin. *Nature* 367: 216–222.

Collier, R.J. (1990) Diphtheria Toxin: Structure and Function of a Cytocidal Protein. *In:* J. Moss and M. Vaugham (eds): *ADP-Ribosylating Toxins and G Proteins.* American Society for Microbiology, Washington D.C., pp. 3–19.

Defrise-Quertain, F., Cabiaux, V., Vandenbranden, M., Wattiez, R., Falmagne, P. and Ruysschaert, J.M. (1989) pH-dependent bilayer destabilization and fusion of phospholipidic large unilamellar vesicles induced by diphtheria toxin and its fragments A and B. *Biochemistry* 28: 3406–3413.

Deleers, M., Beugnier, N., Falmagne, P., Cabiaux, V. and Ruysschaert, J.M. (1993) Localization in diphtheria toxin fragment B of a region that induces pore formation in planar lipid bylayers at low pH. *FEBS Lett.* 160: 82–86.

Demel, R., Schiavo, G., de Kruijff, B. and Montecucco, C. (199) Lipid interaction of diphtheria toxin and mutants: a study with phospholipid and protein monolayers. *Eur. J. Biochem.* 197: 481–486.

de Paiva A., Poulain B., Lawrence, G.W., Shone C.C., Tauc, L. and Dolly, J.O. (1993) A role for the interchain disulfied or its participating thiols in the internalization of botulinum neurotoxin A revealed by a toxin derivative that bind to ecto-acceptors and inhibits transmitter release intracellularly. *J. Biol. Chem.* 268: 20838–20844.

Donovan, J.J., Simon, M.I., Draper, R.K. and Montal, M. (1981) Diphtheria toxin forms transmembrane channels in planar lipid bilayers. *Proc. Natl. Acad. Sci. USA* 78: 172–176.

Donovan, J.J., Simon, M. and Montal, M. (1982) Insertion of diphtheria toxin into and across membranes: role of phosphoinositide asymmetry. *Nature* 298: 669–672.

Donovan, J.J. and Middlebrook, J.L. (1986) Ion-conducting channels produced by botulinum toxin in planar lipid membranes. *Biochemistry* 25: 2872–2876.

Draper, R.K. and Simon, M.I. (1980) The entry of diphtheria toxin into the mammalian cell cytoplasm: evidence for lysosomal involvement. *J. Cell. Biol.* 87: 849–854.

Endo, Y., Tsurugi, K., Yutsudo, T., Takeda, Y., Ogasawara, T. and Igarashi, E. (1988) Site of action of Vero toxin from *Escherichia coli* 0157:H7 and of Shiga toxin on eukaryotic ribosomes. *Eur. J. Biochem.* 171: 45–50.

Friedlander, A.M. (1986) Macrophages are sensitive to anthrax lethal toxin through an acid-dependent process. *J. Biol. Chem.* 261: 7123–7126.

Gambale, F. and Montal, M. (1988) Characterization of the channel properties of tetanus toxin in planar lipid bilayers. *Biophys. J.* 53: 771–783.

Glaser, P., Sakamoto, H., Bellalou, J., Ullmann, A. and Danchin, A. (1988) Secretion of cyclolysin, the calmodulin-sensitive adenylate cyclase-haemolysin bifunctional protein of Bordetella pertussis. *EMBO J.* 7: 3997–4004.

Hoch, D.H., Romero-Mira, M., Ehrlich, B.E., Finkelstein, A., DasGupta, B.R. and Simpson, L.L. (1985) Channels formed by botulinum, tetanus, and diphtheria toxins in planar lipid bilayers: relevance to translocation of proteins across membranes. *Proc. Natl. Acad. Sci. USA* 82: 1692–1696.

Hudson, T.H., Scharff, J., Kimak, A.G. and Neville, D.M. (1988) Energy requirements for diphtheria toxin translocation are coupled to the maintenance of a plasma membrane potential and a proton gradient. *J. Biol. Chem.* 263: 4773–4781.

Jankovic, J. and Brin, M.F. (1991) Therapeutic uses of botulinum toxin. *New Engl. J. Med.* 324: 1186–1194.

Jiang, G., Solow, R. and Hu, V.W. (1989) Characterization of diphtheria toxin-induced lesions of liposomal membranes. *J. Biol. Chem.* 264: 13424–13429.

Jiang, J.X., Abrams, F.S. and London, E. (1991) Folding changes in membrane-inserted diphtheria toxin that may play important roles in its translocation. *Biochemistry* 30: 3857–3864.

Kagan, B.L., Finkelstein, A. and Colombini, M. (1981). Diphtheria toxin fragment forms large pores in phospholipid bilayer membranes. *Proc. Natl. Acad. Sci. USA* 78: 4950–4954.

Kantardjieff, K., Collier, R.J. and Eisenberg, D. (1989) K-ray grade crystals of the enzymatic fragment of diphtheria toxin. *J. Biol. Chem.* 264: 10402–10404.

Kounnas, M.Z., Morris, R.E., Thompson, M.R., FitzGerald, D.J., Strickland, D.K. and Saelinger, C.B. (1992) The α_2-macroglobulin receptor/low density lipoprotein receptor-related protein binds and internalizes *Pseudomonas* exotoxin A. *J. Biol. Chem.* 267: 12420–12423.

Leppla, S.H. (1982) Anthrax toxin edema factor: a bacterial adenylate cyclase that increases cyclic AMP concentrations of eukaryotic cells. *Proc. Natl. Acad. Sci. USA* 79: 3162–3166.

Leppla, S.H. (1991) The anthrax toxin complex. *In:* J.E. Alouf and J.H. Freer (eds): *A Sourcebook of Bacterial Protein Toxins.* Academic Press, London, pp. 303–348.

London, E. (1992) Diphtheria toxin: membrane interaction and membrane translocation. *Biochim. Biophys. Acta* 1113: 25–51.

Matteoli, M. Takei, K., Perin, M., Sudhof, T.C. and De Camilli, P. (1992) Exo-endocytotic recycling of synaptic vesicles in developing processes of cultured hyppocampal neurons. *J. Cell. Biol.* 117: 849–861.

Mellanby, J. and Green J. (1981) How does tetanus toxin act? *Neuroscience* 6: 281–300.

Menestrina, G., Forti, S. and Gambale, F. (1989) Interaction of tetanus toxin with lipid vesicles: effect of pH, surface charge and transmembrane potential on the kinetics of channel formation. *Biophys. J.* 55: 393–405.

Menestrina, G., Schiavo, G. and Montecucco, C. (1994) Molecular Mechanisms of Action of Bacterial Protein Toxins. *In:* H. Baum (ed.): *Molecular Aspects of Medicine*, Pergamon Press, Oxford, Vol. 15, pp. 79–193.

Milne, J.C. and Collier, R.J. (1993) pH-dependent permeabilization of the plasma membrane of mammalian cells by anthrax protective antigen. *Mol. Microbiol.* 10: 647–653.

Montecucco, C. (1986) How do tetanus and botulinum toxins bind to neuronal membranes? *Trends Biochem. Sci.* 11: 314–317.

Montecucco, C., Schiavo, G. and Tomasi, M. (1985). pH-dependence of the phospholipid interaction of diphtheria-toxin fragments. *Biochem. J.* 231: 123–128.

Montecucco, C., Schiavo, G., Brunner, J., Duflot, E., Boquet, P. and Roa, M. (1986) Tetanus toxin is labeled with photoactivatable phospholipids at low pH. *Biochemistry* 25: 919–924.

Montecucco, C., Schiavo, G., Gao, Z., Bauerlein, E., Boquet, P. and DasGupta, B.R. (1988) Interaction of botulinumand tetanus toxins with the lipid bilayer surface. *Biochem. J.* 251: 379–383.

Montecucco, C., Schiavo, G. and DasGupta, B.R. (1989) Effect of pH on the interaction of botulinum neurotoxins A, B and E with liposomes. *Biochem. J.* 259: 47–53.

Montecucco, C., Papini, E. and Schiavo, G. (1991) Molecular models of toxin membrane translocation. *In:* J.E. Alouf and J.H. Freer (eds): *A Sourcebook of Bacterial Protein Toxins.* Academic Press, London, pp. 45–56.

Montecucco, C., Papini, E., Schiavo, G., Padovan, E. and Rossetto, O. (1992) Ion channel and membrane translocation of diphtheria toxin. *FEMS Microbiol. Immunol.* 195: 101–111.

Montecucco, C. and Schiavo, G. (1993) Tetanus and botulism neurotoxins: a new group of zinc proteases. *Trends Biochem. Sci.* 18: 324–327.

Montecucco, C. and Schiavo, G. (1994) The mechanism of action of tetanus and botulism neurotoxins. *Mol. Microbiol.* 13: 1–8.

Montesano, R., Roth, J., Robert, A. and Orci, L. (1982) Non-coated membrane invaginations are involved in binding and internalization of cholera and tetanus toxin. *Nature* 296: 651–653.

Moskaug, J.O., Sandvig, K. and Olsnes, S. (1987) Cell-mediated reduction of the interfragment disulfide in nicked diphtheria toxin. *J. Biol. Chem.* 262: 10339–10345.

Moskaug, J.O., Sandvig, K. and Olsnes, S. (1989) Role of anions in low-pH-induced translocation of diphtheria toxin. *J. Biol. Chem.* 264: 11367–11372.

Moskaug, J.O., Stenmark, H. and Olsnes, S. (1991) Insertion of diphtheria toxin B-fragment into the plasma membrane at low pH. *J. Biol. Chem.* 266: 2652–2659.

Moya, M., Dautry-Varsat, A., Gould, B., Louvard, D. and Boquet, P. (1985) Inhibition of coated pit formation in HepG2 cells blocks the cytotoxicity of diphtheria toxin but not that of ricin toxin. *J. Cell. Biol.* 101: 548–559.

Naglich, J.G., Metherall, J.E., Russell, D.W. and Eidels, L. (1992) Expression cloning of a diphtheria toxin receptor: identity with a heparin-binding EGF-like growth factor precursor. *Cell* 69: 1051–1061.

Nesmeyanova, M.A. (1982) On the possible participation of acid phospholipids in the translocation of secreted proteins through the bacterial cytoplasmic membrane. *FEBS Lett.* 142: 189–193.

Neville, D.M. and Hudson, T.H. (1986) Transmembrane transport of diphtheria toxin, related toxins and colicons. *Annu. Rev. Biochem.* 55: 195–224.

Nishiki, T., Kamata, Y., Nemoti, Y., Omiri, A., Ito, T., Takahashi, M. and Kozaki, S. (1994) Identification of protein receptor for *Clostridium botulinum* type B neurotoxin in rat brain synaptosomes. *J. Biol. Chem.* 269: 10498–10503.

Olsnes, S., Carvajal, E., Sundan, A. and Sandvig, K. (1985) Evidence that membrane phospholipids and protein are required for binding of diphtheria toxin in Vero cells. *Biochim. Biophys. Acta* 846: 334–341.

Osen-Sand, A., Catsicas, M., Staple, J.K., Jones, K.A., Ayala, G., Knowles, J., Grenningloh, G. and Catsicas, S. (1993) Inhibition of axonal growth by SNAP-25 antisense oligonucleotides *in vitro* and *in vivo*. *Nature* 364: 445–448.

Oyler, G.A., Higgins, G.A., Hart, R.A., Battenberg, E., Billingsley, M., Bloom, F.E. and Wilson, M.C. (1989) The identification of a novel synaptosomal-associated protein, SNAP-25, differently expressed by neuronal subpopulations. *J. Cell. Biol.* 109: 3039–3052.

Papini, E., Colonna, R., Schiavo, G., Cusinato, F., Tomasi, M., Rappuoli, C. and Montecucco, C. (1987a) Diphtheria toxin and its mutant *crm* 197 differ in their interaction with lipids. *FEBS Lett.* 215: 73–78.

Papini, E., Colonna, R., Cusinato, F., Montecucco, C., Tomasi, M. and Rappuoli, R. (1987b) Lipid interaction of diphtheria toxin and mutants with altered fragment B: liposome aggregation and fusion. *Eur. J. Biochem.* 169: 629–635.

Papini, E., Schiavo, G., Tomasi, M., Colombatti, M., Rappuoli, R. and Montecucco, C. (1987c) Lipid interaction of diphtheria toxin and mutants with altered fragment B: hydrophobic photolabelling and cell intoxication. *Eur. J. Biochem.* 169: 637–644.

Papini, E., Sandonà, D., Rappuoli, R. and Montecucco, D. (1988) On the membrane translocation of diphtheria toxin: at low pH the toxin induces ion channels on cells. *EMBO J.* 7: 3353–3359.

Papini, E., Rappuoli, R., Murgia, M. and Montecucco, C. (1993) Cell penetration of diphtheria toxin. *J. Biol. Chem.* 268: 1567–1574.

Pappenheimer, A.M. (1982) Diphtheria: studies on the biology of an infectious disease. *Harvey Lect.* 76: 45–73.

Parker, M.W. and Pattus, F. (1992) Rendering a membrane protein soluble in water: a common packing motif in bacterial protein toxins. *Trends Biochem. Sci.* 18: 391–395.

Pelham, H.R.B. (1991) Recycling of proteins between the endoplasmic reticulum and Golgi complex. *Curr. Opin. Cell. Biol.* 3: 585–591.

Perin, M.S., Johnson, P.A., Ozcelik, T., Jahn, R., Franke, U. and Sudhof, T.S. (1991) Structural and functional conservation of synaptotagmin (p65) in Drosophila and humans. *J. Biol. Chem.* 266: 615–622.

Perin, M.S. (1994) The COOH-terminus of synaptotagmin mediates interaction with the neurexins. *J. Biol. Chem.* 269: 8576–8581.

Rappuoli, R. and Pizza, M.G. (1991) Structural and evolutionary aspects of ADP-ribosylating toxins. *In:* J.E. Alouf and J.H. Freer (eds): *Sourcebook of Bacterial Protein Toxins.* Academic Press, London, pp. 1–23.

Robinson, J.P., Schmid, M.F., Morgan, D.G. and Chiu, W. (1988) Three-dimensional structural analysis of tetanus toxin by electron crystallography. *J. Mol. Biol.* 200: 367–375.

Sandvig, K. and Olsnes, S. (1980) Diphtheria toxin entry into cells is facilitated by low pH. *J. Cell. Biol.* 87: 828–832.

Sandvig, K. and Olsnes, S. (1988) Diphtheria toxin-induced channels in Vero cells selective for monovalent cations. *J. Biol. Chem.* 263: 12352–12359.

Sandvig, K. and Olsnes, S. (1991) Membrane translocation of diphtheria toxin. *In:* J.E. Alouf and J.H. Freer (eds): *Sourcebook of Bacterial Protein Toxins.* Academic Press, London, pp. 57–73.

Sandvig, K., Garred, O., Prydz, K., Kozlov, J.V., Hansen, S.H. and van Deurs, B. (1992) Retrograde transport of endocytosed Shiga toxin to the endoplasmic reticulum. *Nature* 358: 510–512.

Schiavo, G., Papini, E., Genna, G. and Montecucco, C. (1990) An intact interchain disulfide bond is required for the neurotoxicity of tetanus toxin. *Infect. Immun.* 58: 4136–4141.

Schiavo, G., Demel, R. and Montecucco, C. (1991a) On the role of polysialoglycosphin-golipids as tetanus toxin receptors: a study with lipid monolayers. *Eur. J. Biochem.* 199: 705–711.

Schiavo, G., Rossetto, O., Ferrari, G. and Montecucco, C. (1991b) Tetanus toxin receptor. Specific cross-linking of tetanus toxin to a protein of NGF-differentiated PC 12 cells. *FEBS lett.* 290: 227–230.

Schiavo, G., Poulain, B., Rossetto, O., Benfenati, F., Tauc, L. and Montecucco, C. (1992a) Tetanus toxin is a zinc protein and its inhibition of neurotransmitter release and protease activity depend on zinc. *EMBO J.* 11: 3577–3583.

Schiavo, G., Benfenati, F., Poulain, B., Rossetto, O., Polverino de Laureto, P., DasGupta, B.R. and Montecucco, C. (1992b) Tetanus and botulinum-B neurotoxins block neuro-transmitter release by a proteolytic cleavage of synaptobrevin. *Nature* 359: 832–835.

Schiavo, G., Rossetto, O., Santucci, A., DasGupta, B.R. and Montecucco, C. (1992c) Botulinum neurotoxins are zinc proteins. *J. Biol. Chem.* 267: 23479–23483.

Schiavo, G., Shone, C.C., Rossetto, O., Alexandre, F.C.G. and Montecucco, C. (1993a) Botulinum neurotoxin serotype F is a zinc endopeptidase specific for VAMP/synaptobre-vin. *J. Biol. Chem.* 268: 11516–11519.

Schiavo, G., Rossetto, O., Catsicas, S., Polverino de Laureto, P., DasGupta, B.R., Benfe-nati, F. and Montecucco, C. (1993b) Identification of the nerve-terminal targets of botulinum neurotoxins serotypes A, D and E. *J. Biol. Chem.* 268: 23784–23787.

Schiavo, G., Santucci, A., DasGupta, B.R., Metha, P.P., Jontes, J., Benfenati, F., Wilson, M.C. and Montecucco, C. (1993b) Botulinum neurotoxins serotypes A and E cleave SNAP-25 at distinct COOH-terminal peptide bonds. *FEBS Lett.* 335: 99–103.

Schiavo, G., Poulain, B., Benfenati, F., DasGupta, B.R. and Montecucco, C. (1993d) Novel targets and catalytic activities of bacterial protein toxins. *Trends Microbiol.* 1: 170–174.

Schiavo, G., Malizio, C., Trimble, W.S., Polverino de Laureto, P., Milan, G., Sugiyama, H., Johnson, E.A. and Montecucco, C. (1994) Botulinum G neurotoxin cleaves VAMP/synap-tobrevin at a single Ala-Ala peptide bond. *J. Biol. Chem.* 269: 20213–20216.

Seetharam, S., Chaudhary, V.K., Fitzgerald, D. and Pastan, I. (1991) Increased cytotoxic activity of *Pseudomonas aeruginosa* exotoxin and two chimeric toxins ending in KDEL. *J. Biol. Chem.* 266: 17376–17381.

Shiver, J.W. and Donovan, J.J. (1987) Interactions of diphtheria toxin with lipid vesicles: determinants of ion channel formation. *Biochim. Biophys. Acta* 903: 48–55.

Shone, C.C., Hambleton, P. and Melling, J. (1987) A 50-kDa fragment from the NH$_2$-termi-nus of the heavy subunit of *Clostridium botulinum* type A neurotoxin forms channels in lipid vesicles. *Eur. J. Biochem.* 167: 175–180.

Simon, S.M. and Blobel, G. (1991) A protein-conducting channel in the endoplasmic reticulum. *Cell* 65: 371–380.

Simon, S.M. and Blobel, G. (1992) Signal peptides open protein-conducting channels in *E. coli. Cell* 89: 677–684.

Simpson, L.L. (1982) The interaction between aminoquinolines and presynaptically acting neurotoxins. *J. Pharmacol. Exp. Ther.* 222: 43–48.

Simpson L.L. (ed.) (1989) *Botulinum Neurotoxins and Tetanus Toxin.* Academic Press, New York.

Sixma, T.K., Pronk, S.E., Kalk, K.H., Wartna, E.S., van Zanten, B.A.M., Witholt, B. and Hol, W.J.J. (1991) Crystal structure of a cholera toxin-related heat-labile enterotoxin from *E. coli. Nature* 351: 371–377.

Sixma, T.K., Stein, P.E., Hol, W.G.J. and Read, R.J. (1993) Comparison of the B-pentamers of heat-labile enterotoxin and verotoxin-1; two structures with remarkable similarity and dissimilarity. *Biochemistry* 32: 191–198.

Söllner, T., Whiteheart, S.W., Brunner, M., Erdjument-Bromage, H., Geromanos, S., Tempst, P. and Rothman, J.E. (1993a) SNAP receptors implicated in vesicle targeting and fusion. *Nature* 362: 318–324.

Söllner, T., Bennett, M., Whiteheart, S.W., Scheller, R.H. and Rothman, J.E. (1993b) A protein assembly-disassembly pathway in vitro that may correspond to sequential steps of synaptic vesicle docking, activation, and fusion. *Cell* 75: 409–418.

Stein, P.E., Boodhoo, A., Tyrrell, G.J., Brunton, J.L. and Read, R.J. (1992) Crystal structure of the cell-binding B oligomer of verotoxin-1 from *E. coli. Nature* 355: 748–750.

Stein, P.E., Boodhoo, A., Armstrong, G.D., Cockle, S.A., Klein, M.H. and Read, R.J. (1994) The crystal structure of pertussis toxin. *Nature Struct. Biol.* 2: 45–56.

Stenmark, H., Moskaug, J.O., Mashus, J.H., Sandvig, K. and Olsnes, S. (1991) Peptides fused to the amino-terminal end of diphtheria toxin are translocated into the cytosol. *J. Cell. Biol.* 113: 1025–1032.

Stirpe, F. and Barbieri, L. (1986) Ribosome-inactivating proteins up to date. *FEBS Lett.* 195: 1–8.

Uchida, T., Moriyama, Y., Futai, M. and Mekada, E. (1990) The cytotoxic action of diphtheria toxin and its degradation in intact Vero cells are inhibited by bafilomycin A1, a specific inhibitor of vacuolar-type H-ATPase. *J. Biol. Chem.* 265: 21940–21945.

van Heyningen, W.E. (1968) Tetanus. *Sci. Am.* 218: 69–77.

Williamson, L.C. and Neale, E.A. (1994) Bafilomycin A1 inhibits the action of tetanus toxin in spinal cord neurons in cell culture. *J. Neurochem.* 63: 2342–2345.

Yamaizumi, M., Mekada, E., Uchida, T. and Okada, Y. (1978) One molecule of diphtheria toxin fragment A introduced into a cell can kill the cell. *Cell* 15: 245–250.

Yavin, E. and Nathan, A. (1986) Tetanus toxin receptors on nerve cells contain a trypsin-sensitive component. *Eur. J. Biochem.* 154: 403–407.

Biochemistry of Cell Membranes
ed. by S. Papa & J. M. Tager
© 1995 Birkhäuser Verlag Basel/Switzerland

Non-clathrin coat proteins in biosynthetic vesicular protein transport

C.L. Harter and F.T. Wieland

Institute of Biochemistry I, University of Heidelberg, Im Neuenheimer Feld 328, D-69120 Heidelberg, Germany

Introduction

Biosynthetic vesicular protein transport defines a pathway from the cotranslational import of proteins into the endoplasmic reticulum (ER) via the Golgi apparatus to various cellular organelles or to the cell surface. During constitutive protein transport soluble luminal cargo proteins are secreted once their transport vesicles fuse with the plasma membrane, and transmembrane cargo proteins are expressed as cell surface proteins.

After their translocation into the ER, proteins are taken up into vesicles that bud off the ER membrane, and are transported by stepwise fusion and budding events through the Golgi apparatus. Vesicular luminal contents are delivered to the lumina of the Golgi cisternae by this mechansim, and are subject to posttranslational modifications in this organelle. Besides its role as a posttranslational modification machinery the Golgi acts as a distribution center: from its exit post, the trans Golgi network (TGN), proteins are sorted to their various destinations.

In this short review we will summarize our present knowledge of the basic components and mechansims that enable a transport vesicle to bud from a donor membrane.

Golgi-derived transport vesicles

The Golgi complex is a polarized structure that is formed by different subcompartments: cis Golgi network (CGN), cis, medial, trans Golgi, and trans Golgi network (TGN). Its best known and predominant functions are posttranslational processing of newly synthesized membrane and secreted proteins, as well as protein sorting and sphingolipid biosynthesis. Transport through the Golgi stacks occurs in vesicles that travel vectorially in the cis to trans direction, allowing orderly and sequential processing like addition and trimming of carbohydrate units or sulfation of cargo proteins. A vesicle carrying cargo from the cis compartment must fuse with the medial compartment in order to ensure the correct processing of its cargo. Transport can easily be assessed by following the above-mentioned post-translational modifications of a selected protein in transit (Farquhar, 1985; Pfeffer and Rothman, 1987).

Progress towards the molecular description of these transport vesicles has been achieved by the development of cell-free assays (Rothman, 1992). *In vitro* systems allow one to dissect the complex transport machinery into single steps, e.g., the movement from one compartment to another. Specific proteins and factors required can be tested and identified by depletion, inactivation or replacement. The best character-ized biosynthetic transport vesicles today are those performing constitu-tive intercisternal transport in the Golgi stack. It was a combination of electron microscopy and biochemistry that shed light on the molecular mechanisms of the formation and composition of vesicles in mammalian cells.

Transport between Golgi cisternae has been reconstituted in a cell-free system by incubation of two different populations of Golgi mem-branes with cytosol and ATP as an energy source at 37°C (Fries and Rothman, 1980; Balch et al., 1984a). A "donor" population containing the cargo to be transported, like for example VSV G protein, is incubated with an "acceptor" population of Golgi that contains a glycosylating enzyme, GlcNAc transferase which is absent in the donor population. After addition of UDP-^{3}H-GlcNAc the radioactive aminosugar is transferred to the cargo protein, and this label serves as a measure for the amount of protein transported from the donor to the acceptor compartment (Braell et al., 1984). The transporting vesicles were visualized by immunoelectron microscopy and were found to bud preferentially from the rims of the Golgi cisternae. The vesicles are uniformly in size, with a diameter of about 75 nm (Orci et al., 1986). They have an electron-dense, fuzzy protein coat, distinct from the clathrin coat of endocytic vesicles. Coated vesicles were predominantly found in budding regions and this was taken as a first indication that the coat drives the budding reaction by deforming the membrane into a spherical shape. Omission of either cytosol or an ATP regenerating system as well as incubation at 0°C completely blocked coated vesicle formation (Balch et al., 1984b).

The observation that GTPγS treatment results in an accumulation of Golgi-derived coated vesicles has allowed their isolation in amounts sufficient for a biochemical characterization of their major protein components (Melancon et al., 1987; Serafini et al., 1991a). Coated vesicles were generated by incubation of isolated Golgi membranes with cytosol and an ATP-generating system in the presence of GTPγS

(Serafini et al., 1991a). Isolation of the vesicles was performed by differential centrifugation steps in the presence of various salt concentrations (Malhotra et al., 1989). Final purification was achieved by isopycnic sucrose gradient centrifugation. Fractions at about 40% sucrose (w/v) contained the putative transport vesicles as has been shown by the following criteria: (i) electron microscopy that revealed the presence of a homogenous population of coated vesicles of about 75 nm in diameter and (ii) immunological analysis that demonstrated these fractions to contain typical cargo proteins: VSV G protein, if VSV infected CHO cells were used as the Golgi donor population, or albumin, if liver cells were used as the Golgi source. On SDS-PAGE these vesicles gave rise to a characteristic protein pattern clearly segregated from the majority of proteins at the top of the gradient and from a few proteins at higher sucrose concentrations. The major part of these proteins turned out to be membrane-associated rather than integral membrane proteins, an indication that they represent the fuzzy coat observed on these vesicles by electron microscopy. This was further substantiated by proteolytic digestion of purified vesicles where these proteins were degraded under conditions that let the cargo proteins intact (Seerafini et al., 1991a). Most of them have been characterized at a molecular level and will be described in the following chapter.

Coat proteins

The proteins forming the coat of Golgi-derived vesicles include the COP family (for COat Proteins) and a small GTP-binding protein, ADP-ribosylation factor (ARF) (Tab. 1) (Serafini et al., 1991a; Serafini et al., 1991b; Orci et al., 1993b). ARF was originally discovered as cofactor in Cholera toxin-catalyzed ADP-ribosylation of stimulatory G protein α-subunits (Kahn and Gilman, 1986). Its putative role in intracellular protein transport to or within the Golgi apparatus was suggested due to a secretion phenotype in yeast and the localization of its membrane-bound form to Golgi stacks (Stearns et al., 1990). In addition to their membrane associated form, COPs were found in the cytosol where they form a soluble complex, termed coatomer, for coat protomer (Waters et al., 1991). The discovery of coatomer allowed the identification of its subunits: α-COP, (160 kD), β-COP (107 kd), β'-COP (102 kD), γ-COP (98 kD), δ-COP (61 kD), ε-COP (35 kD) and ζ-COP (20 kD). All COPs, except α-COP, have been cloned from cDNA libraries and sequenced. The first COP identified was the β-subunit, whose amino acid sequence shows 17% identity to β-adaptin, a coat protein of clathrin-coated vesicles (Duden et al., 1991). This protein is identical to a protein that has been discovered due to a cross-reaction with a monoclonal antibody directed against microtubule-associated proteins,

Table 1. Coat proteins of Golgi-derived transport vesicles

Protein	Mr (Da)	Present in coatomer	Present in vesicles	Monomeric in cytosol	Protein features
α-COP	~160 000	+	+	−	?
β-COP	107 010	+	+	−	homologous to β-adaptin of the AP2 complex
β'-COP	102 041	+	+	−	five WD40 repeated motifs of β-transducin family
γ-COP	97 385	+	+	−	homologous to Sec21p of S. cerevisiae
δ-COP	57 109	+	+	−	−
ε-COP	34 500	+	+	+	−
ξ-COP	20 219	−	+	+	homologous to AP19/17 of the AP1/AP2 complex
ARF	21 000		+	+	ADP-ribosylation factor; small GTP-binding protein

For references see text.

and that has been localized to the Golgi apparatus (Allan and Kreis, 1986; Duden et al., 1991; Serafini et al., 1991a). β'-COP (MW 102 041), not related structurally to β-COP, and so named because this protein comigrates with β-COP in conventional SDS-acrylamide gels, was only discovered upon resolution of the 100 kD family of coatomer in a modified gel system containing 6M urea and a reduced concentration of bisacrylamide in the separating gel (Stenbeck et al., 1993). Independently, the same protein was identified due to its crossreaction with monoclonal antibody that was generated against the cytosolic chaperone TCP1 (Harrison-Lavoie et al., 1993). The N-terminal third of β'-COP consists of five repeated WD40 motifs typically found in β-subunits of trimeric G proteins (van der Voorn and Ploegh, 1992). The third member of the 100 kD family of coatomer, γ-COP, was found to be highly homologous to Sec21p of *S. cerevisiae* (35.1% identity, 55% homology) (Stenbeck et al., 1992). SEC21 represents a secretion mutant in yeast that has been shown to be defective in ER to Golgi transport (Kaiser and Schekman, 1990; Hosobuchi et al., 1992). Analogous to the mammalian system, Sec21p is present in a large cytosolic complex similar in size to mammalian coatomer (Hosobuchi et al., 1992). For δ-COP and ε-COP (57 kD and 36 kD) no homology has been found to any known protein (Auerbach et al., in preparation; Hara-Kuge et al., 1994). ζ-COP, like β-COP, has weak homology to a clathrin adaptor subunit (Duden et al., 1991; Kuge et al., 1993). ζ-COP is the only coatomer subunit that, in addition to its coatomer-bound form, has been shown to occur as a soluble monomer in the cytosol (Kuge et al., 1993). Comparison of peptide sequences obtained from COPs isolated from soluble coatomer and from purified transport vesicles clearly revealed that each individual COP is a constituent of the cytosolic coatomer as well as of the transport vesicles (Waters et al., 1991; Stenbeck et al., 1992; Kuge et al., 1993; Hara-Kuge et al., 1994). Both on vesicles and in coatomer the individual COPs occur in a one-by-one stoichiometry (Serafini et al., 1991a; Waters et al., 1991). The presence in transport vesicles of all COPs has been proven by immunoelectron microscopy (unpublished data; Orci et al., 1993c).

The second protein component of the coat, ARF (21 kD), was detected by blotting with ^{32}P-GTP of isolated Golgi-transport vesicles after SDS-PAGE (Serafini et al., 1991b). Additional evidence that ARF is a true constituent of the coat of vesicular carriers came from immunoelectron microscopy (Serafini et al., 1991b). ARF and COPs accumulate in buds and coated vesicles on Golgi membranes, but not at flattened Golgi cisternae and not in uncoated vesicles (Orci et al., 1993a; Orci et al., 1993b). Like coatomer, the small GTP-binding protein ARF is present both on membranes and in the cytosol. The protein contains an aminoterminal myristic acid that is masked as long as ARF is in its GDP-bound form (Kahn and Gilman, 1986; Kahn et al., 1991). Exchange of GDP leads to a conformational change by which the myristyl residue is exposed. This hydrophobic "tail" is believed to serve as an anchor for membrane binding of ARF-GTP.

Mechanisms of vesicle budding

Functional and structural characterization of the proteins involved as well as genetic evidence mainly derived from transport defective mutants in *S. cerevisiae* have shed light on the principle mechanisms that underlie the budding of a transport vesicle from a donor membrane. This process has been reconstituted *in vitro* for the mammalian Golgi system and for the endoplasmic reticulum from mammals and yeast (Pryer et al., 1992, Rothman and Orci, 1992).

Vesicle budding from mammalian Golgi membranes

In a Golgi-*in vitro* system the crucial role ARF plays in the budding reaction has been elucidated. In its GTP-bound form, ARF binds to membranes in a rather unspecific manner. This binding seems to be a prerequisite for a highly specific interaction of ARF-GTP with a membrane receptor protein in the donor Golgi membrane (Helms et al., 1993; Orci et al., 1993b; Palmer et al., 1993). This receptor has not yet been characterized. Once ARF is bound to the donor membrane, coatomer will be recruited from the cytosol (Donaldson et al., 1992a; Palmer et al., 1993). No curvature of the membranes is observed when ARF alone is bound. In contrast, addition of purified coatomer to Golgi membranes pretreated with ARF leads to the appearance of budding structures as revealed by electron microscopy (Orci et al., 1993a). This finding strongly indicates that binding of coatomer initiates and triggers the formation of membrane protrusions and thus rules the shape (and diameter) of a transport vesicle (Fig. 1).

However, coat assembly is insufficient to complete the pinching off of vesicles. What drives the production of complete vesicles and what are the intermediate steps from the binding of coatomer to the budding of vesicles? After coat assembly at the Golgi membrane a fission event has to occur in order to release the coated vesicle. In contrast to fusion which is initiated at the cytoplasmic side, this membrane fission occurs at the base of the bud on the luminal (cisternal) side. This fission event (periplasmic fusion) is dependent on long chain fatty acyl-CoA (Pfanner et al., 1989; Ostermann et al., 1993). Both coated vesicle formation and transport are blocked by a nonhydrolyzable analogue of palmitoyl-CoA (Pfanner et al., 1989). This might imply that a periplasmic fusion protein is activated by fatty acylation on the cytoplasmic side (Fig. 1).

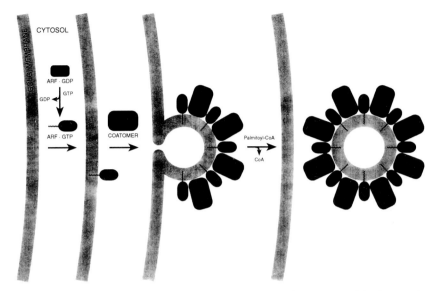

Figure 1. Steps in the budding of a Golgi-derived COP-coated transport vesicle. See text for details.

The observation that treatment with GTPγS and AlF$_4^-$ affect intra-Golgi transport and vesicle formation points to the additional involvement of trimeric G proteins in vesicle budding (Melancon et al., 1987; Donaldson et al., 1991b; Robinson and Kreis, 1992). AlF$_4^-$ (which in the presence of GDP activates trimeric G proteins by mimicking the γ-phosphate group of GTP) does not bind to GTPases of the ras superfamily (e.g., small GTP-binding proteins like ARF), but is selective for trimeric G proteins (Bigay et al., 1987; Kahn, 1991). Indeed, treatment with AlF$_4^-$ only enhances binding of β-COP (which probably represents coatomer) to Golgi membranes, but does not affect ARF-binding (Donaldson et al., 1991a) Accordingly, the β-COP/coatomer binding may be specifically regulated by trimeric G proteins. The involvement of heterotrimeric G proteins (the α subunit of which is mostly affected by GTPγS/AlF$_4^-$) in binding of coatomer to membranes has been demonstrated by the inhibitory effect of exogenous G$_{\beta/\gamma}$ subunits (Donaldson et al., 1991a). A specific G protein, G$_{i\alpha3}$, if overexpressed localizes to Golgi membranes and inhibits secretion of a proteoglycan (Ercolani et al., 1990; Stow et al., 1991). In the study of formation of neuronal constitutive secretory vesicles and immature secretory granules from the TGN, multiple trimeric G proteins of the G$_{i/o}$ and G$_s$ class have been found on the TGN membrane where they appear to regulate the formation of these vesicles both negatively (G$_{i/o}$) and positively (G$_s$) (Leyte et al., 1992). However, the identity of the G protein(s) on the Golgi membrane (and possibly other membranes of the endomembrane system) that regulate(s) coatomer binding and vesicles formation is unknown; it is also not known whether a direct interaction between the G protein and coatomer exists. Alternatively, coatomer might activate a G protein indirectly for instance via its interaction with a GDP-GTP exchange factor. At present, it seems likely that the formation of intra-Golgi transport vesicles is regulated through an interaction between coatomer and components of a machinery associated with heterotrimeric G proteins. The identity and molecular mechanisms underlying this interaction wait to be elucidated.

Vesicle budding from mammalian ER membranes

Studies on protein export from the ER were performed in semi-intact cells and *in vivo*. Similar to Golgi vesicle formation, both ARF and β-COP were found to be required for the export of proteins from the ER (Balch et al., 1992; Peter et al., 1993). Transport of VSV G protein from the ER is efficiently blocked by incubation of semi-intact cells with a peptide identical to the N-terminus of ARF or by transfecting cells with an ARF mutant, ARF1(T31N), that has a preferential affinity for GDP compared to the wildtype protein (Balch et al., 1992; Dascher and Balch, 1994). Similarly, incubation of semi-intact cells with specific anti β-COP antibodies or microinjection of these antibodies into living cells prevents export of VSV G protein from the ER (Pepperkok et al., 1993; Peter et al., 1993). However, the protein coat of ER transport vesicles seems to be distinct from the Golgi coat (Tab. 2).

First, the cytosolic protomer of the ER coat has been suggested to be larger than the cytosolic coatomer complex that binds to Golgi membranes: ≥ 1000 kD for ER procoat as compared to 700 kD for Golgi coatomer (Waters et al., 1991; Peter et al., 1993). In addition to β-COP, the ER procoat contains rab1B and the mammalian homologue of Sec23p, a yeast secretory mutant protein (Orci et al., 1991; Peter et al., 1993). β-COP and rab1B have been shown to be required for the formation of ER derived coated vesicles and thus might be recruited from a common cytosolic pool (Plutner et al., 1991; Peter et al., 1993). The role of Sec23p in mammalian cells is not known, nor is it known whether this ER coat protomer contains the complete set of COPs of the Golgi coatomer.

Second, vesicle formation from the ER is promoted by Sar1a and Sar1b, the mammalian homologues of yeast Sar1p (Kuge et al., 1994). Sar1a and Sar1b were found to be enriched in transitional elements of the ER and on vesicular profiles in the proximal Golgi cisternae. The Sar1p GTPase is needed for export from the ER, but not for intra-Golgi transport. Thus, Sar1p might somehow be involved in the segregation of

Table 2. Cytosolic proteins involved in vesicle formation during biosynthetic protein transport

Intra-Golgi in mammals	ER to Golgi in mammals	Intra-Golgi, ER to Golgi in *S. cerevisiae*
coatomer	ER-specific coatomer?	coatomer?
ARF	ARF	ARF?
	Sar1a	Sar1p
	Sar1b	Sec23p/Sec24p
	rab1B	Sec13p/p150

For references see text.

ER membrane. Once sorted at the exit of the ER, COP-coated Golgi vesicles transport cargo at its bulk concentration from one cisterna to the next in a Sar1p independent manner.

Vesicle budding from yeast membranes

Two distinct sets of coat proteins might play a role in membrane transport in yeast depending on the experimental system considered (Tab. 2).

One protein coat is similar to the ARF-coatomer coat in mammals. Support for a function of ARF in intracellular protein transport in yeast results from the finding that an intact ARF gene is essential for viability and proper glycosylation and secretion (Stearns et al., 1990). Yet, the exact site of ARF function in the secretory pathway in yeast is not known. Furthermore, ER to Golgi transport is blocked in a mutant strain defective in the SEC 21 gene, the yeast homologue of γ-COP (Kaiser and Schekman, 1990; Hosobuchi et al., 1992; Stenbeck et al., 1992). Homologues of the α-, β'- and δ- subunits of mammalian coatomer have also been found in yeast (unpublished data, Harter et al., 1993). Their secretion phenotype, however, has not been characterized. The high degree of structural conservation between yeast and mammalian coatomer strongly suggests a conserved functional role of the coatomer (Hosobuchi et al., 1992; Harter et al., 1993). Yet, the *in vivo* requirement for the COP coat in yeast has not been reproduced in the *in vitro* system.

The coat of ER derived transport vesicles obtained from an *in vitro* system is composed of three cytosolic protein fractions: Sar1p, a complex of Sec23p/Sec24p, and a complex of Sec13p/p150 (Tab. 2) (Nakano and Muramatsu, 1989; Hicke et al., 1992; Salama et al., 1993). The formation of fusion competent transport vesicles from the ER in yeast was reconstituted *in vitro* with these purified proteins. Sar1p, a

GTPase closely related to ARF, has been suggested to play a role analogous to that of ARF, but at an earlier stage of secretion (d'Enfert et al., 1991). The coat would only assemble when Sar-GTP is bound.

The general principle that underlies vesicle formation seems to be the binding of a GTPase to the target membrane which is a prerequisite for the recruitment of cytosolic proteins that exist in the cytosol as a complex. Uncoating and coat formation is regulated by a cycle of GTP hydrolysis and GDP-GTP exchange. For yeast, the GTPase activating protein for Sar1p, Sec23p, was identified, as well as the GDP-GTP exchange factor, Sec12p, a transmembrane glycoprotein (Barlowe and Schekman, 1993; Yoshihisa et al., 1993). In the mammalian systems regulators for the GTP-GDP-cycle are not identified.

Function of the protein coat

The transport of cargo molecules in vesicular carriers that bud from the donor membrane and fuse with the target membrane ensures that the structural and functional organization of the endomembrane system is maintained. At least two sorts of vesicles are likely to exist: the one moving cargo in the forward direction described in this paper, and the other type recycling membrane lipids and proteins that escaped retention (Lippincott-Schwartz, 1993). What are the characteristic features of these two types of vesicles and what regulates their formation and consumption? Constitutive intracellular transport in the forward ("anterograde") direction is mediated by COP-coated vesicles. Budding of vesicles is inititated by the assembly of the protein coat (Fig. 1). Once the budding is completed, the vesicle coat has to be removed in order to allow the vesicle to fuse with the target membrane. For this uncoating reaction, the hydrolysis of GTP is required, since in the presence of GTPγS, a nonhydrolyzable analog of GTP, coated vesicles accumulate (Melancon et al., 1987). The same effect is obtained in cells transfected with a GTPase mutant of ARF (Q71L) (Tanigawa et al., 1993; Dascher and Balch, 1994; Zhang et al., 1994). This mutant protein binds but does not hydrolyze GTP, resulting in the accumulation of coated vesicles that are incompetent for fusion with the acceptor membrane. Therefore, uncoating may be considered as a reversal of coat assembly: binding of coatomer to membranes requires ARF-GTP–when GTP is hydrolyzed ARF undergoes a conformational change by which the myristic acid tail will be retracted, and ARF-GDP will be released into the cytosol followed by coatomer. This cycle allows the protein coat to serve as a device to prevent direct fusion, e.g., it couples fusion to budding (Elazar et al., 1994).

Direct fusion occurs by treatment of cells with the fungal metabolite brefeldin A (BFA). It has been shown that in the presence of BFA secretion is blocked (Misumi et al., 1986). The Golgi complex vesiculates and forms extended tubules that fuse with the ER (Klausner et al., 1992). Thus, BFA inhibits the anterograde membrane traffic while it seems to enhance a retrograde transport from the Golgi to the ER (recycling pathway) (Lippincott-Schwartz et al., 1990). What causes the dramatic rearrangement of the Golgi apparatus upon BFA treatment? One explanation for the translocation of Golgi functions into the ER was that transport into the ER induced by BFA represents enhanced trafficking through the normal retrograde pathway as a result of the absorption of Golgi protein and membrane components into this pathway (Lippincott-Schwartz, 1993). On the other hand, the redistribution of Golgi contents into the ER might be caused by uncontrolled membrane fusion: both ARF and β-COP (probably as part of coatomer) rapidly dissociate from Golgi membranes (Donaldson et al., 1991a, b). BFA blocks binding of coatomer to Golgi membranes and thus the budding of coated vesicles by interfering with the initial binding of ARF (Donaldson et al., 1992a; Palmer et al., 1993). ARF binding requires a nucleotide exchange factor that is sensitive to BFA treatment (Donaldson et al., 1992b; Helms and Rothman, 1992). Accordingly, when coat assembly is blocked by BFA treatment vesicular transport stops and the involved compartments fuse directly. Thus, BFA uncouples fusion from budding (Elazar et al., 1994).

An interaction of coatomer with the consensus motif KKXX for retrieval of ER membrane proteins has been shown using mammalian cells transfected with a fusion protein that contains the cytoplasmic tail of the ER resident protein E19 of adenovirus or with WBP1, a component of the oligosaccharyl transferase of *S. cerevisiae* (Jackson et al., 1990; Cosson and Letourneur, 1994). This finding might be taken as an indication for the involvement of coatomer in a recycling pathway. In this context, it might be of interest that several coatomer subunits have been localized by electron microscopy to special compartments, so-called BFA-bodies that are closely associated with the ER (Orci et al., 1993c). Possibly, BFA bodies represent a membrane subcompartment from which coatomer is recruited to membranes mediating both antero-grade and retrograde trafficking. Further studies are required towards the characterization of putative binding partners of coatomer.

In summary, under physiological conditions, the COP coat regulates vesicular trafficking at distinct levels: (i) it induces vesicle budding, (ii) it prevents direct fusion of membranes, possibly by masking the receptors involved in membrane fusion, and (iii) it might be involved in a recycling pathway.

Perspectives

As as general mechanism, ARF-GTP binding (or the binding of related small GTP-binding proteins) is a prerequisite for recruitment of coatomer from cytosol. In light of the dramatic effects caused by interfering with the GDP-GTP exchange on ARF by BFA, ARF seems to represent a major regulatory component in keeping the steady state of the endomembrane system. Characterization of the guanine nucleotide exchange protein, target of BFA, will be a key to understand the mechanisms that underly ARF-regulation. In addition, nothing is known to date about the donor membrane protein(s) that interact specifically with ARF. Likewise, the membrane partner(s) to which coatomer is recruited during the budding reaction remain to be characterized.

As outlined above, several observations indicate that in addition to small G proteins, trimeric G proteins seem to be involved in vesicles budding, leading to an even higher complexity of regulation of reactions crucial for vesicular transport. Although it is not clear why constitutive vesicular transport should be regulated at all it will be of importance to pin down the molecular mechanism by which trimeric G proteins are involved in budding. They may play a role for the coordinate dissociation and association of organelles during and after mitosis.

Seemingly discrepant results exist as to the components needed for vesicle budding in yeast. In an *in vitro* system, no coatomer is found on transport vesicles. On the other hand, a coatomer closely related to that from mammals exists in yeast cytosol, and from genetic results it was concluded that Sec21p, a subunit of this coatomer is needed for ER budding. One possibility to resolve this problem is that the coat components defined biochemically are a first layer on the vesicles that is then covered by a second layer made from coatomer, and that the kinetics of coatomer uncoating after the generation of a transport vesicle would not allow one to isolate coatomer containing vesicles from the yeast system.

Acknowledgements
This work was supported by the Deutsche Forschungsgemeinschaft (SFB 352) and the Human Frontiers Science Program to F.T.W.

References

Allan, V.J. and Kreis, T.E. (1986) A microtubule-binding protein associated with the membranes of the Golgi apparatus. *J. Cell Biol.* 103: 2229–2239.

Balch, W.E., Dunphy, W.G., Braell, W.A. and Rothman, J.E. (1984a) Reconstitution of the transport of protein between successive compartments of the Golgi measured by the coupled incorporation of N-acetylglucosamine. *Cell* 39: 405–416.

Balch, W.E., Glick, B.S. and Rothman, J.E. (1984b) Sequential intermediates in the pathway of intercompartmental transport in a cell-free system. *Cell* 39: 525–536.

Balch, W.E., Kahn, R.A. and Schwaninger, R. (1992) ADP-ribosylation factor is required for vesicular trafficking between the endoplasmic reticulum and the cis-Golgi compartment. *J. Biol. Chem.* 267: 13053–13061.

Barlowe, C. and Schekman, R. (1993) SEC12 encodes a guanine-nucleotide-exchange factor essential for transport vesicle budding from the ER. *Nature* 365: 347–349.

Bigay, J., Deterre, P., Pfister, C. and Chabre, M. (1987) Fluoride complexes of aluminium and beryllium act on G-proteins are reversibly bound analogues of the γ phosphate group of GTP. *EMBO J.* 6: 2907–2913.

Braell, W.A., Balch, W.E., Dobbertin, D.C. and Rothman, J.E. (1984) The glycoprotein that is transported between successive compartments of the Golgi in a cell-free system resides in stacks of cisternae. *Cell* 39: 511–524.

Cosson, P. and Letourneur, F. (1994) Coatomer interaction with di-lysine endoplasmic reticulum retention motifs. *Science* 263: 1629–1631.

d'Enfert, C., Wuestehube, L.J., Lila, T. and Schekman, R. (1991) Sec12p-dependent membrane binding of the small GTP-binding protein Sar1p promotes formation of transport vesicles from the ER. *J. Cell Biol.* 114: 663–670.

Dascher, C. and Balch, W.E. (1994) Dominant inhibitory mutants of ARF1 block endoplasmic reticulum to Golgi transport and trigger disassembly of the Golgi apparatus. *J. Biol. Chem.* 269: 1437–1448.

Donaldson, J.G., Cassel, D., Kahn, R.A. and Klausner, R.D. (1992a) ADP-ribosylation factor, a small GTP-binding protein, is required for binding of the coatomer protein beta-COP to Golgi membranes. *Proc. Natl. Acad. Sci. USA* 89: 6408–6412.

Donaldson, J.G., Finazzi, D. and Klausner, R.D. (1992b) Brefeldin A inhibits Golgi membrane-catalysed exchange of guanine nucleotide onto ARF protein. *Nature* 360: 350–352.

Donaldson, J.G., Kahn, R.A., Lippincott, S.J. and Klausner, R.D. (1991a) Binding of ARF and beta-COP to Golgi membranes: possible regulation by a trimeric G protein. *Science* 254: 1197–1199.

Donaldson, J.G., Lippincott, S.J. and Klausner, R.D. (1991b) Guanine nucleotides modulate the effects of brefeldin A in semipermeable cells: regulation of the association of a 110-kD peripheral membrane protein with the Golgi apparatus. *J. Cell. Biol.* 112: 579–588.

Duden, R., Griffiths, G., Frank, R., Argos, P. and Kreis, T.E. (1991) Beta-COP, a 110 kd protein associated with non-clathrin-coated vesicles and the Golgi complex, shows homology to beta-adaptin. *Cell* 64: 649–665.

Elazar, Z., Orci, L., Ostermann, J., Amherdt, M., Tanigawa, G. and Rothman, J.E. (1994) ADP-Ribosylation factor and coatomer couple fusion to vesicle budding. *J. Cell Biol.* 124: 415–424.

Ercolani, L., Stow, J.L., Boyle, J.F., Holtzman, E.J., Lin, H., Grove, J.R. and Ausiello, D.A. (1990) Membrane localization of the pertussis toxin-sensitive G-protein subunits alpha i-2 and alpha i-3 and expression of a metallothionein-alpha i-2 fusion gene in LLC-PK1 cells *Proc. Natl. Acad. Sci. USA* 87: 4635–4639.

Farquhar, M.G. (1985) Progress in unraveling pathways of Golgi traffic. *Annu. Rev. Cell Biol.* 1: 447–488.

Fries, E. and Rothman, J.E. (1980) Transport of vesicular stomatitis virus glycoprotein in a cell-free extract. *Proc. Natl. Acad. Sci. USA* 77: 3870–3874.

Hara-Kuge, S., Kuge, O., Orci, L., Amherdt, M., Ravazzola, M., Wieland, F.T. and Rothman, J.E. (1994) En bloc incorporation of coatomer subunits during the assembly of COP-coated vesicles. *J. Cell Biol.* 124: 883–892.

Harrison-Lavoie, K.J., Lewis, V.A., Hynes, G.M., Collison, K.S., Nutland, E. and Willison, K.R. (1993) A 102 kDa subunit of a Golgi-associated particle has homology to beta subunits of trimeric G proteins. *EMBO J.* 12: 2847–2853.

Harter, C., Draken, E., Lottspeich, F. and Wieland, F.T. (1993) Yeast coatomer contains a subunit homologuos to mammalian beta'-COP. *FEBS Lett.* 332: 71–73.

Helms, J.B., Palmer, D.J. and Rothman, J.E. (1993) Two distinct populations of ARF bound to Golgi membranes. *J. Cell Biol.* 121: 751–760.

Helms, J.B. and Rothman, J.E. (1992) Inhibition by brefeldin A of a Golgi membrane enzyme that catalyses exchange of guanine nucleotide bound to ARF. *Nature* 360: 352–354.

Hicke, L., Yoshihisa, T. and Schekman, R. (1992) Sec23p and a novel 105-kDa protein function as a multimeric complex to promote vesicle budding and protein transport from the endoplasmic reticulum. *Mol. Biol. Cell* 3: 667–676.

Hosobuchi, M., Kreis, T. and Schekman, R. (1992) SEC21 is a gene required for ER to Golgi protein transport that encodes a subunit of a yeast coatomer. *Nature* 360: 603–605.

Jackson, M.R., Nilsson, T. and Peterson, P.A. (1990) Identification of a consensus motif for retention of transmembrane proteins in the endoplasmic reticulum. *EMBO J.* 9: 3153–3162.

Kahn, R.A. (1991) Fluoride is not an activator of the smaller (20–25 kDa) GTP-binding proteins. *J. Biol. Chem.* 266: 15595–15597.

Kahn, R.A. and Gilman, A.G. (1986) The protein cofactor necessary for ADP-ribosylation of Gs by cholera toxin is itself a GTP binding protein. *J. Biol. Chem.* 261: 7906–7911.

Kahn, R.A., Kern, F.G., Clark, J., Gelmann, E.P. and Rulka, C. (1991) Human ADP-ribosylation factors. A functionally conserved family of GTP-binding proteins. *J. Biol. Chem.* 266: 2606–2614.

Kaiser, C.A. and Schekman, R. (1990) Distinct sets of SEC genes govern transport vesicle formation and fusion early in the secretory pathway. *Cell* 61: 723–733.

Klausner, R.D., Donaldson, J.G. and Lippincott, S.J. (1992) Brefeldin A: insights into the control of membrane traffic and organelle structure. *J. Cell Biol.* 116: 1071–1080.

Kuge, O., Dascher, C., Orci, L., Rowe, T., Amherdt, M., Plutner, H., Ravazzola, M., Tanigawa, G., Rothman, J.E. and Balch, W.E. (1994) Sar1 promotes vesicle budding from the endoplasmic reticulum but not golgi compartments. *J. Cell Biol.* 125: 51–65.

Kuge, O., Hara, K.S., Orci, L., Ravazzola, M., Amherdt, M., Tanigawa, G., Wieland, F.T. and Rothman, J.E. (1993) zeta-COP, a subunit of coatomer, is required for COP-coated vesicle assembly. *J. Cell Biol.* 123: 1727–1734.

Leyte, A., Barr, F.A., Kehlenbach, H. and Huttner, W.B. (1992) Multiple trimeric G-proteins on the trans-Golgi network exert stimulatory and inhibitory effects on secretory vesicle formation. *EMBO J.* 11: 4795–4804.

Lippincott-Schwartz, J., Donaldson, J.G., Schweizer, A., Berger, E.G., Hauri, H.P., Yuan, L.C. and Klausner, R.D. (1990) Microtubule-dependent retrograde transport of proteins into the ER in the presence of brefeldin A suggests an ER recycling pathway. *Cell* 60: 821–836.

Lippincott-Schwartz, J. (1993) Bidirectional membrane traffic between the endoplasmic reticulum and Golgi apparatus. *Trends Cell Biol.* 3: 81–88.

Malhotra, V., Serafini, T., Orci, L., Shepherd, J.C. and Rothman, J.E. (1989) Purification of a novel class of coated vesicles mediating biosynthetic protein transport through the Golgi stack. *Cell* 58: 329–336.

Melancon, P., Glick, B.S., Malhotra, V., Weidman, P.J., Serafini, T., Gleason, M.L., Orci, L. and Rothman, J.E. (1987) Involvement of GTP-binding "G" proteins in transport through the Golgi stack. *Cell* 51: 1053–1062.

Misumi, Y., Miki, A., Takatsuki, A., Tamura, G. and Ikehara, Y. (1986) Novel blockade by brefeldin A of intracellular transport of secretory proteins in cultured rat hepatocytes. *J. Biol. Chem.* 261: 11398–11403.

Nakano, A. and Muramatsu, M. (1989) A novel GTP-binding protein, Sar1p, is involved in transport from the endoplasmic reticulum to the Golgi apparatus. *J. Cell Biol.* 109: 2677–2691.

Orci, L., Glick, B.S. and Rothman, J.E. (1986) A new type of coated vesicular carrier that appears not to contain clathrin: its possible role in protein transport within the Golgi stack. *Cell* 46: 171–184.

Orci, L., Ravazzola, M., Meda, P., Holcomb, C., Moore, H.P., Hicke, L. and Schekman, R. (1991) Mammalain Sec23p homologue is restricted to the endoplasmic reticulum transitional cytoplasm. *Proc. Natl. Acad. Sci. USA* 88: 8611–8615.

Orci, L., Palmer, D.J., Ravazzola, M., Perrelet, A., Amherdt, M. and Rothman, J.E. (1993a) Budding from Golgi membranes requires the coatomer complex of non-clathrin coat proteins. *Nature* 362: 648–652.

Orci, L., Palmer, D.J., Amherdt, M. and Rothman, J.E. (1993b) Coated vesicle assembly in the Golgi requires only coatomer and ARF proteins from the cytosol. *Nature* 364: 732–734.

Orci, L., Perrelet, A., Ravazzola, M., Wieland, F.T., Schekman, R. and Rothman, J.E. (1993c) "BFA bodies": A subcompartment of the endoplasmic reticulum. *Proc. Natl. Acad. Sci. USA* 90: 11089–11093.

Ostermann, J., Orci, L., Tani, K., Amherdt, M., Ravazzola, M., Elazar, Z. and Rothman, J.E. (1993) Stepwise assembly of functionally active transport vesicles. *Cell* 75:1015–1025.

Palmer, D.J., Helms, J.B., Beckers, C.J., Orci, L. and Rothman, J.E. (1993) Binding of coatomer to Golgi membranes requires ADP-ribosylation factor. *J. Biol. Chem.* 268: 12083–12099.

Pepperkok, R., Scheel, J., Horstmann, H., Hauri, H.P., Griffiths, G. and Kreis, T.E. (1993) Beta-COP is essential for biosynthetic membrane transport from the endoplasmic reticulum to the Golgi complex *in vivo*. *Cell* 74: 71–82.

Peter, F., Plutner, H., Zhu, H., Kreis, T.E. and Balch, W.E. (1993) Beta-COP is essential for transport of protein from the endoplasmic reticulum to the Golgi *in vitro*. *J. Cell Biol.* 122: 1155–1167.

Pfanner, N., Orci, L., Glick, B.S., Amherdt, M., Arden, S.R., Malhotra, V. and Rothman, J.E. (1989) Fatty acyl-coenzyme A is required for budding of transport vesicles from Golgi cisternae. *Cell* 59: 95–102.

Pfeffer, S.R. and Rothman, J.E. (1987) Biosynthetic protein transport and sorting by the endoplasmic reticulum and Golgi. *Annu. Rev. Biochem.* 56: 829–852.

Plutner, H., Cox, A.D., Pind, S., Khosravi, F.R., Bourne, J.R., Schwaninger, R., Der, C.J. and Balch, W.E. (1991) Rab1b regulates vesicular transport between the endoplasmic reticulum and successive Golgi compartments. *J. Cell Biol.* 115: 31–43.

Pryer, N.K., Wuestehube, L.J. and Schekman, R. (1992) Vesicle-mediated protein sorting. *Annu. Rev. Biochem.* 61: 471–516.

Robinson, M.S. and Kreis, T.E. (1992) Recruitment of coat proteins onto Golgi membranes in intact and permeabilized cells: effects of brefeldin A and G protein activators. *Cell* 69: 129–138.

Rothman, J.E. (1992) Reconstitution of intracellular transport. *Methods Enzymol.* 219.

Rothman, J.E. and Orci, L. (1992) Molecular dissection of the secretory pathway. *Nature* 335: 409–415.

Salama, N.R., Yeung, T. and Schekman, R.W. (1993) The Sec13p complex and reconstitution of vesicle budding from the ER with purified cytosolic proteins. *EMBO J.* 12: 4073–4082.

Serafini, T., Stenbeck, G., Brecht, A., Lottspeich, F., Orci, L., Rothman, J.E. and Wieland, F.T. (1991a) A coat subunit of Golgi-derived non-clathrin-coated vesicles with homology to the clathrin-coated vesicle coat protein beta-adaptin. *Nature* 349: 215–220.

Serafini, T., Orci, L., Amherdt, M., Brunner, M., Kahn, R.A. and Rothman, J.E. (1991b) ADP-ribosylation factor is a subunit of the coat of Golgi-derived COP-coated vesicles: a novel role for a GTP-binding protein. *Cell* 67: 239–253.

Stearns, T., Willingham, M.C., Botstein, D. and Kahn, R.A. (1990) ADP-ribosylation factor is functionally and physically associated with the Golgi complex. *Proc. Natl. Acad. Sci. USA* 87: 1238–1242.

Stenbeck, G., Harter, C., Brecht, A., Herrmann, D., Lottspeich, F., Orci, L. and Wieland, F.T. (1993) beta'-COP, a novel subunit of coatomer. *EMBO J.* 12: 2841–2845.

Stenbeck, G., Schreiner, R., Herrmann, D., Auerbach, S., Lottspeich, F., Rothman, J.E. and Wieland, F.T. (1992) Gamma-COP, a coat subunit of non-clathrin-coated vesicles with homology to Sec21p. *FEBS Lett.* 314: 195–198.

Stow, J.L., de, A.J., Narula, N., Holtzman, E.J., Ercolani, L. and Ausiello, D.A. (1991) A heterotrimeric G protein, G. alpha i-3, on Golgi membranes regulates the secretion of a heparan sulfate proteoglycan in LLC-PK1 epithelial cells. *J. Cell Biol.* 114: 1113–1124.

Tanigawa, G., Orci, L., Amherdt, M., Ravazzola, M., Helms, J.B. and Rothman, J.E. (1993) Hydrolysis of bound GTP by ARF protein triggers uncoating of Golgi-derived COP-coated vesicles. *J. Cell Biol.* 123: 1365–1371.

van der Voorn, L. and Ploegh, H.L. (1992) The WD-40 repeat. *FEBS Lett.* 307: 131–134.

Waters, M.G., Serafini, T. and Rothman, J.E. (1991) 'Coatomer': a cytosolic protein complex containing subunits of non-clathrin-coated Golgi transport vesicles. *Nature* 349: 248–251.

Yoshihisa, T., Barlowe, C. and Schekman, R. (1993) Requirement for a GTPase-activating protein in vesicle budding from the endoplasmic reticulum. *Science* 259: 1466–1468.

Zhang, C-j., Rosenwald, A.G., Willingham, M.C., Skuntz, S., Clark, J. and Kahn, R.A. (1994) Expression of a dominant allele of human ARF1 inhibits traffic *in vivo*. *J. Cell Biol.* 124: 289–300.

Biochemistry of Cell Membranes
ed. by S. Papa & J. M. Tager
© 1995 Birkhäuser Verlag Basel/Switzerland

Role of the GTP/GDP switch for the function of rab5 in endocytosis

H. Stenmark and M. Zerial

Cell Biology Programme, European Molecular Biology Laboratory, D-69012 Heidelberg, Germany

Summary. Rab GTPases are specific regulators of vesicular transport. They are predicted to exist in two distinct conformations, depending on whether GDP or GTP is bound. Here, we discuss the role of the nucleotide state for the function of rab5, which operates in the early endocytic pathway. Regions involved in nucleotide dependent conformational changes contribute to functionally distinguish rab5 from rab6. Furthermore, mutational analysis of rab5 suggests that the GTP-bound form of the molecule is the active one in vesicle docking or fusion. This leads to propose a model for rab5 function in endocytosis.

Introduction

Distinct members of the GTPase superfamily are involved in fundamental biological processes, such as protein synthesis, membrane translocation of proteins, signal transduction, cytoskeletal organization, and vesicular transport (Bourne et al., 1991). Crystallographic studies on two GTPases, elongation factor-Tu (involved in prokaryote protein synthesis) and p21-ras (involved in signal transduction) have established a crucial feature of this class of proteins: They exist in two distinct conformations, depending on whether GDP or GTP is bound. This allows the GTPases to interact with other molecules in a nucleotide-dependent fashion. The conversion from GTP- to GDP-bound form occurs through hydrolysis of GTP, whereas the switch from GDP- to GTP-bound form is dependent on nucleotide exchange. Both these processes occur in general slowly *in vitro*, but their rate is greatly increased by cellular factors, termed GAPs (GTPase activating proteins) and GEFs (Guanine nucleotide exchange factors), respectively (Bourne et al., 1991).

Small GTPases of the rab family have received much attention recently, due to their involvement in vesicular transport between separate membrane compartments. The more than 30 different rab GTPases identified so far shown striking differences in intracellular localization, in agreement with the view that they serve as specific regulators of membrane traffic (Simons and Zerial, 1993; Zerial and Stenmark, 1993).

For instance, rab2 is localized to an intermediate compartment between the endoplasmic reticulum and the Golgi apparatus (Chavrier et al., 1990), rab5 is found on the cytoplasmic side of the plasma membrane and on early endosomes (Chavrier et al., 1990), whereas rab6 is located to the Golgi complex (Antony et al., 1992). Rab proteins are present in two pools, one membrane-bound and the other cytosolic. The membrane association is reversible and is regulated by rab GDI (GDP dissociation inhibitor) (Araki et al., 1990; Soldati et al., 1993; Ullrich et al., 1993). While geranylgeranylated rab proteins are hydrophobic and bind nonspecifically to various surfaces, the complex between rab GDI and the rab GTPase is soluble, suggesting that rab GDI may serve as a solubilizing factor and prevent mistargeting of rab proteins to the inappropriate membranes (Ullrich et al., 1994; Zerial and Stenmark, 1993).

Although biochemical analysis, *in vitro* transport assays, overexpression studies and yeast genetics have provided important information about the function of rab GTPases in membrane traffic, their mode of action remains unknown. To begin with, it is not clear what role their GTP/GDP switch plays (Simons and Zerial, 1993). In order to address this issue we have chosen to study rab5, a rate-limiting factor in the early endocytic pathway. The function of rab5 can be readily studied *in vivo*, in transfected cells, as well as *in vitro*, in an endosome fusion assay (Bucci et al., 1992; Gorvel et al., 1991). While increased levels of rab5 have a stimulatory effect in both systems, interference with rab5 function is inhibitory. It is therefore possible to score for negative and positive effects in a way that facilitates a detailed functional analysis.

Results and Discussion

The function of different rab GTPases in distinct transport processes is paradoxed by the high similarity between them (Valencia et al., 1991) Obviously, specific structural elements must be responsible for this functional diversity. Possible candidates are the N- and C-termini, since these are the most divergent parts of rab GTPases. In addition, however, one would expect the switch regions (those that change conformation depending on the nucleotide bound) to be involved in specific interactions with accessory proteins. Molecular modeling based on crystallographic data from related GTPases identify the effector loop (loop 2) as well as α-helices 2 and 3 as switch regions (Stouten et al., 1993). Also in these regions, there are sequence variations among different rab GTPases. To test the role of the N- and C-termini and the switch regions we took advantage of the fact that rab5, when overexpressed, strongly stimulates the rate of endocytosis (Bucci et al., 1992), whereas other rab GTPases do not. This allowed us to conduct a series

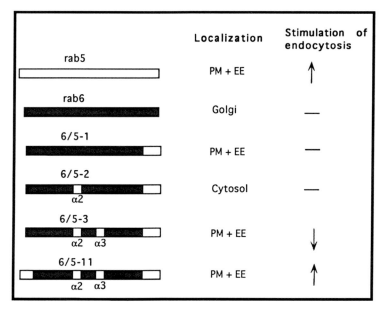

Figure 1. Functional properties of rab5, rab6 and chimeras. An upward arrow means stimulation, a downward arrow inhibition. PM, plasma membrane; EE, early endosomes.

of experiments where parts of rab5 were transplanted onto rab6, to determine whether structural elements could be identified that imposed rab5-like effect on endocytosis onto rab6 (Stenmark et al., 1994b).

The main results of these experiments are represented schematically in Figure 1. We first demonstrated that the C-terminus of rab5 was sufficient to retarget rab6 from its normal location in the Golgi complex to rab5-containing compartments (plasma membrane and early endosomes), in agreement with the view that the C-terminus functions as a localization signal (Chavrier et al., 1991; Brennwald and Novick, 1993). However, this hybrid protein was without effect on endocytosis, indicating that localization of a rab protein to a certain membrane is not sufficient to specify its function. Additional replacement of one of the switch regions, helix $\alpha 2$, led to a protein that was cytosolic, despite having the ability to be geranylgeranylated. Although the reason for this is not clear, it became important to substitute helix $\alpha 3$ as well, since this is predicted to interact with $\alpha 2$. Indeed, the resulting protein was membrane localized again. However, instead of stimulating endocytosis like rab5, it rather inhibited it. This suggested that the hybrid protein interacted with rab5 accessory proteins in a nonproductive fashion. Assuming that another critical part was still missing, we then replaced in addition the third region involved in conformational changes, loop 2. However, this was without effect. In contrast, when we exchanged the divergent N-termini of the proteins, the resultant hybrid had the ability to stimulate endocytosis (see Fig. 1). Other combinations were without effect. We thus found that regions involved in conformational changes, together with the divergent N- and C-termini define the functional specificity of rab5. Most likely, they do so by interacting with specific regulatory or target molecules.

The experiments with rab6/rab5 chimeras emphasized the crucial role of the switch regions for rab5 function. They did not address, however, how the GTP/GDP switch mediates the function of rab5. To investigate this, we constructed two rab5 mutants with alterations in their GTP/GDP cycle (Stenmark et al., 1994a). Rab5 S34N corresponds to the inhibitory S17N mutant of p21-ras, which inhibits a GEF (Feig and Cooper, 1988; Medema et al., 1993). Like p21-ras S17N, rab5 S34N has preferential affinity for GDP. Expression of this protein would be expected to inhibit nucleotide exchange of endogenous rab5, thereby leading to a decreased level of GTP-bound rab5 on the membrane. Interestingly, expression of this mutant led to the accumulation in the cell of fragmented early endosomes and early endocytic transport intermediates. Furthermore, endocytosis of transferrin occured at a slower rate than in control cells, and the protein inhibited homotypic fusion between early endosomes in cell-free assay (see Tab. 1). It thus appears that budding of endocytic vesicles, but not docking to or fusion with early endosomes, can occur in the absence of GTP-bound rab5.

The other mutant we tested, rab5 Q79L is analogous to the strongly activating Q61L mutant of p21-ras (Der et al., 1986) and displays a very low intrinsic GTPase activity. Furthermore, when expressed in cells, it was found mainly in the GTP-bound form, unlike wild-type rab5. Overexpression of this mutant had essentially the opposite effect as the S34N mutant: It expanded the size of early endosomes dramatically, and, like overexpressed wild-type rab5, it increased the rate of endocytosis. Unlike wild-type rab5, however, it caused a partial inhibition of recycling at high expression levels, leading to a lowered number of transferrin receptors on the cell surface. In the cell-free assay, cytosol

Table 1. Properties of wild-type and mutant rab5 when overexpressed in BHK cells using the T7 RNA polymerase recombinant vaccina virus system

Protein	% GTP	EE size	Endocytosis	EE fusion
rab5-wt	21	large	stimulated	stimulated
rab5 S34N	5	small	inhibited	inhibited
rab5 Q79L	63	extra large	stimulated	stimulated

The table summarizes the nucleotide state of the different overexpressed proteins *in vivo* (% GTP) and the effects of overexpressed proteins on the size of early endosomes, the rate of endocytosis and the extent of early endosome fusion *in vitro*.

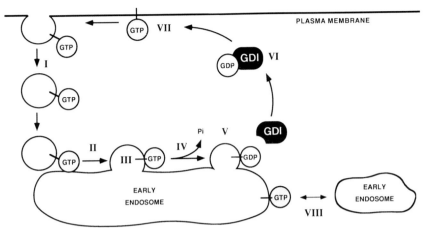

Figure 2. Hypothetical model for the role of rab5 in the early endocytic pathway. Rab5:GTP is present on clathrin-coated vesicles (I) that bud off from the plasma membrane. Rab5:GTP promotes docking (II) between the endocytic vesicle and the early endosome. This is a prerequisite for fusion (III) to occur. Subsequently, rab5 hydrolyzes GTP (IV), and rab5:GDP is removed from the membrane by GDI (V). GDI recycles rab5:GDP back to the plasma membrane via the cytosol (VI). Here, rab5:GDP associates with the membrane, and GDP is exchanged with GTP (VII). Rab5:GTP can now be recruited to a coated pit to enter a new round of transport. A mutant with decreased GTPase activity would reside longer in the early endosome membrane in the GTP-bound form, and might therefore increase attachment to, and thereby fusion with, neighboring early endosomes (VIII).

containing rab5 Q79L stimulated fusion strongly, although the mutant was found in lower concentration than wild-type rab5 in the cytosol. The combined results with this mutant (Stenmark et al., 1994a)(see Tab. 1) indicate that the GTP-bound form of rab5 is capable of promoting vesicle fusion, conceivably by enhancing the association between two compatible membranes (*i.e.*, two early endosomes or an endocytic vesicle and an early endosome). Importantly, these results suggest that GTP hydrolysis by rab5 is not required prior to membrane fusion, in opposition to a previous hypothesis (Bourne, 1988). We thus propose the following model for the function of rab5 (Fig. 2): Rab5 is present on clathrin-coated vesicles (Bucci et al., 1992) that bud from the plasma membrane (I). Docking of the uncoated vesicle to the early endosome requires rab5 in the GTP-bound form (II) and is a prerequi-site for membrane fusion to occur (III). GTP hydrolysis (IV) then causes a conformational change in rab5, switching off its interaction with target proteins and allowing the removal from the membrane by GDI (V)(Ullrich et al., 1993). In this way, rab5 can be recycled back to the plasma membrane via the cytosol (VI). Membrane insertion is accompanied by nucleotide exchange (Ullrich et al., 1994) (VII). Ac-cording to this model, a mutant with reduced GTPase activity would reside longer in the early endosome membrane and might therefore promote interaction with neighbouring early endosomes. This might cause the formation of the giant early endosomes observed in the presence of rab5 Q79L.

Conclusion

The experiments discussed here emphasize the role of nucleotide-dependent conformational changes for the function of rab5 in endocytosis. Regions involved in these changes contribute to the functional specificity of rab5 and are therefore likely to interact with specific accessory proteins. As it appears that the GTP-bound form is the active one in vesicle docking and fusion, the helices $\alpha2$ and $\alpha3$, possibly together with the N-terminus, may interact with and activate a target at the vesicle or early endosome membrane when GTP is bound. Upon GTP hydrolysis, this interaction would be terminated. Although no target molecules for rab5 have been identified so far, interesting candidates are members of the VAMP and syntaxin families of vesicle targeting molecules, or molecules regulating their activity (Novick and Brennwald, 1993; Sollner et al., 1993; Zerial and Stenmark, 1993). Future work will aim at clarifying this issue.

References

Antony, C., Cibert, C., Geraud, G., Santa-Maria, A., Maro, B., Mayau, V. and Goud, B. (1992) The small GTP-binding protein Rab6p is distributed from medial Golgi to the *trans*-Golgi network as determined by a confocal microscopy approach. *J. Cell Science* 103: 785–796.

Araki, S., Kikuchi, A., Hata, Y., Isomura, M. and Takai, Y. (1990) Regulation of reversible binding of *smg* p25A, a *ras* p21-like GTP-binding protein, to synaptic plasma membranes and vesicles by its specific regulatory protein, GDP dissociation inhibitor. *J. Biol. Chem.* 265: 13007–13015.

Bourne, H., Sanders, D. and McCormick, F. (1991) The GTPase superfamily; conserved structure and molecular mechanism. *Nature* 349: 117–127.

Bourne, H.R. (1988) Do GTPases direct membrane traffic in secretion? *Cell* 53: 669–671.

Brennwald, P. and Novick, P. (1993) Interactions of three domains distinguishing the Ras-related GTP-binding proteins Ypt1 and Sec4. *Nature* 362: 560–563.

Bucci, C., Parton, R.G., Mather, I.H., Stunnenberg, H., Simons, K., Hoflack, B. and Zerial, M. (1992) The small GTPase rab5 functions as a regulatory factor in the early endocytic pathway. *Cell* 70: 715–728.

Chavrier, P., Gorvel, J.-P., Steltzer, E., Simons, K., Gruenberg, J. and Zerial, M. (1991) Hypervariable C-terminal domain of rab proteins acts as a targeting signal. *Nature* 353: 769–772.

Chavrier, P., Parton, R.G., Hauri, H.P., Simons, K. and Zerial, M. (1990) Localization of low molecular weight GTP binding proteins to exocytic and endocytic compartments. *Cell* 62: 317–329.

Der, C.J., Finkel, T. and Cooper, G.M. (1986) Biologial and biochemical properties of human *ras*H genes mutated at codon 61. *Cell* 44: 167–176.

Feig, L.A. and Cooper, G.M. (1988) Inhibition of NIH 3T3 cell proliferation by a mutant *ras* protein with preferential affinity for GDP. *Mol. Cell. Biol.* 8: 3235–3243.

Gorvel, J.-P., Chavrier, P. Zerial, M. and Gruenberg, J. (1991) Rab5 controls early endosome fusion *in vitro*. *Cell* 64: 915–925.

Medema, R.H., de Vries-Smits, A.M., van der Zon, G.C., Maassen, J.A. and Bos, J.L. (1993) Ras activation by insulin and epidermal growth factor through enhanced exchange of guanine nucleotides on p21ras. *Mol. Cell. Biol.* 13: 155–162.

Novick, P. and Brennwald, P. (1993) Friends and family: The role of the Rab GTPases in vesicular transport. *Cell* 75: 597–601.

Simons, K. and Zerial, M. (1993) Rab proteins and the road maps for intracellular transport. *Neuron* 11: 789–799.

Soldati, T., Riederer, M.A. and Pfeffer, S.R. (1993) Rab GDI: A solubilizing and recycling factor for rab9 protein. *Mol. Biol. Cell* 4: 425–434.

Söllner, T., Whiteheart, S.W., Brunner, M., Erdjument-Bromage, H., Geromanos, S., Tempst, P. and Rothman, J.E. (1993) SNAP receptors implicated in vesicle targeting and fusion. *Nature* 362: 318–324.

Stenmark, H., Parton, R.G., Steele-Mortimer, O., Lütcke, A., Gruenberg, J. and Zerial, M. (1994a) Inhibition of rab5 GTPase activity stimulates membrane fusion in endocytosis. *EMBO J.* 13: 1287–1296.

Stenmark, H., Valencia, A., Martinez, O., Ullrich O., Goud, B. and Zerial, M. (1994b) Distinct structural elements of rab5 define its functional specificity. *EMBO J.* 13: 575–583.

Stouten, P.F., Sander, C. Wittinghofer, S.W. and Valencia, A. (1993) How does the switch II region of G-domains work? *FEBS Lett.* 320:1–6.

Ullrich, O., Horiuchi, H., Bucci, C. and Zerial, M. (1994) Membrane association of Rab5 mediated by GDP-dissociation inhibitor and accompanied by GDP/GTP exchange. *Nature* 368: 157–160.

Ullrich, O., Stenmark, H., Alexandrov, K., Huber, L.A., Kaibuchi, K., Sasaki, T., Takai, Y. and Zerial, M. (1993) Rab GDP dissociation inhibitor as a general regulator for the membrane association of rab proteins. *J. Biol. Chem.* 268: 18143–18150.

Valencia, A., Chardin, P., Wittinghofer, A. and Sander, C. (1991) The *ras* protein family: evolutionary tree and role of conserved amino acids. *Biochem.* 30: 4637–4648.

Zerial, M. and Stenmark, H. (1993) Rab GTPases in vesicular transport. *Cur. Op. Cell Biol.* 5: 613–620.

Biochemistry of Cell Membranes
ed. by S. Papa & J. M. Tager
© 1995 Birkhäuser Verlag Basel/Switzerland

Regulation of autophagy

E.F.C. Blommaart and A.J. Meijer

E.C. Slater Institute, University of Amsterdam, Academic Medical Centre, Meibergdreef 15, NL-1105 AZ Amsterdam, The Netherlands

Summary. The protein content of a cell is determined not only by the rate of protein synthesis but also by the rate of proteolysis. Part of the degradation of intracellular protein occurs in the lysosomes and is mediated by autophagy. Factors controlling autophagy, including amino acids, insulin, glucagon and cell volume, have opposite effects on protein synthesis. Recent evidence indicates that lysosomal protein degradation can be selective and occurs via ubiquitin dependent and independent pathways.

Introduction

The amount of cellular protein is determined by the relative rates of synthesis and degradation. At constant intracellular protein concentration, i.e. at steady state, these rates are equal. Although turnover of protein results in energy dissipation, regulation at the level of protein degradation is a very effective means of controlling protein levels (Paskin and Mayer, 1977).

Both intra- and extralysosomal pathways are responsible for the breakdown of cellular proteins. Extralysosomal proteolytic pathways include the 26S protease complex with multiple proteolytic activities, which is also part of the ATP- and ubiquitin-dependent proteolytic system (Hershko and Ciechanover, 1992), the Ca^{++}-dependent proteases (calpains) (Mellgren, 1987) and other, presumably ATP-independent, proteases. In liver, these systems are responsible for the breakdown of a small pool of proteins with a high turnover rate, whereas the degradation of long-lived protein predominantly occurs in the lysosomes by autophagy (Mortimore et al., 1989; Seglen and Bohley, 1992). In muscle, however, the contribution of extralysosomal proteolytic systems to overall protein degradation is greater and accelerated breakdown of myofibrillar protein, e.g. during starvation, proceeds via the ubiquitin system (Wing and Goldberg, 1993).

In this short review, we will highlight recent developments in the field of autophagy. Since the majority of the studies on autophagy has been carried out in liver, presumably because of the quantitative importance of autophagy in this tissue and because of the easy experimental accessibility of it, most of our discussion will be based on results obtained with this tissue. However, where necessary, data obtained with other cell types will also be discussed.

Lysosomal protein degradation

Three mechanisms of lysosomal degradation of intracellular proteins can be distinguished. These include direct targeting of proteins to the lysosomes by certain signal sequences, microautophagy and macroautophagy. Each of these mechanisms will briefly be described, followed by a detailed discussion of the mechanisms controlling the macroautophagic pathway which is, from a quantitative point of view, the most important lysosomal proteolytic system.

Uptake of specific cytosolic proteins by lysosomes

Under stress conditions such as starvation or serum withdrawal from cell culture media, in many cell types (e.g. liver, kidney, heart, fibroblasts but not in skeletal muscle (Wing et al., 1991; Goldberg and Dice, 1991)), specific cytosolic proteins can be directly targeted to the lysosomes for degradation. Proteins following this route possess peptide sequences similar to Lys-Phe-Glu-Arg-Gln (KFERQ) (Dice, 1990). This peptide signal is recognized by the cytosolic heat shock protein hsc73 which binds to the KFERQ region and helps in the recognition process by the lysosomes (Terlecki et al., 1992). After binding of the complex to a lysosomal receptor protein, the KFERQ containing protein is translocated across the lysosomal membrane and subsequently degraded (Terlecki and Dice, 1993). Serum withdrawal from cultured fibroblasts results in an activation of the hsc73 protein (rather than in an increase of the amount of hsc73) so that the recognition of KFERQ-like peptide motifs by the lysosomes becomes more effective (Terlecki and Dice, 1993).

Another example of targeting of a cytosolic protein to the lysosome was discovered in the yeast *Saccharomyces cerevisiae* (Chiang and Schekman, 1991). Shifting yeast from a medium containing ethanol to glucose as a carbon source resulted in the breakdown of the gluconeogenic enzyme fructose 1,6-diphosphatase in the yeast vacuole, which is the equivalent of the lysosome in mammalian cells. Mutants deficient in vacuolar proteases were unable to degrade fructose 1,6-diphosphatase. Presumably glucose stimulates a receptor protein in the membrane of the vacuole.

Direct specific lysosomal uptake of glyceraldehyde-3-phosphate dehydrogenase in rat liver has recently been described, but the mechanism by which the lysosomes recognize this protein is not known yet (Aniento et al., 1993).

Microautophagy

During microautophagy, portions of the cytoplasm are directly taken up by the lysosomes by invagination of their membranes (Ahlberg et al., 1982; Mortimore et al., 1983; De Waal et al., 1986; Mortimore et al., 1988a). Micro-autophagosomes are formed, but they are inside the lysosomes where they are rapidly degraded. In contrast to macroautophagy, microautophagy is insensitive to amino acids (Mortimore et al., 1983, 1988a). It is also assumed in the literature that microautophagy is ATP independent (Ahlberg et al., 1982). In our opinion however, it is difficult to understand that this complicated process of membrane flow does not require an input of energy.

There are indications that microautophagy declines in long-term starvation, although the signal responsible for this is unknown (Mortimore et al., 1983, 1988a). An attractive idea, although never tested, is that ketone bodies, which can replace glucose as a fuel for the brain under those conditions, are able to control this process.

Macroautophagy

During macroautophagy, small portions of the cytoplasm, sometimes even containing whole organelles (Fig. 1), are surrounded by a sequestering membrane which originates from the ribosome-free part of the rough endoplasmic reticulum (Dunn, 1990a). These structures, so-called autophagosomes, mature into degradative autophagosomes in a stepwise process: acquisition of lysosomal membrane proteins by fusion with vesicles deficient in hydrolytic enzymes (prelysosomes?), followed by acidification and fusion with existing lysosomes to acquire hydrolytic enzymes (Dunn, 1990b). Degradation of protein and other macromolecular material occurs in these mature autophagosomes. In liver, autophagosomes are synthesized and degraded with a half life of 8 min (Pfeifer et al., 1978; Schworer and Mortimore, 1979). Because of this high turnover rate of autophagosomes, the steady-state volume of autophagosomes in the cell is low even when autophagic flux is high. In hepatocytes, the volume of autophagosomes varies between 0.2% in the fed state to 1–1.5% of the total cell volume in the fasted state (Schworer and Mortimore, 1979; Kovács et al., 1982; Pfeifer, 1987).

Macroautophagy is responsible for the acceleration of proteolysis in many cell types when the concentration of amino acids and insulin falls. In rat liver *in vivo*, for example, when rates of protein synthesis and degradation are equal, protein turnover (which is a mixture of macroautophagy, microautophagy and extralysosomal proteolysis) is 1.5–2% of total intracellular protein per hour (Mortimore et al., 1989). During fasting, this balance is disturbed primarily due to an increase in

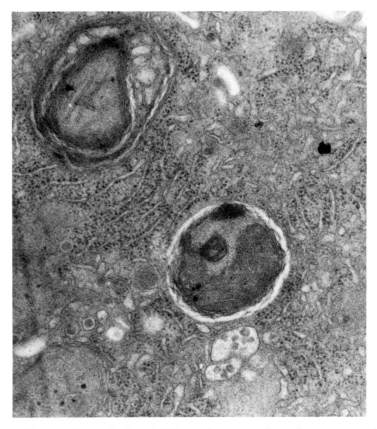

Figure 1. Electron micrograph of two initial autophagosomes of a rat hepatocyte containing still recognizable mitochondria (magnification 35 000 ×).

macroautophagy. In rats and mice there is a net loss of 20–25% of liver cell protein after 24 h fasting (Mortimore et al., 1989), equivalent to a rate of 1% per hour. Under extreme experimental conditions, i.e. in the absence of amino acids as can be imposed in a perfused liver system or in isolated hepatocytes, macroautophagy can reach values as high as 4–5% protein loss per hour (Mortimore et al., 1989; Seglen and Bohley, 1992).

Control of macroautophagy

Energy requirement

Macroautophagy is ATP-dependent (Plomp et al., 1988; Schellens and Meijer, 1991) with ATP being required for each of the three steps involved in the process: sequestration, fusion of autophagosomes with lysomes and intralysosomal proteolysis (Plomp et al., 1989). The ATP dependence of the last step presumably reflects the ATP requirement of the lysosomal H^+ pump. It is remarkable how sensitive hepatic autophagic sequestration, as measured by the sequestration of electro-injected cytosolic [^{14}C]sucrose (Plomp et al., 1988), is towards small changes in the intracellular concentration of ATP. Although under normal conditions the intrahepatic ATP concentration is rather strongly buffered, it is postulated that under certain pathological conditions, e.g. in alcoholic liver disease, the decline in autophagy resulting from decreasing ATP concentrations contributes to the accumulation of fat and protein (Plomp et al., 1988).

Control by amino acids and hormones

Since one of the major functions of macroautophagy is to produce amino acids for survival of the organism when nutrients fall short, it is not surprising that amino acids are effective product inhibitors of this process. The system is also hormonally controlled: insulin inhibits while glucagon stimulates (Mortimore et al., 1989; Seglen and Bohley, 1992). The relevant parameter is presumably the glucagon/insulin ratio (Parilla et al., 1974). In vitro, in the isolated perfused liver, the hormones are only effective as modulators of autophagy at intermediate amino acid concentrations. Under these experimental conditions insulin and glucagon have no effect in the absence of amino acids when autophagic flux is maximal nor do they act in the presence of high concentrations of amino acids when macroautophagy is maximally inhibited (Mortimore et al., 1989).

Electronmicroscopic studies (Schworer and Mortimore, 1979; Kovács et al., 1981) and data obtained with electro-injected cytosolic [^{14}C]sucrose and other sugars (Seglen and Gordon, 1984; Seglen and Bohley, 1992) led to the conclusion that inhibition of macroautophagy by amino acids primarily occurs at the sequestration step. High concentrations of amino acids do not, however, completely inhibit autophagic sequestration of [^{14}C]sucrose (Seglen and Gordon, 1984), even though they completely inhibit overall autophagic proteolytic flux. This indicates that amino acids also interfere with post-sequestrational steps. This is, indeed, the case. At high concentrations, the amino acid asparagine, for example, is also able to inhibit the fusion between autophagosomes and lysosomes (Høyvik et al., 1991). A direct effect of some amino acids on the lysosomes, resulting in a rise of the intralysosomal pH, can also contribute to the inhibition of proteolysis (Völkl et al., 1993).

It has also become clear that not all amino acids are equally effective
as inhibitors of macroautophagy. From studies with perfused rat liver,
Mortimore et al. (Pösö and Mortimore, 1984; Mortimore et al., 1988b,
1989, 1991) defined a regulatory group of eight amino acids consisting
of leucine, phenylalanine, tyrosine, glutamine, proline, histidine, tryptophan, and methionine with alanine (which by itself does not affect
macroautophagy) as a synergistic coregulator. A similar group of regulatory amino acids (except that asparagine replaced methionine) was
defined by Seglen and coworkers in experiments carried out with isolated rat hepatocytes (Seglen et al., 1980). In perifused rat hepatocytes,
under true steady state conditions, combinations of near-physiological
concentrations of leucine with either alanine, proline, glutamine or
asparagine were particularly effective in inhibiting autophagic proteolysis (Leverve et al., 1987; Caro et al., 1989). This was also found in the
perfused liver (Mortimore et al., 1988b).

An interesting system to obtain information about the nature of amino
acids involved in the control of macroautophagy is the liver in the perinatal
period. In this period, the liver grows rapidly, which is in part due to
suppression of the process (as compared to the adult situation) by high
plasma amino acid concentrations combined with the fact that, in this
period, glucagon is not yet able to exert its catabolic effects (Blommaart
et al., 1993). From a detailed study on the relationship between extracellular and intracellular amino acid concentrations in cultured hepatocytes
obtained from rats at various stages of development, it was concluded that
leucine, phenylanine and tyrosine are important regulators of macroautophagy in vivo (Blommaart et al., 1993).

To summarize, the data obtained with liver suggest that leucine,
phenylalanine and tyrosine, in combination with a few other amino
acids like alanine and glutamine, are the most important amino acids
involved in the control of autophagic proteolysis.

Interestingly, leucine and glutamine have also been mentioned as
regulators of lysosomal proteolysis in skeletal muscle (cf. Meijer et al.,
1990 for review). Under many conditions, both in vitro and in vivo there
is an inverse relationship between the intramuscular glutamine concentration and the rate of proteolysis (Rennie et al., 1986). Parenteral
administration of the dipeptide alanylglutamine (glutamine is chemically more stable as a dipeptide) to patients after surgery, appears to be
able to correct, at least in part, the negative nitrogen balance (Stehle et
al., 1989). However, it has to be borne in mind that the situation in
muscle is different from that in liver in that stress-induced proteolysis in
muscle is not only lysosomal, but also occurs by an extralysosomal
process in which myofibrillar protein is degraded via the ubiquitin-dependent proteolytic pathway (Wing and Goldberg, 1993).

An exciting new development has been the discovery that cell swelling
has anabolic effects on many metabolic pathways, closely resembling the

effects of insulin (Häussinger and Lang, 1991). For example, in liver an increase in cell volume stimulates the synthesis of glycogen (Baquet et al., 1990), fat (Baquet et al., 1991) and protein (Stoll et al., 1992) and has inhibitory effects on glycogenolysis (Lang et al., 1989) and autophagic protein degradation (Häussinger et al., 1990). Hypo-osmotic cell swelling, like insulin, increases the sensitivity of macroautophagic proteolysis to inhibition by low concentrations of amino acids (Meijer et al., 1993; Luiken et al., 1994). It has even been proposed that the antiproteolytic effect of insulin proceeds via an increase in cell volume, caused by activation of the $Na^+K^+/2Cl^-$ cotransporter with massive influx of KCl and NaCl (Hallbrucker et al., 1991a).

An increase in cell volume can also be caused by an influx of amino acids via Na^+ coupled amino acid transport systems. Due to intracellular amino acid accumulation, intracellular osmolarity rises and the cells swell. Alanine and glutamine are present in plasma at relatively high concentrations and both amino acids are transported via Na^+ coupled transporters. It is likely that these two amino acids exert their antiproteolytic effect via cell swelling (Häussinger and Lang, 1991; Hallbrucker et al., 1991b). Leucine, tyrosine and phenylalanine, on the other hand, are not transported via Na^+ coupled systems and these amino acids do not accumulate inside cells, but rather equilibrate across the plasma membrane. The antiproteolytic effects of these amino acids are probably specifically exerted at the macroautophagic sequestration step (Mortimore et al., 1989; Seglen and Bohley, 1992) and are independent of cell swelling. Synergism between, for example, alanine, glutamine and leucine (Leverve et al., 1987; Mortimore et al., 1988b; Caro et al., 1989) can now be explained by the fact that cell swelling caused by glutamine or alanine influx potentiates the inhibition of macroautophagic sequestration by leucine.

The importance of cell volume in the control of proteolysis *in vivo* is highlighted by the observation that in a large number of clinical patients the cellular hydration state of skeletal muscle was inversely correlated with the degree of nitrogen loss from this tissue (Häussinger et al., 1993).

Hardly anything is known about the mechanism by which amino acids and changes in cell volume affect autophagic sequestration. In fact, this question can only be properly answered if information becomes available about the molecular machinery involved in the process. Complementation studies with yeast mutants deficient in autophagy have shown the participation of at least 15 different protein components (Tsukada and Ohsumi, 1993). However, the nature of these proteins is, as yet, unknown. Undoubtedly, cytoskeleton proteins will be among them. Cytochalasins B and D, compounds that cause depolymerization of microfilaments, inhibit autophagosome formation, while nocodazole, which causes depolymerization of microtubuli, inhibits the fusion between autophagosomes and lysosomes (Aplin et al., 1992). Other compounds interfering with the cytoskeleton also inhibit macroautophagy (Mayer and Doherty, 1986; Holen et al., 1992, 1993).

Recent experiments with isolated hepatocytes carried out in our own laboratory have yielded results that may throw light on the mechanism by which amino acids control macroautophagy. Under conditions where protein synthesis was blocked we observed that the degree of inhibition of macroautophagic proteolysis by single amino acids or by amino acid combinations strictly correlated with the phosphorylation of a 31 kD protein which we have identified as ribosomal protein S6. Moreover, hypo-osmolarity, while having no effect on its own, not only increased the sensitivity of proteolysis to inhibition by amino acids, but also promoted the effect of amino acids on the phosphorylation of S6. Likewise, in the presence of low concentrations of amino acids, glucagon stimulated proteolysis and reduced S6 phosphorylation, whereas the opposite effect, inhibition of proteolysis and increased S6 phosphorylation, was found with insulin (Luiken et al., 1994). Significantly, a combination of leucine, phenylalanine and tyrosine under hypo-osmotic incubation conditions largely mimicked the effect of a complete, physiological mixture of amino acids on both proteolysis and S6 phosphorylation (Luiken et al., 1994).

The effect of amino acids is due to an activation of S6 kinase and is not due to an inhibition of S6 phosphatase. Inhibition of S6 kinase relieves the inhibitory effect of amino acids on macroautophagic sequestration [Blommaart, Luiken and Meijer, unpublished].

S6 can be phosphorylated at five different serine residues (Wettenhall et al., 1982). Although the function of these phosphorylations is not known (Sturgill and Wu, 1991; Kozma and Thomas, 1992), it is assumed that S6 phosphorylation enhances the efficiency of mRNA translation (Palen and Traugh, 1987; Morley and Traugh, 1993), the more so because S6 phosphorylation occurs under conditions which promote cell growth and cell division (Sturgill and Wu, 1991; Kozma and Thomas, 1992). In our experiments, however, protein synthesis was blocked by low concentrations of cycloheximide so that the effect of S6 phosphorylation on proteolysis could not be caused by an increase in the rate of protein synthesis.

Of course, it is possible that the same signal, i.e. a protein kinase, increases the phosphorylation of S6 to increase protein synthesis and at the same time phosphorylates another, as yet unknown, protein to simultaneously inhibit macroautophagic proteolysis. However, an alternative view is that S6 phosphorylation is directly involved in the control of the autophagic process. The problem then is to understand the mechanism by which a ribosomal protein can control autophagy. The key to the answer of this problem is perhaps the fact that the autophagosomal membrane is derived from ribosome-free regions of the rough endoplasmic reticulum (Dunn, 1990a). A high degree of phosphorylation of S6 may promote the binding of ribosomes to such regions so that less membrane becomes available for autophagosome formation. In this way, the same signal, i.e. S6 phosphorylation, would then contribute to the control of both protein synthesis and of protein degradation. One can even envisage a scenario in which autophagosomes are continuously formed on the endoplasmic reticulum unless ribosomal attachment prevents this. There are indications in the literature supporting this view. In the pancreas, for example, with its highly developed rough endoplasmic reticulum, degranulation of the ER caused by administration of puromycin or other compounds *in vivo* is accompanied by enhanced autophagosome formation (Réz et al., 1976). In cultured hepatocytes, synthesis of export proteins, synthesized on the ER, but not of house-keeping proteins, which are largely synthesized on free ribosomes in the cytosol, decreases with decreasing concentration of amino acids in the culture medium (Tanaka and Ichihara, 1983; Dahn et al., 1993).

These observations strongly suggest that phosphorylation of S6 not only controls protein synthesis but also controls macroautophagic proteolysis. This is an attractive mechanism to explain control of these opposing processes.

The next question concerns the mechanism by which amino acids increase phosphorylation of S6. As indicated above, the effect of amino acids proceeds via activation of S6 kinase rather than through inhibition of S6 phosphatase. Recent experiments of Miotto et al. (1993) with isolated hepatocytes have provided direct evidence that the initial effect of amino acids is at the level of the plasma membrane, in line with an earlier proposal that this might be the case (Miotto et al., 1992). They synthesized an isopeptide (H-Leu8-Lys4-Lys2-Lys-ßAla-OH), consisting of a branched tree of seven lysines, to which eight leucines are coupled via their carboxylgroup to the α and ε groups of the terminal lysines. This peptide did not permeate the plasma membrane nor was it hydrolysed. Nevertheless, the peptide effectively inhibited autophagic proteolysis at low concentration, in analogy with the effect of free leucine. Upon replacement of the terminal leucines by isoleucine, which is a non-regulating amino acid, the antiproteolytic action of the peptide was lost. The simplest explanation is that the plasma membrane contains a receptor for amino acids which, in analogy with hormone and growth factor receptors, is coupled to S6 via a protein phosphorylation cascade. According to this, as yet hypothetical, mechanism the inhibitory effect of cell swelling on proteolysis can now be explained by assuming that stretching of the plasma membrane increases the affinity of the amino acid receptor for amino acids.

In Figure 2, we have schematically summarized the interaction of amino acids with the entire autophagic proteolytic pathway. The

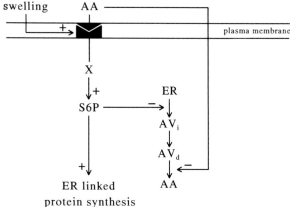

Figure 2. Schematic overview of the co-regulation of macroautophagic proteolysis and protein synthesis by amino acids and cell swelling. Amino acids (AA) stimulate a protein kinase cascade (X) via a plasma membrane receptor; this causes phosphorylation of ribosomal protein S6 (S6P). As a result of this, the first step of the macroautophagic pathway is inhibited and endoplasmic reticulum (ER)-linked protein synthesis is stimulated. Also, there is a direct, receptor-independent, inhibitory effect of amino acids on the last step of the process via an increase in the lysosomal pH. Cell swelling potentiates the effect of amino acids via a change in the receptor. AV_i, initial autophagic vacuole; AV_d, degradative autophagic vacuole.

scheme includes interaction with the amino acid receptor, the protein kinase pathway ("factor X") leading to S6 phosphorylation, and the direct effect of amino acids on the intralysosomal pH. The latter effect is independent of the amino acid receptor, since it can also be demonstrated in permeabilized hepatocytes (Luiken and Meijer, unpublished).

An important new development is the finding that hepatocytes permeabilized with *Staphylococcus aureus* α-toxin, which creates stable plasma membrane channels allowing only small molecules (<1000 Da) to pass through, are still capable of autophagy provided ATP is added (Kadowaki et al., 1994). Under these conditions, the process is sensitive to inhibition by the sequestration inhibitor 3-methyladenine but the inhibitory effect of amino acids is lost. The latter observation suggests that the amino acid signal transduction pathway, starting at the plasma membrane level (cf. Fig. 2) is interrupted in this system. Also of interest is that autophagy in these permeabilized cells is inhibited by low concentrations of GTPγS. This indicates that one or more GTP-binding proteins are involved in the formation of autophagosomes (Kadowaki et al., 1994). Perhaps these proteins belong to the GTP-binding proteins of the *rab* family (Olkkonen et al., 1993).

Specificity of macroautophagy

It has been assumed for a long time that macroautophagy is a non-selective process in which macromolecules are randomly degraded in the same ratio as they occur in the cytoplasm (Kominami et al., 1983; Kopitz et al., 1990). However, recent observations strongly suggest that this may not always be the case, and that macroautophagy can be selective under some conditions. For example, in the perfused liver, although autophagic breakdown of protein and RNA (mainly ribosomal RNA) is sensitive to inhibition by amino acids and insulin, glucagon accelerates proteolysis but has no effect on RNA degradation (Lardeux and Mortimore, 1987). The mechanism by which ribosomes are excluded from autophagic degradation under these conditions is unknown. One may speculate that in the hepatocyte during starvation, this is a means to prevent excessive degradation of ribosomes since these structures are required for the synthesis of essential proteins.

Another example of selective autophagy is the degradation of super-fluous peroxisomes in hepatocytes from clofibrate-treated rats. When hepatocytes from these rats, in which the number of peroxisomes is greatly increased, are incubated in the absence of amino acids to ensure maximal flux through the macroautophagic pathway, peroxisomes are degraded at a relative rate that exceeds that of any other component in the liver cell (Luiken et al., 1992). The accelerated degradation of peroxisomes was sensitive to inhibition by 3-methyladenine, a specific autophagic sequestration inhibitor. Interestingly, the accelerated removal of peroxisomes was prevented by long-chain but not short-chain fatty acids (Luiken et al., 1992). Since long-chain fatty acids are substrates for peroxisomal ß-oxidation this indicates that these organelles are removed by autophagy when they are functionally redundant. Our hypothesis is that acylation (palmitoylation?) of a peroxisomal membrane protein protects the peroxisome against autophagic sequestration. Selective degradation of peroxisomes has also been demonstrated *in vivo*, in rats treated with di-(2-ethylhexyl)phtalate, another peroxisome proliferator, followed by the removal of this drug (Yokota, 1993; Yokota et al., 1993). Remnants of peroxisomes have been found inside autophago-lysosomal structures in fibroblasts from Zellweger patients (Heikoop et al., 1992). In these patients, there is a defect in the import system for many peroxisomal proteins so that the remnants of these organelles, often referred to as peroxisomal "ghosts", are not functional and must be removed.

Selective degradation of peroxisomes has also been observed in methylotrophic yeasts. In these cells, methanol is a peroxisomal substrate and growth on methanol results in induction of peroxisomes. Switching these cells from methanol to a glucose containing growth medium results in a specific removal of these organelles by autophagy, while mitochondria, for example, are retained under these conditions (Tuttle et al., 1993).

Selective elimination of smooth endoplasmic reticulum, after prior induction with phenobarbital, followed by removal of the drug has also been reported (Masaki et al., 1987). This is due to increased autophagy rather than to decreased synthesis of SER components (Luiken and Meijer, unpublished).

Indications for selective degradation, of damaged mitochondria in this case, were found in liver biopsies of patients with Reye's syndrome (Partin et al., 1971) and in an Influenza B virus model of Reye's syndrome in the mouse (Woodfin and Davis, 1986).

From all these observations, we conclude that under normal conditions macroautophagy may be largely unselective and serves to produce amino acids for gluconeogenesis, for example in starvation. However, when there is an excess of a certain cell structure or when cell structures are damaged, the macroautophagic system is able to recognize this and to degrade the structure concerned. Nothing is known about the recognition signals. A possibility is that ubiquitination of membrane proteins is required to mark the structure to be degraded for autophagic sequestration (see also next section).

Ubiquitin may be involved in macroautophagy

According to recent evidence, ubiquitin not only contributes to extralysosomal proteolysis but is also involved in autophagic protein degradation. Thus, in fibroblasts ubiquitin-protein conjugates can be found in the lysosomes as shown by immunohistochemistry and immunogold electronmicroscopy (Doherty et al., 1989; Laszlo et al., 1990). Free ubiquitin can also be found inside lysosomes (Schwartz et al., 1988). Accumulation of ubiquitin-protein conjugates in filamentous, presumably lysosomal structures are also found in a large number of neurodegenerative diseases, e.g. in hippocampus neurones in Alzheimer's disease, in neuronal inclusions ("Lewy bodies"), in the brainstem in Parkinson's disease, in Pick's disease and in Down Syndrome (Mayer et al., 1991). Mallory bodies in the liver of alcoholics also contain ubiquitin-protein conjugates (Mayer et al., 1991).

This presence of ubiquitin-protein conjugates in filamentous inclusions in neurons and other cells can be due to a defect in the extralysosomal ubiquitin-dependent proteolytic pathway. However, it is also possible that these filamentous inclusions represent an attempt of the cell to get rid of unwanted material (proteins, organelles) via autophagy (Mayer et al., 1991).

Direct evidence that ubiquitin may be involved in the control of macroautophagy came from experiments with CHO cells with a temperature-sensitive mutation in the ubiquitin-activating enzyme E1 (Gropper et al., 1991). Wild type cells increased their rate of proteolysis in response to stress (amino acid depletion, increased temperature). This was prevented by the acidotropic agent ammonia or by the autophagic sequestration inhibitor 3-methyladenine, indicating that the accelerated proteolysis occurred by autophagy. In the mutant cells, there was no such increase in proteolysis in response to stress at the restrictive temperature. Apparently, functional E1 is required for autophagic proteolysis. However, formation of initial autophagosomes was not affected in the mutant cells (Lenk et al., 1992; Schwartz et al., 1992). Perhaps these autophagosomes are not functional because ubiquitin-protein conjugates are not available as autophagic substrates. An alternative interpretation is that the ubiquitin system is required for maturation of autophagosomes and/or their fusion with lysosomes. Another possibility is that a short-lived protein in the cell specifically inhibits the fusion step and that ubiquitination and degradation of this protein is required for macroautophagy to proceed (Gropper et al., 1991).

Conclusions

It is surprising that autophagy has received so little attention in the past. After all, the level of proteins and other macromolecular structures in the cell is determined by the balance between synthesis and degradation so that, from a quantitative point of view, degradation is as important as protein synthesis. We hope to have made clear in this review that autophagy is an intriguing process. To understand the process in all its details presents a formidable challenge for the future. Problems to be solved are the unraveling of the molecular machinery involved in this complicated process, the elucidation of the signals responsible for the recognition of structures to be degraded by autophagy and, likewise, elucidation of the signals that prevent the degradation of structures. An exciting possibility is that *rab* proteins, which are so important in endocytosis and exocytosis, are also involved in autophagy (Olkkonen et al., 1993). Does a plasma membrane receptor for amino acids indeed exist and how is binding of specific amino acids to this receptor transmitted to intracellular targets (like S6)? Are other targets, like cytoskeleton proteins, also involved in the control of autophagy by amino acids?

References

Ahlberg, J., Marzella, L. and Glaumann, H. (1982) Uptake and degradation of proteins by isolated rat liver lysosomes. Suggestion of a microautophagic pathway of proteolysis. *Lab. Invest.* 47: 523–532.

Aniento, F., Roche, E., Cuerro, A.M. and Knecht, E. (1993) Uptake and degradation of glyceraldehyde 3-phosphate dehydrogenase by rat liver lysosomes. *J. Biol. Chem.* 268: 10463–10470.

Aplin, A., Jasianowski, T., Tuttle, D.L., Lenk, S.E. and Dunn, W.A. (1992) Cytoskeletal elements are required for the formation and maturation of autophagic vacuoles. *J. Cell. Physiol.* 152: 458–466.

Baquet, A., Hue, L., Meijer, A.J., van Woerkom, G.M. and Plomp, P.J.A.M. (1990) Swelling of hepatocytes stimulates glycogen synthesis. *J. Biol. Chem.* 265: 955–959.

Baquet, A., Maisin, L. and Hue, L. (1991) Swelling of rat hepatocytes activates acetylCoA carboxylase in parallel to glycogen synthase. *Biochem. J.* 278: 887–890.

Blommaart, P.J.E., Zonneveld, D., Meijer, A.J. and Lamers, W.H. (1993) Effects of intracellular amino acid concentrations, cyclic AMP, and dexamethasone on lysosomal proteolysis in primary cultures of perinatal rat hepatocytes. *J. Biol. Chem.* 268: 1610–1617.

Caro, L.H.P., Plomp, P.J.A.M., Leverve, X.M. and Meijer, A.J. (1989) A combination of intracellular leucine with either glutamate or aspartate inhibits autophagic proteolysis in isolated rat hepatocytes. *Eur. J. Biochem.* 181: 717–720.

Chiang, H.L. and Schekman, R. (1991) Regulated import and degradation of a cytosolic protein in the yeast vacuole. *Nature* 350: 313–318.

Dahn, M.S., Hsu, C.J., Lange, M.P., Kimball, S.R. and Jefferson, L.S. (1993) Factors affecting secretory protein production in primary cultures of rat hepatocytes. *Proc. Soc. Exptl. Biol. Med.* 203: 38–44.

Dice, J.F. (1990) Peptide sequences that target cytosolic proteins for lysosomal proteolysis. *Trends in Biochem. Sci.* 15: 305–309.

Doherty, F.J., Osborne, N.U., Wassell, J.A., Heggie, P.E., Laszlo, L. and Mayer, R.J. (1989) Ubiquitin-protein conjugates accumulate in the lysosomal system of fibroblasts treated with cysteine proteinase inhibitors. *Biochem. J.* 263: 47–55.

Dunn, W.A. (1990a) Studies on the mechanisms of autophagy: formation of the autophagic vacuole. *J. Cell Biol.* 110: 1923–1933.

Dunn, W.A. (1990b) Studies on the mechanisms of autophagy: maturation of the autophagic vacuole. *J. Cell Biol.* 110: 1935–1945.

Gropper, R., Brandt, R.A., Elias, S., Bearer, C.F., Mayer, A., Schwartz, A.L. and Ciechanover, A. (1991) The ubiquitin-activating enzyme, E1, is required for stress-induced lysosomal degradation of cellular proteins. *J. Biol. Chem.* 266: 3602–3610.

Hallbrucker, C., vom Dahl, S., Lang, F., Gerok, W. and Häussinger, D. (1991a) Inhibition of hepatic proteolysis by insulin. Role of hormone-induced alterations of the cellular K^+ balance. *Eur. J. Biochem.* 199: 467–474.

Hallbrucker, C., vom Dahl, S., Lang, F. and Häussinger, D. (1991b) Control of hepatic proteolysis by amino acids. The role of cell volume. *Eur. J. Biochem.* 197: 717–724.

Häussinger, D., Hallbrucker, C., vom Dahl, S., Lang, F. and Gerok, W. (1990) Cell swelling inhibits proteolysis in perfused rat liver. *Biochem. J.* 272: 239–242.

Häussinger, D. and Lang, F. (1991) Cell volume in the regulation of hepatic function: a mechanism for metabolic control. *Biochim. Biophys. Acta* 1071: 331–350.

Häussinger, D., Roth, E., Lang, F. and Gerok, W. (1993) Cellular hydration state, an important determinant of protein catabolism in health and disease. *Lancet* 341: 1330–1332.

Heikoop, J.C., van den Berg, M., Strijland, A., Weijers, P.J., Just, W.W., Meijer, A.J. and Tager, J.M. (1992) Turnover of peroxisomal vesicles by autophagic proteolysis in cultured fibroblasts from Zellweger patients. *Eur. J. Cell Biol.* 57: 165–171.

Hershko, H. and Ciechanover, A. (1992) The ubiquitin system for protein degradation. *Ann. Rev. Biochem.* 61: 761–807.

Holen, I., Gordon, P.B. and Seglen, P.O. (1992) Protein kinase-dependent effects of okadaic acid on hepatocytic autophagy and cytoskeletal integrity. *Biochem. J.* 284: 633–636.

Holen, I., Gordon, P.B. and Seglen, P.O. (1993) Inhibition of hepatic autophagy by okadaic acid and other protein phosphatase inhibitors. *Eur. J. Biochem.* 215: 113–122.

Høyvik, H., Gordon, P.B., Berg, T.O., Strømhaug, P.E. and Seglen, P.O. (1991) Inhibition of autophagic-lysosomal delivery and autophagic lactolysis by asparagine. *J. Cell Biol.* 113: 1305–1312.

Kadowaki, M., Venerando, R., Miotto, G. and Mortimore, G.E. (1994) De novo autophagic vacuole formation in hepatocytes permeabilized by *Staphylococcus aureus* α-toxin. Inhibition by nonhydrolysable GTP analogs. *J. Biol. Chem.* 269: 3703–3710.

Kominami, E., Hashida, S., Khairallah, E.A. and Katunuma, N. (1983) Sequestration of cytoplasmic enzymes in autophagic vacuole-lysosomal system induced by injection with leupeptin. *J. Biol. Chem.* 258: 6093–6100.

Kopitz, J., Kisen, G.Ø., Gordon, P.B., Bohley, P. and Seglen, P.O. (1990) Nonselective autophagy of cytosolic enzymes by isolated rat hepatocytes. *J. Cell Biol.* 111: 941–953.

Kovács, A.L., Grinde, B. and Seglen, P.O. (1981) Inhibition of autophagic vacuole formation and protein degradation by amino acids in isolated rat hepatocytes. *Exptl. Cell Res.* 133: 431–436.

Kovács, A.L., Reith, A. and Seglen, P.O. (1982) Accumulation of autophagosomes after inhibition of hepatocyte protein degradation by vinblastin, leupeptin or a lysosomotropic amine. *Exptl. Cell Res.* 137: 191–201.

Kozma, S.C. and Thomas, G. (1992) Serine/threonine kinases in the propagation of the early mitogenic response. *Rev. Physiol. Biochem. Pharmacol.* 119: 123–155.

Lang, F., Stehle, T. and Häussinger, D. (1989) Water, K^+, H^+, lactate and glucose fluxes during cell volume regulation in perfused rat liver. *Pflügers Arch.* 413: 209–216.

Lardeux, B.R. and Mortimore, G.E. (1987) Amino acid and hormonal control of macromolecular turnover in perfused rat liver. Evidence for selective autophagy. *J. Biol. Chem.* 262: 14514–14519.

Laszlo, L., Doherty, F.J., Osborne, N.U. and Mayer, R.J. (1990) Ubiquitinated protein conjugates are specifically enriched in the lysosomal system of fibroblasts. *FEBS Lett.* 261: 365–368.

Lenk, S.E., Dunn, W.A., Trausch, J.S., Ciechanover, A. and Schwartz, A.L. (1992) Ubiquitin-activating enzyme, E1, is associated with maturation of autophagic vacuoles. *J. Cell Biol.* 118: 301–308.

Leverve, X.M., Caro, L.H.P., Plomp, P.J.A.M. and Meijer, A.J. (1987) Control of proteolysis in perifused hepatocytes. *FEBS Lett.* 219: 455–458.

Luiken, J.J.F.P., van den Berg, M., Heikoop, J.C. and Meijer, A.J. (1992) Autophagic degradation of peroxisomes in isolated hepatocytes. *FEBS Lett.* 304: 93–97.

Luiken, J.J.F.P., Blommaart, E.F.C., Boon, L., van Woerkom, G.M. and Meijer, A.J. (1994) Cell swelling and the control of autophagic proteolysis in hepatocytes: involvement of phosphorylation of ribosomal protein S6? *Biochem. Soc. Trans.* 22: 508–511.

Masaki, R., Yamamoto, A. and Tashiro, Y. (1987) Cytochrome P-450 and NADPH-cytochrome P-450 reductase are degraded in the autolysosomes in rat liver. *J. Cell Biol.* 104: 1207–1215.

Mayer, R.J., Arnold, J., Laszlo, L., Landon, M. and Lowe, L. (1991) Ubiquitin in health and disease. *Biochim. Biophys. Acta* 1089: 141–157.

Mayer, R.J. and Doherty, F. (1986) Intracellular protein catabolism: state of the art. *FEBS Lett.* 198: 181–193.

Meijer, A.J., Gustafson, L.A., Luiken, J.J.F.P., Blommaart, P.J.E., Caro, L.H.P., van Woerkom, G.M., Spronk, C. and Boon, L. (1993) Cell swelling and the sensitivity of autophagic proteolysis to inhibition by amino acids in isolated rat hepatocytes. *Eur. J. Biochem.* 215: 449–454.

Meijer, A.J., Lamers, W.H. and Chamuleau, R.A.F.M. (1990) Nitrogen metabolism and ornithine cycle function. *Physiol. Revs.* 70: 701–748.

Mellgren, R.L. (1987) Calcium-dependent proteases: an enzyme system active at cellular membranes? *FASEB J.* 1: 110–115.

Miotto, G., Venerando, R., Khurana, K.K., Siliprandi, N. and Mortimore, G.E. (1992) Control of hepatic proteolysis by leucine and isovaleryl-L-carnitine through a common locus. Evidence for a possible mechanism of recognition at the plasma membrane. *J. Biol. Chem.* 267: 22066–22072.

Miotto, G., Venerando, R., Marin, O., Piutti, C. and Siliprandi, N. (1993) Inhibition of hepatic autophagy by MAP-8-LEU, a synthetic isopeptide impermeant to the plasma membrane. *2nd IUBMB Conference, Biochemistry of Cell Membranes,* Sept. 29–Oct. 3, Bari, Italy, abstract p. 87.

Morley, S.J. and Traugh, J.A. (1993) Stimulation of translation in 3T3-L1 cells in response to insulin and phorbolester is directly correlated with increased phosphate labelling of initiation factor (eiF-)4F and ribosomal protein S6. *Biochimie* 75: 985–989.

Mortimore, G.E., Hutson, N.J. and Surmacz, C.A. (1983) Quantitative correlation between proteolysis and macro- and microautophagy in mouse hepatocytes during starvation and refeeding. *Proc. Natl. Acad. Sci. USA* 80: 2179–2183.

Mortimore, G.E., Khurana, K.K. and Miotto, G. (1991) Amino acid control of proteolysis in perfused livers of synchronously fed rats. *J. Biol. Chem.* 266: 1021–1028.

Mortimore, G.E., Lardeux, B.R. and Adams, C.E. (1988a) Regulation of microautophagy and basal protein turnover in rat liver. Effects of short-term starvation. *J. Biol. Chem.* 263: 2506–2512.

Mortimore, G.E., Pösö, A.R. and Lardeux, B.R. (1989) Mechanism and regulation of protein degradation in liver. *Diabetes/Metabolism Revs.* 5: 49–70.

Mortimore, G.E., Wert, J.J. and Adams, C.E. (1988b) Modulation of the amino acid control of hepatic protein degradation by caloric deprivation. *J. Biol. Chem.* 263: 19545–19551.

Olkkonen, V.M., Dupree, P., Killisch, I., Lutcke, A., Zerial, M. and Simons, K. (1993) Molecular cloning and subcellular localizations of three GTP-binding proteins of the Rab subfamily. *J. Cell Sci.* 106: 1249–1261.

Palen, E. and Traugh, J.A. (1987) Phosphorylation of ribosomal protein S6 by cAMP dependent protein kinase and mitogen-stimulated S6 kinase differentially alters translation of globin mRNA. *J. Biol. Chem.* 262: 3518–3523.

Parilla, R., Goodman, M.N. and Toews, C.J. (1974) Effect of glucagon:insulin ratios on hepatic metabolism. *Diabetes* 23: 725–731.

Partin, J., Schubert, W.K. and Partin, J.S. (1971) Mitochondrial ultrastructure in Reye's syndrome (encephalopathy and fatty degeneration of the viscera). *New Engl. J. Med* 285: 1339–1343.

Paskin, N. and Mayer, R.J. (1977) The role of enzyme degradation in enzyme turnover during tissue differentiation. *Biochim. Biophys. Acta* 474: 1–10.

Pfeifer, U. (1987) Functional morphology of the lysosomal apparatus. *In:* H. Glaumann and F.J. Ballard (eds): *Lysosomes, Their Role in Protein Breakdown*, Academic Press London, pp. 4–59.

Pfeifer, U., Werde, E. and Bergeest, H. (1978) Inhibition by insulin of the formation of autophagic vacuoles. A morphometric approach to the kinetics of intracellular degradation by autophagy. *J. Cell Biol.* 78: 152–167.

Plomp, P.J.A.M., Gordon, P.B., Meijer, A.J., Høyvik, H. and Seglen, P.O. (1989) Energy dependence of different steps in the autophagic-lysosomal pathway. *J. Biol. Chem.* 254: 6699–6704.

Plomp, P.J.A.M., Wolvetang, E.J., Groen, A.K., Meijer, A.J., Gordon, P.B. and Seglen, P.O. (1988) Energy dependence of autophagic protein degradation in isolated rat hepatocytes. *Eur. J. Biochem.* 164: 197–203.

Pösö, A.R. and Mortimore, G.E. (1984) Requirement for alanine in the amino acid control of deprivation induced protein degradation in liver. *Proc. Natl. Acad. Sci. USA* 81: 4270–4274.

Rennie, M.J., Babij, P., Taylor, P.M., Hundal, H.S., McLennan, P., Watt, P.W., Jepson, M.M. and Millward, R.J. (1986) Characteristics of a glutamine carrier in skeletal muscle have important consequences for nitrogen loss in injury, infection and chronic disease. *Lancet* 1: 1008–1011.

Réz, G., Kiss, A., Buseck, M.J. and Kovács, J. (1976) Attachment of ribosomes to endoplasmic membranes in mouse pancreas. Degranulation *in vivo* caused by the inducers of autophagocytosis neutral red, vinblastine, puromycin, and cadmium ions, and prevention by cycloheximide. *Chem. Biol. Interactions* 13: 77–87.

Schellens, J.P.M. and Meijer, A.J. (1991) Energy depletion and autophagy. Cytochemical and biochemical studies in isolated rat hepatocytes. *Histochem. J.* 23: 460–466.

Schwartz, A.L., Brandt, R.A., Geuze, H. and Ciechanover, A. (1992) Stress-induced alterations in autophagic pathway: relationship to ubiquitin system. *Am. J. Physiol.* 262: C1031–C1038.

Schwartz, A.L., Ciechanover, A., Brandt, R.A. and Geuze, H.J. (1988) Immunoelectron microscopic localization of ubiquitin in hepatoma cells. *EMBO J.* 7: 2961–2966.

Schworer, C.M. and Mortimore, G.E. (1979) Glucagon induced autophagy and proteolysis in rat liver: mediation by selective deprivation of intracellular amino acids. *Proc. Natl. Acad. Sci. USA* 76: 3169–3173.

Seglen, P.O. and Bohley, P. (1992) Autophagy and other vacuolar protein degradation mechanisms. *Experientia* 48: 158–172.

Seglen, P.O. and Gordon, P.B. (1984) Amino acid control of autophagic sequestration and protein degradation. *J. Cell Biol.* 99: 435–444.

Seglen, P.O., Gordon, P.B. and Poli, A. (1980) Amino acid inhibition of the autophagic/lysosomal pathway of protein degradation in isolated rat hepatocytes. *Biochim. Biophys. Acta* 630: 103–118.

Stehle, P., Mertens, N., Puchstein, C.H., Zander, J., Albers, S., Lawin, P. and Fürst, P. (1989) Effect of parenteral glutamine peptide supplements on muscle glutamine loss and nitrogen balance after major surgery. *Lancet* 1: 231–233.

Stoll, B., Gerok, W., Lang, F. and Häussinger, D. (1992) Liver cell volume and protein synthesis. *Biochem. J.* 287: 217–222.

Sturgill, T.W. and Wu, J. (1991) Recent progress in characterization of protein kinase cascades for phosphorylation of ribosomal protein S6. *Biochim. Biophys. Acta* 1092: 350–357.

Tanaka, K. and Ichihara, A. (1983) Different effects of amino acid deprivation on synthesis of intra- and extracellular proteins in rat hepatocytes in primary culture. *J. Biochem.* 94: 1339–1348.

Terlecki, S.R., Chiang, H.L., Olson, T.S. and Dice, J.F. (1992) Protein and peptide binding and stimulation of *in vitro* lysosomal proteolysis by the 73-kDa heat shock cognate protein. *J. Biol. Chem.* 267: 9202–9209.

Terlecki, S.R. and Dice, J.F. (1993) Polypeptide import and degradation by isolated lysosomes. *J. Biol. Chem.* 268: 23490–23495.

Tsukada, M. and Ohsumi, Y. (1993) Isolation and characterization of autophagy-defective mutants of *Saccharomyces cerevisiae*. *FEBS Lett.* 333: 169–174.

Tuttle, D.L., Lewin, A.S. and Dunn, W.A. (1993) Selective autophagy of peroxisomes in methylotrophic yeasts. *Eur. J. Cell Biol.* 60: 283–290.

Völkl, H., Friedrich, F., Häussinger, D. and Lang, F. (1993) Effect of cell volume on acridine orange fluorescence in hepatocytes. *Biochem. J.* 295: 11–14.

De Waal, E.J., Vreeling-Sindelárová, H., Schellens, J.P.M., Houtkoper, J.M. and James, J. (1986) Quantitative changes in the lysosomal vacuolar system of rat hepatocytes during short-term starvation. A morphometric analysis with special reference to macro- and micro-autophagy. *Cell Tissue Res.* 234: 641–648.

Wettenhall, R.E.H., Cohen, P., Caudwell, B. and Holland, R. (1982) Differential phosphorylation of ribosomal protein S6 in isolated rat hepatocytes after incubation with insulin and glucagon. *FEBS Lett.* 148: 207–213.

Wing, S.S., Chiang, H.L., Goldberg, A.L. and Dice, J.F. (1991) Proteins containing peptide sequences related to Lys-Phe-Glu-Arg-Gln are selectively depleted in liver and heart, but not skeletal muscle, of fasted rats. *Biochem. J.* 275: 165–169.

Wing, S.S. and Goldberg, A.L. (1993) Glucocorticoids activate the ATP-ubiquitin-dependent proteolytic system in skeletal muscle during fasting. *Am. J. Physiol. (Endocrinol. Metab. 27)* 264: E668–676.

Woodfin, B.M. and Davis, L.E. (1986) Liver autophagy in the influenza B virus model of Reye's syndrome in mice. *J. Cell. Biochem.* 31: 271–275.

Yokota, S. (1993) Formation of autophagosomes during degradation of excess peroxisomes induced by administration of dioctyl phtalate. *Eur. J. Cell Biol.* 61: 67–80.

Yokota, S. Himeno, M., Roth, J., Brada, D. and Kato, K. (1993) Formation of autophagosomes during degradation of excess peroxisomes induced by di-(2-ethylhexyl)phtalate treatment. 2. Immunocytochemical analysis of early and late autophagosomes. *Eur. J. Cell Biol.* 62: 372–383.

Biochemistry of Cell Membranes
ed. by S. Papa & J. M. Tager
© 1995 Birkhäuser Verlag Basel/Switzerland

Glycosphingolipids and sphingolipid activator proteins: Cell biology, biochemistry and molecular genetics

K. Suzuki[1], G. van Echten-Deckert[2], A. Klein[2] and K. Sandhoff[2]

[1]*Brain and Development Research Center, Department of Neurology and Psychiatry, University of North Carolina School of Medicine, Chapel Hill, NC 27599, USA*
[2]*Institut für Organische Chemie und Biochemie, Universität Bonn, D-53121 Bonn, Germany*

Summary. This review focuses, after a short description of the biosynthesis, topology and intracellular traffic of glycosphinogolipids (GSLs), on a new model of the topology of endocytosis and lysosomal digestion of GSLs. The essential role of the sphingolipid activator proteins (SAPs) in the catabolism of vesicle-bound GSLs with short hydrophilic head groups is described. The new model and the proposed function of SAPs are strongly supported by the pathobiochemistry of sphingolipid storage diseases caused by the inherited deficiency of either the SAPs or the lysosomal hydrolases.

Introduction

Glycosphingolipids are a group of complex lipids which contain a long-chain base, sphingosine, as the basic building block. In almost all naturally-occurring sphingolipids, sphingosine is acylated by a long-chain fatty acid to ceramide. A complex hydrophilic side chain, consisting of carbohydrate, sialic acid and other constituents is attached to the terminal hydroxyl group of sphingosine. These lipids are characteristic integral amphiphilic constituents of the plasma membrane in vertebrates (Ledeen and Yu, 1982; Svennerholm, 1984; van Echten and Sandhoff, 1989). Their composition varies in different cell types and in different developmental stages. The pattern can be further altered by viral transformation and oncogenesis (for review see Markwell et al., 1981; Hakomori, 1984). The biochemical pathways of biosynthesis and degradation of glycosphingolipids have been largely clarified but the cell biological mechanism by which they are transported from sites of synthesis to other membranes and from the membranes to the lysosome, the site of their degradation, has not yet been well defined. It is generally assumed (Fishman and Brady, 1976; Morré et al., 1979) that, as is the case in glycoproteins, routing of glycosphingolipids is coupled with a vesicular membrane flow, from the endoplasmatic reticulum (ER) through the cisternae of the Golgi complex and finally to the plasma membrane (for review see Schwarzmann and Sandhoff, 1990). There is also general agreement that a portion of the plasma membrane destined for degradation is endocytosed and carried by endosomes to the lysosome, where the constituent glycosphingolipids are degraded. However, several glycosphingolipid-binding or transfer proteins have been described in the literature and their involvement in the intracellular trafficking of glycosphingolipids cannot yet be excluded (Thompson et al., 1986; Tiemeyer et al., 1989). Inherited deficiencies of protein components essential for degradation of glycosphingolipids cause serious metabolic disorders. The category of genetic disorders that has attracted attention in recent years includes defective sphingolipid activator proteins, which are small, non-enzymatic lysosomal glycoproteins essential for *in vivo* degradation of glycosphingolipids with short hydrophilic head groups.

Biosynthesis and intracellular topology

Glycosphingolipid biosynthesis starts with condensation of serine and palmitoyl-CoA, yielding 3-ketosphinganine (Mandon et al., 1991). This ketone is then rapidly reduced to sphinganine by an NADPH-dependent reductase (Stoffel et al., 1968). Enzymatic addition of amide-linked fatty acid to sphinganine to form dihydro-ceramide appears to take place on the cytosolic face of the endoplasmic reticulum (ER) (Mandon et al., 1992). Introduction of the 4-trans double bond occurs only after the addition of fatty acid (reviewed by van Echten and Sandhoff, 1993). Thus, sphingosine is not an intermediate of the *de novo* biosynthetic pathway but can be generated only by degradation of ceramide. The next step, glucosylation of ceramide, also appears to occur on the cytosolic side of the Golgi membrane (Coste et al., 1986; Trinchera et al., 1991). However, the topological localization of the following step, galactosylation of glucosylceramide to form lactosylceramide (LacCer), the common precursor of most glycosphingolipids, including gangliosides, is not definitive, although the luminal side has been suggested as being the most likely site (van Echten and Sandhoff, 1993).

   The subsequent sequential addition of monosaccharide or sialic acid residues to the growing oligosaccharide chain, yielding complex glycosphingolipids and gangliosides, all occurs on the luminal side of the Golgi apparatus and is catalyzed by a series of membrane-bound glycosyltransferases (reviewed by Schwarzmann and Sandhoff, 1990). Therefore, transfer of either glucosylceramide (GlcCer) or lactosylceramide (LacCer) from the cytosolic to the luminal side of the Golgi membranes must occur during biosynthesis of complex glycosphingolipids. Such a transfer, however, has not been experimentally demonstrated.

While many of the glycosyltransferases involved are specific for the respective steps, analogous glycolipid acceptors, differing only in the number of sialic acid residues on the internal galactose of the oligosaccharide chain, share the same glycosyltransferase for transfer of the same sugar in the Golgi apparatus of rat liver (reviewed by van Echten and Sandhoff, 1993).

Topology and mechanism of degradation and the role of sphingolipid activator proteins

Fragments of the plasma membrane and their constituents reach the lysosomal compartment mainly by an endocytic membrane flow through the early and late endocytic reticulum (Griffiths et al., 1988). During this vesicular membrane flow molecules are subjected to a sorting process which directs some of the molecules to the lysosomal compartment, some others to the Golgi, and some even back to the plasma membrane (Koval and Pagano, 1989, 1990; Wessling-Resnick and Braell, 1990; Kok et al., 1991). It remains, however, an open question whether components of the plasma membrane are integrated into the lysosomal membrane after successive steps of vesicle budding and fusion along the endocytotic pathway before they are catabolized. It seems unlikely that the components of the lysosomal membrane originating from the plasma membrane can be more or less selectively degraded by the lysosomal enzymes. It would require another mechanism to distinguish intrinsic constituents of the lysosomal membrane from those that are derived from the plasma membrane.

An alternative possibility is suggested by the observation of multivesicular bodies at the level of the early and late endosomal reticulum (McKanna et al., 1979; Hopkins et al., 1990; Kok et al., 1991; Zachgo et al., 1992). Portions of the endosomal membranes, possibly those enriched in components derived from the plasma membranes, bud off into the endosomal lumen and thus form the intra-endosomal vesicles. These vesicles enriched in plasma membrane constituents could be delivered into the lysosol for final degradation of their constituents by successive processes of membrane fission and fusion along the endocytic pathways (Fig. 1). Thus, glycoconjugates, originating from the outer leaflet of the plasma membrane would enter the lysosol on the outer leaflet of endocytic vesicles, facing the lysosol containing digestive enzymes. This hypothesis (Fürst and Sandhoff, 1992) is supported by the accumulation of multivesicular storage bodies in cells, e.g., the Kupffer cells and fibroblasts, of patients with a complete deficiency of the SAP (sphingolipid activator protein) precursor protein and therefore with a combined activator protein deficiency (Harzer et al., 1989; Schnabel et al., 1992). Also, the observation that the epidermal growth

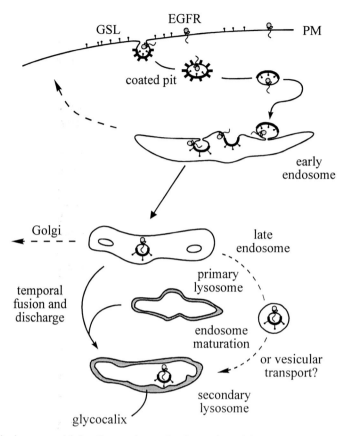

Figure 1. A new model for the topology of endocytosis and lysosomal digestion of GSLs derived from the plasma membrane (modified from Fürst and Sandhoff, 1992). During endocytosis glycolipids of the plasma membrane are supposed to end up in intraendosomal vesicles (multivesicular bodies) from where they are discharged into the lysosomal compartment. GSL, glycosphingolipid; PM, plasma membrane; EGFR, epidermal growth factor receptor; ♀, glycolipid; ⟶ proposed pathway of endocytosis of GSLs derived from the plasma membrane into the lysosomal compartment; −−→ other intracellular routes for GSLs derived from the plasma membrane.

factor receptor derived from the plasma membrane and internalized into lysosomes of hepatocytes is not integrated into the lysosomal membrane (Renfrew and Hubbard, 1991) gives support to our above hypothesis. Furthermore, the above view is in accordance with recent observations by van Deurs et al. (1993), who found that spherical multivesicular bodies were the predominant endocytic compartments in HEp-2 cells and that they entered into lysosomes within 60 to 90 min.

It is not yet clear whether the early endosomes are gradually transformed, first to late endosomes, and then to lysosomes. This would require addition of several organelle-specific components to the trans-forming and maturing compartment (Stoorvogel, 1993; Murphy et al., 1993). It is also possible that early endosomes, late endosomes and lysosomes pre-exist and that intra-endosomal carrier vesicles are trans-ported successively from one compartment to another. Recently, Mullock and Luzio (1993) demonstrated association and fusion of endosomes with pre-existing lysosomes in a cell-free rat liver prepara-tion. Since the experimental conditions did not permit maturation of endosomes to lysosomes, the results indicated some vesicular transport between pre-existing compartments, but did not rule out the first possi-bility.

Finally, components of the plasma membrane, having reached the lysosomes, are digested by a mixture of hydrolases most active in the acidic environment of the lysosol. On the other hand, the lysosomal membrane itself must be protected against premature digestion by the massive glycocalix on the inner surface of the lysosomal membrane contributed by the extensive carbohydrate chains of the lysosomal associated membrane proteins (LAMPs) and the lysosomal integral membrane proteins (LIMPs) (Carlsson and Fukuda, 1990). Degrada-tion of glycosphingolipids occurs by stepwise action of specific acid hydrolases. Several of these enzymes need assistance of small glyco-protein cofactors, the so-called "sphingolipid activator proteins" (SAPs) (Fürst and Sandhoff, 1992) in order to hydrolyze membrane- or vesicle-bound glycosphingolipid substrates with short oligosaccharide head groups.

Since the discovery of sulfatide activator protein by Mehl and Jatzke-witz (1964), several additional factors were described but their identity, specificity and function often remained unclear. When sequence data became available, it turned out that only two genes code for the five known or putative SAPs (Fürst and Sandhoff, 1992). One gene carries the genetic information for the GM2-activator and the second for the *sap*-precursor which is processed to four homologous proteins, includ-ing sulfatide activator protein (*sap*-B) and glucosylceramidase activator protein (*sap*-C). The precursor generates two other homologous proteins which are putative sphingolipid activator proteins, *sap*-A and *sap*-D.

Experimental data by Meier et al. (1991) suggest the mechanism of action of GM2-activator. Hexosaminidase A is a water-soluble hydro-lase for degradation of GM2-ganglioside. The enzyme can recognize and degrade substrates (e.g. GD1a-GalNAc) which extend far enough into the aqueous space (Fig. 2). However, the enzyme cannot degrade GM2-ganglioside which has an oligosaccharide head group too short to be reached by the water-soluble enzyme. Its degradation requires the second component, the GM2-activator, a specialized glycosphingolipid-binding protein, which complexes with GM2-ganglioside, lifts it and

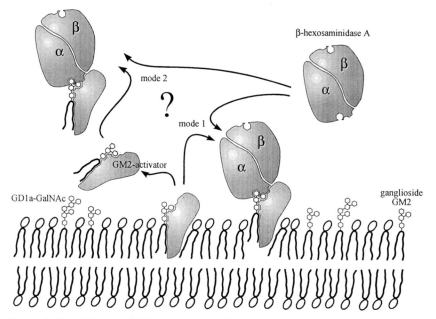

Figure 2. Model for the GM2-activator stimulated degradation of ganglioside GM2 by human hexosaminidase A (modified from Fürst and Sandhoff, 1992). Water-soluble hexosaminidase A does not degrade membrane-bound ganglioside GM2 which has a short carbohydrate chain in the absence of GM2-activator or appropriate detergents. But it degrades ganglioside GD1a-GalNAc which has an extended carbodydrate chain and also analogues of ganglioside GM2 which contain a short acyl residue or no acyl residue (lysoganglioside GM2). They are less firmly bound to the lipid bilayer and more water-soluble than GM2. Ganglioside GM2 bound to a lipid-bilayer, e.g., of an intralysosomal vesicle (see Fig. 1), is hydrolyzed in the presence of the GM2-activator. The GM2-activator binds one ganglioside GM2 molecule and lifts it a few Å out of the membrane. This activator/lipid complex can be reached and recognized by water-soluble hexosaminidase A which cleaves the substrate (mode 1). However, it is also possible that the water-soluble activator/lipid complex leaves the membrane and the enzymatic reaction takes place in free solution (mode 2). The terminal GalNAc-residue of membrane-incorporated ganglioside GD1a-GalNAc protrudes from the membrane far enough to be accessible to hexosaminidase A without an activator.

even extracts it from the membrane to make it accessible for the hexosaminidase A for degradation. While GM2-activator and hexosaminidase A represent a selective and precisely tuned machinery for the degradation of only few structurally similar membrane-bound sphingolipids, *sap*-B stimulates degradation of many lipids by several enzymes from human, plant and even bacterial origin in the test tube (Li et al., 1988). Thus, *sap*-B seems to act more like a physiological detergent with broad specificity at least in *in vitro* conditions (Vogel et al., 1991). On the other hand, unlike GM2-activator and *sap*-B, *sap*-C is reported to activate membrane-associated hydrolytic enzymes by forming complexes with the enzymes and acid phospholipids rather than with the substrates (Berent and Radin, 1981; Ho and Light, 1973; Ho, 1975; Ho and Rigby, 1975; Vaccaro et al., 1993). While *in vitro* data are available regarding the activator functions of *sap*-A and *sap*-D*, the physiological substrates for these putative activator proteins have not been elucidated and thus their reaction mechanisms are not known.

Inherited enzyme- and activator protein deficiencies involving glycosphingolipids

As already mentioned, final degradation of sphingolipids occurs in the lysosome. Here they are degraded by hydrolases in a stepwise manner starting at the hydrophilic end of the molecules. The inherited deficiency of one of these ubiquitously occurring enzymes causes the lysosomal storage of its substrates. The diseases resulting from these defects are rather heterogeneous from the biochemical as well as from the clinical point of view (for reviews see Suzuki, 1993 and appropriate chapters in the new edition of Scriver et al., 1994). The analysis of sphingolipid storage diseases without detectable hydrolase deficiency resulted in identification of a series of genetic disorders due to abnormalities in the

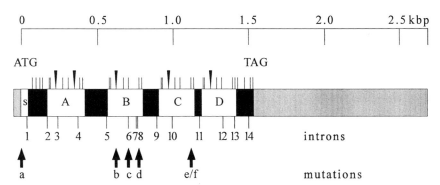

Figure 3. Structure of the *sap*-precursor cDNA (according to Fürst and Sandhoff, 1992, supplemented by a further mutation). The cDNA of *sap*-precursor codes for a sequence of 524 amino acids (or of 527 amino acids, see Holtschmidt et al., 1991) including a signal peptide of 16 amino acids (termed s, for the entry into ER) (Nakano et al., 1989; Fürst and Sandhoff, 1992). The four domains on the precursor, termed saposins A–D by O'Brien et al. (1988), correspond to the mature proteins found in human tissues: A = *sap*-A or saposin A, B = *sap*-B or saposin B or SAP-1 or sulfatide-activator, C = *sap*-C or SAP-2 or saposin C or glucosylceramidase activator protein and D = *sap*-D or saposin D or component C. The positions of cysteine residues are marked by vertical bars and those of the N-glycosylation sites by arrow heads. The positions of the 14 introns and of the known mutations leading to diseases are also given: (a) A1 → T (Met1 → Leu), (Schnabel et al., 1992); (b) C650 → T (Thr217 → Ile), (Rafi et al., 1990; Kretz et al., 1990); (c) 33 bp insertion after G777 (11 additional amino acids after Met259), (Zhang et al., 1990, 1991); (d) G722 → C (Cys241 → Ser), (Holtschmidt et al., 1991); (e) G1154 → T (Cys385 → Phe), (Schnabel et al., 1991); (f) T1155 → G (Cys385 → Gly), (Rafi et al., 1993).

Table 1. Major sphingolipidoses*

Disease	Clinicopathological manifestations	Affected lipids	Enzymatic defects
Farber disease (lipogranulomatosis)	Mostly infantile disease, tender swollen joints, multiple subcutaneous nodules, progressive hoarseness, later flaccid paralysis and mental impairment	Ceramide	Acid ceramidase
Niemann-Pick disease	Neuropathic and non-neuropathic forms, hepatosplenomegaly, foamy cells in bone marrow, severe neurological signs in the neuropathic form (type A)	Sphingomyelin	Sphingomyelinase
Globoid cell leukodystrophy (Krabbe disease)	Almost always infantile disease, white matter signs, peripheral neuropathy, high spinal fluid protein loss of myelin, globoid cells in white matter	Galactosylceramide Galactosylsphingosine	Galactosylceramidase
Metachromatic leukodystrophy	Late infantile, juvenile and adult forms, white matter signs, peripheral neuropathy, high spinal fluid protein, metachromatic granules in the brain, nerves, kidney and urine	Sulfatide	Arylsulfatase A ("sulfatidase")
Multiple sulfatase deficiency	Similar to metachromatic leukodystrophy, but additional gray matter signs, facial and skeletal abnormalities, organomegaly similar to mucopolysaccharidoses	Sulfatide, other sulfated compounds (see text)	Arylsulfatase A, B, C,
Sulfatidase activator deficiency (sap-B deficiency)	Similar to later onset forms of metachromatic leukodystrophy	Sulfatide, Globotriaosylceramide, Digalactosylceramide, GM3-ganglioside	Sulfatidase activator (SAP-1, sap-B)
Gaucher disease	Neuropathic and non-neuropathic forms, hepatosplenomegaly, "Gaucher cells" in bone marrow, severe neurological signs in the neuropathic form, also an intermediate form (type III)	Glucosylceramide Glucosylsphingosine	Glucosylceramidase
SAP-2 deficiency (sap-C deficiency)	Gaucher-like clinical phenotype	Glucosylceramide	SAP-2 (sap-C)
Fabry disease	Primarily adult and non-neurologic, angiokeratoma around buttocks, renal damage, X-linked	Globotriaosylceramide Digalactosylceramide	α-Galactosidase A (trihexosylceramidase)

Disease	Clinical features	Stored material	Defective protein/enzyme
GM1-gangliosidosis	Slow growth, motor weakness, gray matter signs, infantile form with additional facial and skeletal abnormalities and organomegaly, swollen neurons	GM1-ganglioside Galactose-rich fragments of glycoproteins	GM1-ganglioside β-galactosidase
Galactosialidosis	Similar to late-onset form of GM1-gangliosidosis	(Unknown)	"Protective protein" (secondary defect in GM1-ganglioside β-galactosidase and sialidase)
Tay-Sachs disease	Severe gray matter signs, slow growth, motor weakness, hyperacusis, cherry-red spot, head enlargement, swollen neurons	GM2-ganglioside	β-Hexosaminidase A
GM2-gangliosidosis B1 variant	Commonly later onset, slower progression but otherwise similar to Tay-Sachs disease, milder pathology	GM2-ganglioside	β-Hexosaminidase A (normal against non-sulfated artificial substrates but deficient against GM2-ganglioside and sulfated artificial substrate)
GM2-gangliosidosis AB variant	Similar to Tay-Sachs disease	GM2-ganglioside	GM2 activator protein
Sandhoff disease	Panethnic but otherwise virtually indistinguishable from Tay-Sachs disease	GM2-ganglioside, Asialo-GM2-ganglioside, Globoside	β-Hexosaminidase A and B
sap-precursor deficiency	Hyperkinetic behavior, hepatosplenomegaly, storage macrophages in bone marrow, contracted hand joints and membranous storage inclusions	All glycolipids with short sugarchains, e.g. Cer, GlcCer, LacCer, GalCer, DigalCer, Sulfatide, GM3 etc.	sap-precursor, sap-A, B, C, D

*Modified from Suzuki, 1993.

sphingolipid activator proteins, rather than in the hydrolytic enzymes (Fürst and Sandhoff, 1992; Sandhoff et al., 1994; Suzuki, 1994). Known genetic disorders affecting degradation of glycosphingolipids are summarized in Table 1.

Point mutations in the GM2-activator gene result in the clinical and biochemical phenotype similar to Tay-Sachs disease, as expected (Schröder et al., 1991, 1993). However, mutations affecting the *sap*-C domain (Gaucher factor) result in a variant form of Gaucher disease, and mutations affecting the *sap*-B domain (sulfatide-activator) result in variant forms of metachromatic leukodystrophy (Fig. 3 and Tab. 1). These findings indicate that, although these activator proteins have been demonstrated to activate degradation of many other glycosphingolipids in the test tube, they have physiological significance only in a small subset of the lipids *in vivo*. The *in vitro* findings do not necessarily reflect their physiological functions *in vivo*. Recently, two patients in a single family were found to have a mutation in the initiation codon, ATG of the *sap*-precursor gene. This resulted in a complete absence of the *sap*-precursor and consequently of all four SAPs A, B, C, D (Schnabel et al., 1992). The patients exhibited complex clinical and pathological phenotypes, and analytical studies demonstrated simultaneous storage of ceramide, and glycosphingolipids with short oligosaccharide head groups, including glucosylceramide, lactosylceramide, galactosylceramide, diagalactosylceramide, sulfatide and ganglioside GM3 (Harzer et al., 1989; Paton et al., 1992; Bradova et al., 1993). These findings are highly informative in that some of the stored lipids in these patients may require one or more of these sphingolipid activator proteins for their degradation, although they may not have been implicated so far. On the other hand, the level of sphingomyelin in these patients was entirely normal. It is clear that physiological degradation of sphingomyelin does not require any of the four activator proteins even through *in vitro* data have been presented to show that some of them activate its degradation *in vitro* (Tayama et al., 1993).

In the absence of the SAPs, the glycosphingolipids with short carbohydrate chains are not degraded in the total SAP deficiency. As detected by electron microscopy, the storage occurs predominantly within multivesicular bodies and/or within intralysosomal vesicles (Bradova et al., 1993). It appears that these lipids also stabilize the intralysosomal vesicles against degradation by lysosomal enzymes. This observation suggests an additional possible function of glycosphingolipids (and other glycoconjugates) on the plasma membrane. They may protect the cell surface against premature digestion by extracellular hydrolases, e.g., phospholipases. Bianco (1989) reported that a molar fraction of 0.2 or more of GM1-ganglioside in the lipid bilayer completely protected phospholipids within the bilayer against digestion by phospholipase A2 isolated from pig pancreas.

Acknowledgments
We thank Professor Harun Yusuf (Dhaka, Bangladesh) for helpful discussions. Studies from the authors' laboratories were supported in part by research grants, Deutsche Forschungsgemeinschaft, SFB 284, and RO1 NS24289, RO1-NS-28997 and a Mental Retardation Research Center Core grant, P30-HD03110 from the United States Public Health Service. The collaboration between our laboratories has been facilitated by the Senior Scientist Award from the Alexander von Humboldt Stiftung to K. Suzuki.

* *Note added in proof:* Recently we have shown that *sap*-D stimulates the degradation of ceramide by acid ceramidase *in vivo*. The fact that *sap*-D co-purifies with acid ceramidase indicates that the activator shares the same mechanism as *sap*-C (Klein et al., 1994, Azuma et al., 1994).

References

Azuma, N., O'Brien, J.S., Moser, H.W. and Kishimoto, Y. (1994) Stimulation of acid ceramidase activity by saposin D. *Arch. Biochem. Biophys.* 311: 354–357.

Berent, S.L. and Radin, N.S. (1981) Mechanism of activation of glucocerebrosidase by co-β-glucosidase (glucosidase activator protein). *Biochim. Biophys. Acta* 664: 572–582.

Bianco, I.D., Fidelio, G.D. and Maggio, B. (1989) Modulation of phospholipase A2 activity by neutral and anionic glycosphingolipids in monolayers. *Biochem. J.* 258: 95–99.

Bradova, V., Smid, F., Ulrich-Bott, B., Roggendorf, W., Paton, B.C. and Harzer, K. (1993) Prosaposin deficiency: further characterization of the sphingolipid activator protein-deficient sibs. *Human Genet.* 92: 143–152.

O'Brien, J.S., Kretz, K.A., Dewji, N., Wenger, D.A., Esch, F. and Fluharty, A.L. (1988) Coding of two sphingolipid activator proteins (SAP-1 and SAP-2) by same genetic locus. *Science* 241: 1098–1101.

Carlsson, S.R. and Fukuda, M. (1990) The polylactosaminoglycans of human lysosomal membrane glycoproteins Lamp-1 ad Lamp-2. *J. Biol. Chem.* 265: 20488–20495.

Coste, H., Martel, M.-B. and Got, R. (1986) Topology of glucosylceramide synthesis in Golgi membranes from porcine submaxillary glands. *Biochim. Biophys. Acta* 858: 6–12.

Fishman, P.H. and Brady, R.O. (1976) Biosynthesis and function of gangliosides. *Science* 194: 906–915.

Fürst, W. and Sandhoff, K. (1992) Activator proteins and topology of lysosomal sphingolipid catabolism. Review. *Biochim. Biophys. Acta* 1126: 1–16.

Griffiths, G., Hoflack, B., Simons, K., Mellman, I. and Kornfeld, S. (1988) The mannose-6-phosphate receptor and the biogenesis of lysosomes. *Cell* 52: 329–341.

Hakomori, S.I. (1984) Glycosphingolipids as differentiation-dependent, tumor-associated markers and as regulators of cell proliferation. *Trends Biochem. Sci.* 9: 453–458.

Harzer, K., Paton, B.C., Poulos, A., Kustermann-Kuhn, B., Roggendorf, W., Grisar, T. and Popp, M. (1989) Sphingolipid activator protein (SAP) deficiency in a 16-week-old atypical Gaucher disease patient and his fetal sibling: biochemical signs of combined sphingolipidoses. *Eur. J. Pediatr.* 149: 31–39.

Ho, M.W. and Light, N.D. (1973) Glucocerebrosidase: Reconstitution from macromolecular components depends on acidic phospholipids. *Biochem. J.* 136: 821–823.

Ho, M.W. (1975) Specificity of low molecular weight glycoprotein effector of lipid glycosidase. *FEBS Lett.* 53: 243–247.

Ho, M.W. and Rigby, M. (1975) Glucocerebrosidase. Stoichiometry of association between effector and catalytic proteins. *Biochim. Biophys. Acta* 397: 267–273.

Holtschmidt, H., Sandhoff, K., Kwon, H.Y., Harzer, K., Nakano, T. and Suzuki, K. (1991) Sulfatide activator protein: Alternative splicing generates three mRNAs and a newly found mutation responsible for a clinical disease. *J. Biol. Chem.* 266: 7556–7560.

Hopkins, C.R., Gibson, A., Shipman, M. and Miller, K. (1990) Movement of internalized ligand-receptor complexes along a continuous endosomal reticulum. *Nature* 346: 335–339.

Klein, A., Henseler, M., Klein, C., Suzuki, K., Harzer, K. and Sandhoff, K. (1994) Sphingolipid activator protein D (sap-D) stimulates the lysosomal degradation of ceramide in vivo. *Biochem. Biophys. Res. Commun.* 200: 1440–1448.

Kok, J.W., Babia, T. and Hoekstra, D. (1991) Sorting of sphingolipids in the endocytic pathway of HT 29 cells. *J. Cell Biol.* 114: 231–239.

Koval, M. and Pagano, R.E. (1989) Lipid recycling between the plasma membrane and intracellular compartments: transport and metabolism of fluorescent sphingomyelin analogs in cultured fibroblasts. *J. Cell Biol.* 108: 2169–2181.

Koval, M. and Pagano, R.E. (1990) Sorting of an internalized plasma membrane lipid between recycling and degradative pathways in normal and Niemann-Pick, type A fibroblasts. *J. Cell Biol.* 111: 429–442.

Kretz, K.A. Carson, G.S., Morimoto, S., Kishimoto, Y., Fluharty, A.L. and O'Brien, J.S. (1990) Characterization of a mutation in a family with saposin B deficiency: A glucosylation site defect. *Proc. Natl. Acad. Sci. USA* 87: 2541–2544.

Ledeen, R.W. and Yu, R.K. (1982) New strategies for detection and resolution of minor gangliosides as applied to brain fucogangliosides. *Methods Enzymol.* 83: 139–189.

Li, S.-C., Sonnino, S., Tettamanti, G. and Li, Y.-T. (1988) Characterization of a nonspecific activator protein for the enzymatic hydrolysis of glycolipids. *J. Biol. Chem.* 263: 6588–6591.

Mandon, E.C., van Echten, G., Birk, R., Schmidt, R.R. and Sandhoff, K. (1991) Sphingolipid biosynthesis in cultured neurons. Down-regulation of serine palmitoyltransferase by sphingoid bases. *Eur. J. Biochem.* 198: 667–674.

Mandon, E., Ehses, I., Rother, J., van Echten, G. and Sandhoff, K. (1992) Subcellular localization and membrane topology of serine palmitoyltransferase, 3-dehydrosphinganine reductase and sphinganine N-acyltransferase in mouse liver. *J. Biol. Chem.* 267: 11144–11148.

Markwell, M.A.K., Svennerholm, L. and Paulson, J.C. (1981) Specific gangliosides function as host cell receptors for Sendai virus. *Proc. Natl. Acad. Sci. USA* 78: 5406–5410.

McKanna, J.A., Haigler, H.T. and Cohen, S. (1979) Hormone receptor topology and dynamics: Morphological analysis using ferritin-labeled epidermal growth factor. *Proc. Natl. Acad. Sci. USA* 76: 5689–5693.

Mehl, E. and Jatzkewitz, H. (1964) Eine Cerebrosidsulfatase aus Schweineniere. *Hoppe-Seylers Z. Physiol. Chem.* 339: 250–276.

Meier, E.M., Schwarzmann, G., Fürst, W. and Sandhoff, K. (1991) The human GM2 activator protein: A substrate specific cofactor of hexosaminidase A. *J. Biol. Chem.* 266: 1879–1887.

Morré, D.J., Kartenbeck, J. and Franke, W.W. (1979) Membrane flow and interconversions among endomembranes. *Biochim. Biophys. Acta* 559: 71–152.

Mullock, B.M. and Luzio, J.P. (1993) Cell-free interactions between rat-liver endosomes and lysosomes. *Trans. Biochem. Soc.* 21: 721–722.

Murphy, R.F., Schmid, J. and Fuchs, R. (1993) Endosome maturation: Insights from somatic cell genetics and cell-free analysis. *Trans. Biochem. Soc.* 21: 716–720.

Nakano, T., Sandhoff, K., Stümper, J., Christomanou, H. and Suzuki, K. (1989) Structure of full-length cDNA coding for sulfatide activator, a co-β-glucosidase and two other homologous proteins: Two alternate forms of the sulfatide activator. *J. Biochem.* 105: 162–154.

Paton, B.C., Schmid, B., Kustermann-Kuhn, B., Poulos, A. and Harzer, K. (1992) Additional biochemical findings in a patient and fetal sibling with genetic defect in the sphingolipid activator protein (SAP) precursor, prosaposin. *Biochem. J.* 185: 481–488.

Rafi, M.A., Zhang, X.-L., de Gala, G. and Wenger, D.A. (1990) Detection of a point mutation in sphingolipid activator protein-1 mRNA in patients with a variant form of metachromatic leukodystrophy. *Biochem. Biophys. Res. Commun.* 166: 1017–1023.

Rafi, M.A., de Gala, G., Zhang, X. and Wenger, D.A. (1993) Mutational analysis in a patient with a variant form of Gaucher disease caused by SAP-2 deficiency. *Somat. Cell Molec. Genet.* 19: 1–7.

Renfrew, C.A., and Hubbard, A.L. (1991) Degradation of epidermal growth factor receptor in rat liver. Membrane topology through the lysosomal pathway. *J. Biol. Chem.* 266: 21265–21273.

Sandhoff, K., Harzer, K. and Fürst, W. (1995) Sphingolipid activator proteins. In: C.R. Scriver, A.L. Beaudet, W.S. Sly and D. Valle (eds): *The Metabolic Basis of Inherited Disease*, 7th edition. McGraw-Hill, New York, Part 12, 2427–2441.

Schnabel, D., Schröder, M. and Sandhoff, K. (1991) Mutation in the sphingolipid activator protein 2 in a patient with a variant of Gaucher disease. *FEBS Lett.* 284: 57–59.

Schnabel, D., Schröder, M., Fürst, W., Klein, A., Hurwitz, R., Zenk, T., Weber, G., Harzer, K., Paton, B., Poulos, A., Suzuki, K. and Sandhoff, K. (1992) Simultaneous deficiency of sphingolipid activator proteins 1 and 2 is caused by a mutation in the initiation codon of their common gene. *J. Biol. Chem.* 267: 3312–3315.

Schröder, M., Schnabel, D., Suzuki, K. and Sandhoff, K. (1991) A mutation in the gene of a glycolipid-binding protein (GM2-activator) that causes GM2-gangliosidosis variant AB. *FEBS Lett.* 290: 1–3.

Schröder, M., Schnabel, D., Hurwitz, R., Young, E., Suzuki, K. and Sandhoff, K. (1993) Molecular genetics of GM2-gangliosidosis AB variant: A novel mutation and expression in BHK cells. *Hum. Genet.* 92: 437–440.

Schwarzmann, G. and Sandhoff, K. (1990) Metabolism and intracellular transport of glycosphingolipids. "Perspectives in Biochemistry" *Biochemistry* 29: 10865–10871.

Scriver, C.R., Beaudet, A.L., Sly, W.S. and Valle, D. (1995) *The Metabolic Basis of Inherited Disease*, 7th edition. McGraw-Hill, New York.

Stoffel, W., Le Kim, D. and Sticht, G. (1968) Metabolism of sphingosine bases: Biosynthesis of dihydrosphingosine *in vitro*. *Hoppe-Seylers Z. Physiol. Chem.* 349: 664–670.

Stoorvogel, W. (1993) Arguments in favour of endosome maturation. *Trans. Biochem. Soc.* 21: 711–715.

Suzuki, K. (1993) Genetic disorders of lipid, glycoprotein and mucopolysaccharide metabolism. *In:* G. Siegel, B.W. Agranoff, R.W. Albers and P. Molinoff (eds): *Basic Neurochemistry*, 5th edition. Raven Press, New York, pp. 793–812.

Suzuki, K. (1995) Sphingolipid activator proteins. *Essays in Biochemistry*; in press.

Svennerholm, L. (1984) Biological significance of gangliosides. *In:* H. Dreyfus, R. Massarelli, L. Freysz and G. Rebel (eds): *Cellular and Pathological Aspects of Glycoconjugate Metabolism*, Vol. 126. INSERM, France, pp. 21–44.

Tayama, M., Soeda, S., Kishimoto, Y., Martin, B.M., Callahan, J.W., Hiraiwa, M. and O'Brien, J.S. (1993) Effect of saposins on acid sphingomyelinase. *Biochem. J.* 290: 401–404.

Thompson, L.K., Horowitz, P.M., Bently, K.L., Thomas, D.D., Alderete, J.F. and Klebe, R.J. (1986) Localization of the ganglioside-binding site of fibronectin. *J. Biol. Chem.* 261: 5209–5214.

Tiemeyer, M., Yasuda,Y. and Schnaar, R.L. (1989) Ganglioside-specific binding protein on rat brain membranes. *J. Biol. Chem.* 264: 1671–1681.

Trinchera, M., Fabbri, M. and Ghidoni, R. (1991) Topography of glycosyltransferases involved in the initial glycosylations of gangliosides. *J. Biol. Chem.* 266: 20907–20912.

Vaccaro, A.M., Tatti, M., Ciaffoni, F., Salvioli, R., Maras, B. and Barca, A. (1993) Function of saposin C in the reconstitution of glucosylceramidase by phosphatidylserine liposomes. *FEBS Lett.* 336: 159–162.

van Deurs, B., Holm, P.K., Kayser, L., Sandvig, K. and Hansen, S.H. (1993) Multivesicular bodies in HEp-2 cells are maturing endosomes. *Eur. J. Cell Biol.* 61: 208–224.

van Echten, G. and Sandhoff, K. (1989) Modulation of ganglioside biosynthesis in primary cultured neurons. *J. Neurochem.* 52: 207–214.

van Echten, G. and Sandhoff, K. (1993) Ganglioside metabolism. *J. Biol. Chem.* 268: 6341–5344.

Vogel, A., Schwarzmann, G. and Sandhoff, K. (1991) Glycosphingolipid specificity of the human sulfatide activator protein. *Eur. J. Biochem.* 200: 591–597.

Wessling-Resnick, M. and Braell, W.A. (1990) The sorting and segregation mechanism of the endocytic pathway is functional in a cell-free system. *J. Biol. Chem.* 265: 690–699.

Zachgo, S., Dobberstein, B. and Griffiths, G. (1992) A block in degradation of MHC class II-associated invariant chain correlates with a reduction in transport from endosome carrier vesicles to the prelysosome compartment. *J. Cell Sci.* 103: 811–822.

Zhang, X.-L., Rafi, M.A., de Gala, G. and Wenger, D.A. (1990) Insertion in the mRNA of a metachromatic leukodystrophy patient with sphingolipid activator protein-1 deficiency. *Proc. Natl. Acad. Sci. USA* 87: 1426–1430.

Zhang, X.-L., Rafi, M.A., de Gala, G. and Wenger, D.A. (1991) The mechanism for a 33-nucleotide insertion in messenger RNA causing sphingolipid activator protein (SAP-1) – Deficient metachromatic leukodystrophy. *Human Genet.* 87: 211–215.

Biochemistry of Cell Membranes
ed. by S. Papa & J. M. Tager
© 1995 Birkhäuser Verlag Basel/Switzerland

On the mechanism of proton pumps in respiratory chains

S. Papa, M. Lorusso and N. Capitanio

Institute of Medical Biochemistry and Chemistry, University of Bari, I-70124, Bari, Italy

Introduction

While it is generally accepted that a transmembrane proton current (proticity) mediates energy transfer in oxidative phosphorylation (Boyer et al., 1977), the molecular mechanism by which redox membrane proteins act as proticity generators (Mitchell, 1966, 1987a; Papa, 1976; Wikström and Saraste, 1984; Malmström, 1989; Rich, 1991) and the ATP synthase as proticity utilizer (Scarpa et al., 1992) is still not fully understood.

Mitchell (1966) originally proposed the protonmotive activity of electron transfer chains to be a direct consequence of primary catalysis by the redox centers. The respiratory chain of mitochondria was conceived as consisting of three consecutive redox loops (alternating hydrogen and electron carriers) each resulting in the effective translocation of $2H^+$ per $2 e^-$ from the N to the P side of the membrane. Later, to explain the actual H^+/e^- stoichiometries (Papa et al., 1975; Lawford and Garland, 1973; Wikström, 1977; Wikström and Krab, 1979), which turned out to be different from what was initially predicted, Mitchell introduced the protonmotive quinone cycle for the bc_1 complex (Mitchell, 1976) and the oxygen cycle for cytochrome c oxidase (Mitchell et al., 1985), subsequently replaced by the Cu loop (Mitchell, 1987b) (Fig. 1a).

Based on the principle of co-operative thermodynamic linkage of solute binding at separate sites in allosteric proteins (Bohr effects) (Wyman, 1968) and the finding that linkage between electron transfer at the metal and protolytic events does occur in cytochromes (redox Bohr effects) (Dutton and Wilson, 1974; Papa, 1976; Papa et al., 1979), Papa et al. proposed an indirect co-operative model for proton pumping by cytochromes (vectorial Bohr mechanism) (Papa, 1976; Papa et al., 1973; see also Von Jagow and Sebald, 1980). The co-operative mechanism, applied to the bc_1 complex, was later developed into a combined model in which redox Bohr effects were conceived to operate in series with

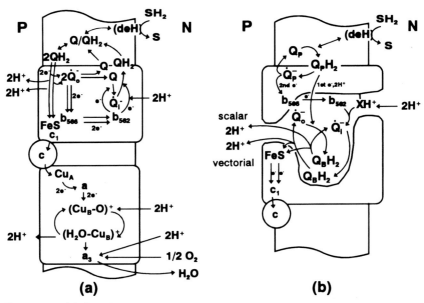

Figure 1. Models for redox and protonmotive activity of the cytochrome system of mitochondria. (a) Ubiquinone cycle for the bc_1 complex and Cu_B loop for cytochrome c oxidase complex. (b) Q-gated H^+ pump for the bc_1 complex. The ubiquinone cycle is based on re-cycling of one of the two electrons donated by ubiquinol of the pool to the bc_1 complex and might involve exchange of protein-bound quinone with the pool. The Q-gated pump envisages a linear split pathway for electron transfer mediated by non-exchangeable protein stabilized quinol/semiquinone, which provides pumping of $2H^+/e^-$. In the Cu_B loop proton translocation from the N to the P side is mediated by reorientation around Cu_B of OH^- (or O^-) and H_2O.

protonmotive redox catalysis by a protein bound quinol/semiquinone couple (Q-gated proton pump) (Papa et al., 1983b, 1990) (Fig. 1b).

A commonly used minimal model for the redox proton pump is the eight-state "cubic" formalism developed by Wikström et al. (1981). This describes the essential steps of the pump and can be generally applied to both direct and indirect coupling mechanisms. The cubic model emphasizes the role of re-orientation of the electron and proton transfer centers (alternating access of the centers to their reactants (Malmström, 1985) and the feasibility of kinetic linkage (Malmström, 1985; Blair et al., 1986)).

In the kinetic-linkage mechanism, the transition-state of an electron transfer step is proposed to be stabilized by protonation, thus promoting, through reorientation of the redox and protolytic center(s), electron transfer from a donor to an acceptor and proton translocation from the input (N) to the output (P) sides. The pump can exhibit variable slipping with respect to electrons or protons (Wilkström et al., 1981; Blair et al., 1986; Malmström, 1989). A drawback of the transition state mechanism may be represented by a low incidence of the protonated "transition state" in the coupled state in which the N phase experiences an alkaline pH. This difficulty can be minimized by redox-Bohr effects which would provide a kinetic advantage by promoting protonation of the transition state when, in its reduced state, it is accessible to protons from the N phase (Malström, 1985; Blair et al., 1986). It has been pointed out that, conversely, redox-Bohr effects without kinetic barrier, such as that envisaged in the transition-state mechanism, might be ineffective for pumping protons against an opposing gradient (Blair et al., 1986).

The present authors favor models for redox proton pumps in which proton transfer reactions taking place directly at the catalytic redox center are associated with kinetic and/or thermodynamic linkage. In this case the anisotropy of protonation of the catalytic/coupling system from the N side and its deprotonation at the P side, rather than deriving directly from the mobility of the primary chemical groups participating in the catalysis (Mitchell, 1966), is imposed on them by redox-Bohr effects in the enzyme, i.e., pK shifts thermodynamically-linked to redox catalysis and properly tuned in time and space (Papa and Lorusso, 1984).

Redox Bohr effects

Redox Bohr effects result in pH dependence of the midpoint redox potential of the electron carriers involved (Urban and Klingenberg, 1969; Dutton and Wilson, 1974; van Gelder et al., 1977; Blair et al., 1985) and in scalar-proton transfer associated to their oxido-reduction (Papa et al., 1979, 1986). In the cytochrome c oxidase isolated from beef-heart mitochondria the number of scalar H^+/COX taken up upon full reduction of hemes a, a_3, Cu_A and Cu_B, and released per mole of soluble oxidase upon oxidation of these centers, increases with pH from 1 at acidic to 2.5 at alkaline pH (Fig. 2) (cf. Hallen and Nilsson, 1992; Mitchell and Rich, 1994). In the cyanide-liganded oxidase with heme a_3 blocked in the oxidized state, the scalar H^+/COX transfer associated to oxido-reduction of heme a, Cu_A and Cu_B (Wrigglesworth et al., 1988; Moody and Rich, 1990), when plotted as a function of pH, fits a bell-shaped curve resulting from combination of those obtained for two hypothetical redox-linked protolytic groups (Capitanio et al., 1990). Since the Em of Cu_A is pH independent (Erecinska et al., 1971) the pK shifts of the two protolytic groups have to be linked to oxido-reduction of heme a (Artzatbanov et al., 1978) and/or Cu_B whose Em in the cyanide-liganded oxidase are apparently pH dependent (Moody and Rich, 1990). Evidence for pH-dependence in the cyanide-liganded oxidase of negative cooperativity between heme a and Cu_B, apparently

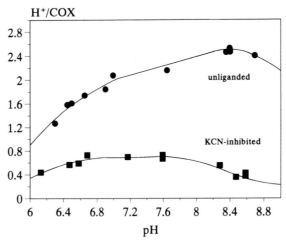

Figure 2. pH dependence of redox-Bohr effects (H$^+$/COX coupling number) in unliganded and cyanide-liganded soluble cytochrome c oxidase purified from beef-heart mitochondria. Mathematical analysis shows that the H$^+$/COX ratios in the unliganded oxidase are best-fitted by a curve derived from at least three protolytic groups undergoing redox-linked pK shifts, the H$^+$/COX ratios for the cyanide-liganded oxidase are best fitted by a curve with two protolytic groups.

linked to (de)protonation of a common protolytic group, has been presented (Moody and Rich, 1990). This would be consistent with the observation that in the CO-liganded oxidase, where heme a$_3$ and Cu$_B$ are held in the reduced state, the Em of heme a is practically pH independent (Ellis et al., 1986).

Redox Bohr effects in the unliganded oxidase apparently derive from three or more protolytic groups undergoing pK shift upon oxido-reduction of the unliganded oxidase. Two are associated to heme a (and possibly CuB) (see also Mitchell and Rich, 1994), the other protolytic group(s) might be associated with heme a$_3$. One group could be represented by H$_2$O$_2$ ligated to the heme-iron (Rich, 1991; Konstantinov et al., 1992), the others represented by protolytic residues in the protein. Genetic analysis of bacterial a,a$_3$ cytochrome c oxidase has shown that there are in subunit I, in the proximity of heme a and the binuclear heme a$_3$–Cu$_B$ center (Fig. 3), a number of conserved protolytic residues whose replacement by site-directed mutagenesis affects spectral and functional characteristics of the redox centers (Hosler et al., 1993; Brown et al., 1993).

Variability of the H$^+$/e$^-$ stoichiometry of proton pumps

This issue of variability of H$^+$/e$^-$ ratio in proton pumps has been extensively addressed by examining the steady-state relationship be-

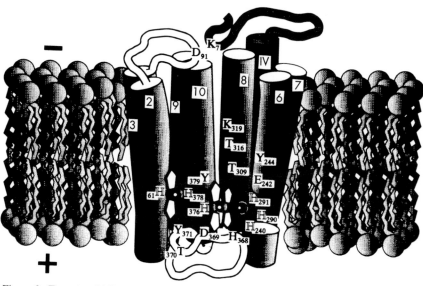

Figure 3. Tentative folding and membrane topology of putative transmembrane α-helices of subunit I of cytochrome c oxidase with conserved residues (outlined letters) serving as ligands for heme a, heme a_3 and Cu_B. Conserved residues possibly involved in intramolecular electron transfer and in proton translocation (full letters) are shown. A conserved Asp at position 91 (in the bovine sequence, 135 in E. Coli) of the second loop connecting transmembrane helices 2 and 3 at the N side might be involved in the proton input channel of the pump (Thomas et al., 1993). The transmembrane helix of subunit IV and the hydrophilic NH_2^- terminal segment exposed at the N surface and apparently involved with its Lys7 in proton translocation is also shown (Planques et al., 1989; Capitanio et al., 1994).

tween respiratory rate and transmembrane proton motive force (Δp) in mitochondria and other membranes. The observed non-linearity of this relationship (Nicholls, 1974; Pietrobon et al., 1983; Brown and Brand, 1986; Murphy, 1989) is taken by some authors as evidence of slip in proton pumps (Pietrobon et al., 1983; Luvisetto et al., 1991; see also Murphy and Brand, 1988). Others, however, argue against this interpretation and consider the non-linearity as due to non-ohmic increase of membrane proton conductance at high Δp (leak) (Brown and Brand, 1986; Brown, 1989).

Recently, Papa et al. (1991) and Capitanio et al. (1991), using a rate method, have shown that the intrinsic $\leftarrow H^+/e^-$ stoichiometry of the cytochrome system of mitochondria varies under the influence of kinetic and thermodynamic factors. A systematic study of the protonmotive activity of cytochrome c oxidase and cytochrome c reductase in intact mitochondria and in the purified enzymes reconstituted in phospholipid vesicles was carried out by these authors. Measurements were carried out in the presence of valinomycin so that K^+ migration prevented build up of a membrane potential. The $\leftarrow H^+/e^-$ stoichiometry in

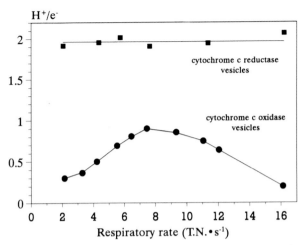

Figure 4. Dependence of the $\leftarrow H^+/e^-$ ratio on the rate of electron flow in reconstituted cytochrome c oxidase and cytochrome c reductase. Cytochrome c oxidase liposomes were supplemented with cytochrome c and ascorbate plus N,N,N',N'-tetramethyl-p-phenylenediamine (TMPD). The respiratory rate was varied changing the concentration of TMPD. Cytochrome c reductase vesicles were supplemented with ferricytochrome c, a trace of soluble cytochrome c oxidase and duroquinol. The respiratory rate was varied by titrating with antimycin A. In all the experiments respiration was initiated by the addition of ferricytochrome c and the $\leftarrow H^+/e^-$ ratios were obtained from the initial rates of electron flow and proton translocation measured spectrophotometrically. For other details see Papa et al. (1991).

cytochrome c oxidase vesicles measured from initial rates at level flow, i.e., under conditions of negligible Δp, first increases with the rate of electron flow to about 1, then decreases again upon further enhancement of the rate of electron flow (Fig. 4) (Capitanio et al., 1991). In cytochrome c reductase vesicles the $\leftarrow H^+/e^-$ stoichiometry is, at level flow, 2 independent of the actual rate of e^- flow (Fig. 4) (Cocco et al., 1992).

The rate dependence of the $\leftarrow H^+/e^-$ stoichiometry in the oxidase can result from slips in e^- transfer. It is thought that, to be coupled to proton pumping, e^- have to follow the sequence cyt.c $\rightarrow Cu_A \rightarrow$ heme $a \rightarrow Cu_B \rightarrow$ heme a_3 (Babcock and Wikström, 1992; see however Brown et al., 1994). Direct e^- slip from Cu_A to Cu_B/a_3, bypassing heme a, will result in decoupling of proton pumping. Electron slip could depend on the redox pressure imposed on the enzyme, negative co-operativity between the high and low potential redox centers (Nicholls and Pedersen, 1974; Wikström et al., 1981; Mitchell et al., 1992) and the relative adjustment of the kinetics of the various redox steps. The low efficiency of proton pumping observed in oxygen pulses of the fully reduced enzyme (Papa et al., 1987a; Oliveberg et al., 1991) as compared to that obtained in reductant pulses of the oxidized enzyme (Wilkström and Krab, 1979; Papa et al., 1987a) could provide an example of how the activity of the pump can be affected by these and other factors (Papa, 1988).

Measurements in intact mitochondria show that under level flow conditions the $\leftarrow H^+/e^-$ ratio for succinate respiration varies from minima of 2 at extreme high (Papa et al., 1980a,b, 1983a) and low respiratory rates (cf. Lorusso et al., 1979) (the same low ratios were found at low respiratory rates in oxygen pulse experiments (Papa et al., 1987b)), to about 3 at intermediate rates, with a rate dependence similar and directly attributed to that exhibited by the $\leftarrow H^+/e^-$ ratio for cytochrome c oxidase (Fig. 4) (Papa et al., 1991; Capitanio et al., 1991). The $\leftarrow H^+/e^-$ ratio for the span succinate to ferricyanide measured at level flow is, on the other hand, constantly 2, independent of the rate of electron flow (Capitanio et al., 1991; Cocco et al., 1992).

At steady-state the highest ratios attainable at intermediate flow rates are significantly lower than those observed for the same rates at level flow, this providing evidence that, besides the flow rate also Δp, in fact its ΔpH component ($\Delta \psi$ was collapsed by valinomycin in these experiments), affects the H^+/e^- stoichiometry (cf. Murphy and Brand, 1988). This is confirmed by other experiments showing that ΔpH but not $\Delta \psi$ depresses the efficiency of proton pumping (Cocco et al., 1992; see also Murphy, 1989). Δp by specifically depressing proton-coupled electron flow can enhance the contribution of decoupled e^- slips. Furthermore, alkalinization, at steady-state, of the N phase can result in proton slip because of loss of protonation asymmetry of the critical proton translocating center in the pump in the input state (Capitanio et al., 1990; Cocco et al., 1992). Consistent with this view is the finding that the H^+/e^- ratio for succinate respiration at steady-state in "inside out" submitochondrial vesicles, which expose the input state of the pump to the outer space, approaches the value of 3 at slightly acidic pH and in the presence of NaSCN to collapse Δp (Papa et al., 1973).

It is interesting to note that the marked rate dependence of the $\leftarrow H^+/e^-$ stoichiometry observed for cytochrome c oxidase at level flow is attenuated at the steady-state (Papa et al., 1991). Under the latter conditions the critical factor for proton pumping seems to be represented for both the oxidase and the reductase by transmembrane ΔpH which imposes a limitation to protonation of the pump in the input state. It is, however, possible that under physiological phosphorylating conditions, proton-coupled uptake of phosphate and respiratory substrates and proton-influx for ATP synthesis contribute to prevent establishment of a large ΔpH (and low H^+ concentration at the N input side of pumps) so that proton pumping by the cytochrome system can approach maximal efficiency.

Rate-dependent slips in redox proton pumps could represent a mean to optimize the intrinsic power output of the pump (Blair et al., 1986),

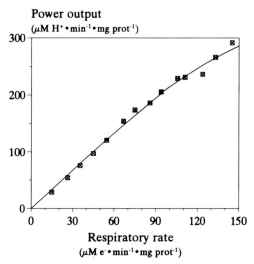

Figure 5. Power output of the protonmotive activity of the mitrochondrial cytochrome system. The power output was obtained multiplying the $\leftarrow H^+/e^-$ ratio by the actual respiratory rate both measured during steady-state succinate respiration in mitochondria.

this being given by the product of the actual electron flow and the corresponding $\leftarrow H^+/e^-$ stoichiometry. The situation for the overall respiratory activity of the mitochondrial cytochrome chain with succinate as substrate is illustrated in Figure 5. This shows that the overall

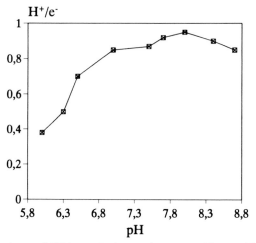

Figure 6 pH-dependence of H^+/e^- ratio in cytochrome c oxidase vesicles. Redox linked proton translocation was measured electrometrically and elicited by addition of an amount of ferrocytochrome c able to substain a limited number (2–3) of turnovers. For further details see Papa et al., 1987a.

net power output at steady-state increases with the respiratory rate, approaching saturation at the highest rates probably set up by dehydrogenase and/or transporter activity.

The $\leftarrow H^+/e^-$ stoichiometry in vesicles reconstituted cytochrome c oxidase was also affected by the pH of the external medium. The $\leftarrow H^+/e^-$ ratio increases in the pH range 6.0 to $7.0 - 7.5$ (Papa et al., 1987a) (Fig. 6). Low H^+/e^- ratios at acidic pH could result from proton slip caused by depression of deprotonation of the pump in the output state at the P side (Maison Peteri and Malmström, 1989). A pH dependence of the $\leftarrow H^+/e^-$ ratio similar to that observed in cytochrome c oxidase vesicles has also been observed in the cytochrome *bo* quinol oxidase (Verkhovskaya et al., 1992).

The mechanism of proton pumping in the redox complexes of respiratory chain

In Figure 7 (a and b) a minimal scheme for the four electron reduction of dioxygen to H_2O by heme-copper oxidases is presented (Babcock and Wikström, 1992; Mitchell et al., 1992; Konstantinov et al., 1992; Malmström, 1990). It consists of: (1) two electron reduction of Cu_B and heme a_3; (2) oxygen binding, and (3) its reduction to heme a_3-bound peroxide (*P*); (4) delivery of the 3rd electron with formation of the first H_2O molecule and the oxyferryl derivative of heme a_3 (*F*); (5) delivery of the 4th electron with formation of the second H_2O molecule and reconversion of the binuclear center to the fully oxidased state (*O*). Based on recent findings of Konstantinov et al. (1992) and Mitchell et al. (1992) (see also Guerrieri et al., 1981) the first two protons consumed in the reduction of dioxygen to H_2O are shown to be taken up by the enzyme reduction of heme a_3 and Cu_B, and the 3rd and 4th protons are taken up separately in reaction (4) and (5) respectively. Wikström (1989) has proposed that the four electrons consumed in the reduction of dioxygen are not all equivalent for proton pumping as only the transfer of the 3rd (reaction 4) and 4th electron (reaction 5) would be associated to pumping of two protons per each electron.

As regards the molecular mechanism by which proton pumping would be associated to electron transfer, various models have been proposed. These can be grouped into two types: exchange of protolytic ligands at the metal centres (A) and redox-linked pK shifts of protolytic groups in the enzyme (B).

In Figure 7 two of the possible models by which the oxygen chemistry at the binuclear center could be directly involved in the protonmotive activity of the heme copper oxidases are presented (cf. Rich, 1991; Mitchell, 1987b, Mitchell, 1988). Both models (a) and (b) incorporate the various oxygen intermediate derivatives of heme a_3 resolved by

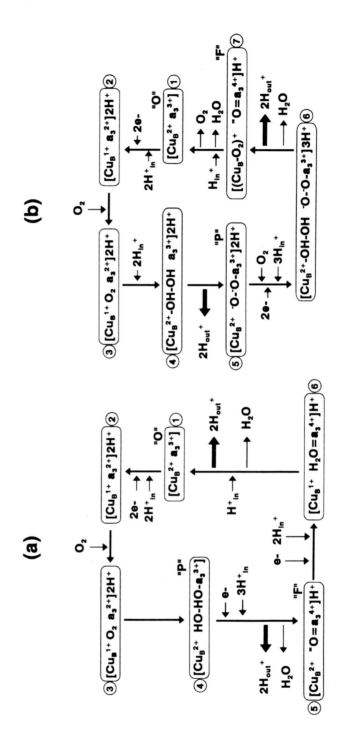

Figure 7. Tentative models for protonmotive dioxygen reduction in heme-cooper oxidases. Model (a) is based on anisotropic protonation of the "P" and "F" intermediates of heme a_3; model (b) on initial production of hydrogen peroxide on Cu_B and its anisotropic (de)protonation. For details see text.

resonance-Raman spectropscopy (Varotsis et al., 1993). Model (a) is a minimal scheme assuming anisotropic protonation from the N phase and deprotonation to the P phase of the peroxy (P) and oxy-ferryl (F) derivative of heme a_3, both protonmotive steps resulting in the translocation of $2H^+$ from the N to the P phase per e^- transferred in the conversion of "P" to "F" and "F" to "O" (Wikström, 1989). Model (b) introduces the proposal that a central process in proton pumping is the production of hydrogen peroxide on Cu_B. It is conceived that the first two electrons, which reduce Cu_B and heme a_3, are transferred to dioxygen bound to the binuclear center with formation of Cu^{2+}-peroxide protonated from the N phase (cf. Rich, 1991). This step is followed by exchange of H_2O_2 from Cu_B^{2+} to a_3^{3+} (via a bridged intermediate?) where it finds an environment favoring its deprotonation in the P phase, with pumping of the first two protons.

The third and fourth electrons, accompanied by binding of a second O_2 and three N protons, produce a second $Cu_B^{2+}-H_2O_2$. One electron of $Cu_B^{2+}-H_2O_2$ is donated to $^-O-^-O-a_3$ with formation of compound F, the first H_2O molecule and pumping in the P phase of the other two protons. The remaining one electron species $(Cu_B-O_2)^+$ transfers the fourth electron to the F compound with formation, with the uptake of a N proton, of the second H_2O and reconversion of the binuclear center to the fully oxidazed state.

The present model makes use of the concept introduced by Mitchell (1987b) of ligation and redox-linked reorientation of oxygen species on Cu_B (see also Rich, 1991). In Mitchell's model the ligands to Cu_B are OH^- (or O^-) and H_2O, in the present model O_2, $(\dot{O}_2^-)$ and H_2O_2.

Bohr effects, apparently shared by Cu_B and heme a (Moody and Rich, 1990), could represent a co-operative device by which heme a participates in the pump by coupling electron delivery to Cu_B with proton translocation from the N phase to the $Cu_B^{2+}-O^--O^-$ compound.

The maximal $\leftarrow H^+/e^-$ stoichiometry and $\leftarrow q^+/e^-$ stoichiometries observed for complex I (Weiss et al., 1991), III (Papa and Lorusso, 1984) and IV (Wikström and Saraste, 1984) of the mitochondrial respiratory chain are 2 and 2, 2 and 1, 1 and 2 respectively (see Fig. 8). A detailed tentative mechanism for proton pumping in the oxidase (complex IV) has just been described. For the pumping activity of complex III two models are available: the Q-cycle (Mitchell, 1976) and the Q-gated proton pump (Papa et al., 1983b, 1990). The relative merits of the two models are discussed by Papa et al. (1990). For the Q cycle many reviews have been published on the issue (Lenaz et al., 1990). As regards the Q-gated proton pump it can be noted that the two electrons transferred from ubiquinol to cytochrome c are not equivalent for proton pumping (Papa et al., 1990). The transfer of the first electron of the quinol of the pool to cytochrome c is associated with release in the

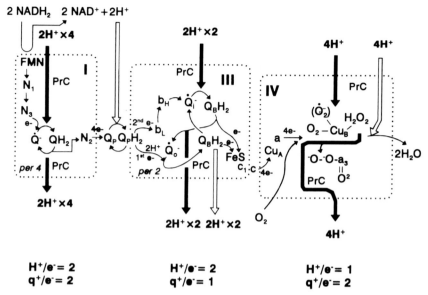

Figure 8. Simplified general scheme for proton pumping at the three coupling sites of the mitochondrial respiratory chain in complex I, III, and IV respectively. The maximal $\leftarrow H^+/e^-$ and $\leftarrow q^+/e^-$ ratios attainable for the three sites are given at the bottom of the scheme. Proton pumping in complex I is based on the information reviewed in (Chan and Li, 1990) and translocation of $2H^+/e^-$ by protein-bound $\dot{Q}^-/QH_2$. For proton translocation in complex III the Q-gated proton pump (Papa et al., 1983b, 1990) is shown. For complex IV a simplified version of the model of Figure 7 is presented.

P phase of two scalar protons, the transfer of the second electron results in vectorial transport of $2H^+$ and $2q^+$ from the N to the P phase (Papa et al., 1990).

Of the $\leftarrow 2\,H^+/e^-$ pumped by the NADH-ubiquinone oxidoreductase, one is considered to be associated to redox cycling of the intrinsic quinone molecule, apparently located between two electron carrying Fe-S centers (N_3 and N_2). The second is attributed to an as yet unidentified redox process on the substrate side of N_3 (Weiss et al., 1991). Recently, Kotylar et al. (1990) have produced evidence showing the existence in the complex of a protein-stabilized, rotenone sensitive, $g = 2.00$ ubisemiquinone similar to the antimycin-sensitive ubisemiquinone of complex III (Ohnishi and Trumpover, 1980), the central element in the Q-gated pump (Papa et al., 1983b, 1990; Lorusso et al., 1989).

It is proposed that, as in complex III, also in complex I electron transfer by the protein stabilized ubisemiquinone/quinol couple from the N_3 to N_2 FeS centers results in vectorial transport across the osmotic barrier of the membrane from the N to the P phase of $2H^+$ per electron.

Thus, in each of the complexes the central element in redox-linked proton pumping is provided by protonmotive redox catalysis at the primary reaction centers: $U\dot{Q}^-/UQH_2$ in complex I and III and $\dot{O}_2^-/H_2O_2$ in complex IV. Transmembrane proton pumping in the three complexes will result in each of the complexes from combination of these protonmotive redox events at the catalytic centers and co-operative proton-input pathways in the protein from the N phase to the catalytic center and proton-output pathways from the center to the P phase, much as in the case of bacteriorhodopsin.

Finally, another important element of proton pumps seems to be represented by hydrophylic extension in the N phase of the transmembrane helices of the protein(s) which by means of polarizable hydrogen-bonded acidic and basic residues (Zundel and Brezinsky, 1992) might capture protons from the relatively alkaline N aqueous phase and mediate their selective access into the proton input channel of the pumps (mouth and filter of the channel) (Capitanio et al., 1994).

References

Artzatbanov, V.Y., Konstantinov, A.A. and Skulachev, V.P. (1978) Involvement of intra-mitochrondrial protons in redox reaction of cytochrome a. *FEBS Lett.* 87: 180–185.

Babcock, G.T. and Wikström, M.K.F. (1992) Oxygen activation and the conservation of energy in cell respiration. *Nature* 356: 301–309.

Blair, D.F., Ellis, W.R., Wang, H., Gray, H.B. and Chan, S.I. (1985) Spectroelectrochemical study of cytochrome c oxidase: pH and temperature dependences of the cytochrome potentials. *J. Biol. Chem.* 261: 11524–11537.

Blair, D.F., Gelles, J. and Chan, S.I. (1986) Redox-linked proton translocation in cytochrome oxidase: The importance of gating electron flow. *Biophys. J.* 50: 713–733.

Boyer, P.D., Chance, B., Ernster, L., Mitchell, P., Racker, E. and Slater, E.C. (1977) Oxidative phosphorylation and photophosphorylation. *Annu. Rev. Biochem.* 46: 955–1026.

Brown, G.C. and Brand, M.D. (1986) Changes in permeability to protons and other cations at high protonmotive force in rat-liver mitochondria. *Biochem. J.* 234: 75–81.

Brown, G.C. (1989) The relative proton stoichiometry of the mitochondrial proton pumps are independent of the protonmotive force. *J. Biol. Chem.* 264: 14704–14709.

Brown, S., Moody, A.J., Mitchell, R. and Rich, P.R. (1993) Binuclear centre structure of terminal protonmotive force. *FEBS Lett.* 316: 216–223.

Brown, S., Rumbley, J.N., Moody, A.J., Thomas, J.W., Gennis, R.B. and Rich, P.R. (1994) Flash photolysis of the carbon monoxide compounds of wild-type and mutant variants of cytochrome *bo* of *Escherichia Coli. Biochim. Biophys. Acta* 1183:521–532.

Capitanio, N., De Nitto, E., Villani, G., Capitanio, G. and Papa, S. (1990) Protonmotive activity of cytochrome c oxidase: Control of oxidoreduction of the heme centers by the protonmotive force in the reconstituted beef heart enzyme. *Biochemistry* 29: 2939–2945.

Capitanio, N., Capitanio, G., De Nitto, E., Villani, G. and Papa, S. (1991) H^+/e^- stoichiometry of mitochondrial cytochrome complexes reconstituted in liposomes. *FEBS Lett.* 288: 179–182.

Capitanio, N., Peccarisi, R., Capitanio, G., Villani, G., De Nitto, E., Scacco, and Papa, S. (1994) Role of nuclear-encoded subunits of mitochondrial cytochrome c oxidase revealed by limited enzymatic proteolysis. *Biochemistry* 33: 12521–12526.

Chan, S.I. and Li, P.M. (1990) Cytochrome c oxidase: Understanding nature's design of a proton pump. *Biochemistry* 29: 1–12.

Cocco, T., Lorusso, M., Di Paola, M., Minuto, M. and Papa, S. (1992) Characteristic of energy-linked proton translocation in liposome reconstituted bovine cytochrome bc_1 complex. Influence of the protonmotive force on the H^+/e^- stoichiometry. *Eur. J. Biochem.* 209: 475–481.

Dutton, P.L. and Wilson, D.F. (1974) Redox potentiometry in mitochondrial and photosynthetic bioenergetics. *Biochim. Biophys. Acta* 346: 165–212.

Ellis, W.R., Wang, H., Blair, D.F., Gray, H.B. and Chan, S.I. (1986) Spectroelectrochemical study of the cytochrome a site in carbon monoxide inhibited cytochrome c oxidase. *Biochemistry* 25: 161–167.

Erecinska, M., Chance, B. and Wilson, D.F. (1971) The oxidation-reduction potential of the copper signal in pigeon heart mitochondria. *FEBS Lett.* 16: 284–286.

Guerrieri, F., Maida, I. and Papa, S. (1981) Redox-Bohr effects in isolated cytochrome bc_1 complex and cytochrome c oxidase from beef-heart mitochondria. *FEBS Lett.* 125: 261–265.

Hallen, S. and Nilsson, T. (1992) Proton transfer during the reaction between fully reduced cytochrome c oxidase and dioxygen: pH and deuterium isotope effects. *Biochemistry* 31: 11853–11859.

Hosler, J.P., Ferguson-Miller, S., Calhoun, W.M., Thomas, J.W., Hill, J., Lemieux, L., Ma, J., Georgia, C., Fetter, J., Shapleigh, J., Tecklenburg, M.M.J., Babcock, G.T. and Gennis, R.B. (1993) Insight into the active-site structure and function of cytochrome oxidase by analysis of bacterial cytochrome aa_3 cytochrome bo. *J. Bioenerg. Biomembr.* 25: 121–136.

Konstantinov, A.A., Capitanio, N., Vygodina, T.V. and Papa, S. (1992) pH changes associated with cytochrome c oxidase reaction with H_2O_2. Protonation state of the peroxy and oxoferryl intermediates. *FEBS Lett.* 312: 71–74.

Kotylar, A.B., Sled, V.D., Moroz, I.A. and Vinogradov, A.D. (1990) Coupling site I and the rotenone sensitive ubisemiquinone in tightly coupled submitochondrial particles. *FEBS Lett.* 264: 17–20.

Lawford, H.G. and Garland, P.B. (1973) Proton translocation coupled to quinol oxidation in ox heart mitochondria. *Biochem. J.* 136: 711–720.

Lenaz, G., Barnabei, O., Rabbi, A. and Battino, M. (1990) *Highlights in Ubiquinone Research*, Taylor and Francis, London.

Lorusso, M., Capuano, F., Boffoli, D., Stefanelli, R. and Papa, S. (1979) The mechanism of transmembrane $\Delta\mu H^+$ generation in mitochondria by cytochrome c oxidase. *Biochem. J.* 182: 133–147.

Lorusso, M., Cocco, T., Boffoli, D., Gatti, D., Meinhardt, S. and Ohnishi, T. (1989) Effect of papain digestion on polypeptide subunits and electron-transfer pathways in mitochondrial bc_1 complex. *Eur. J. Biochem.* 179: 535–540.

Luvisetto, S., Conti, E., Buso, M. and Azzone, G.F. (1991) Flux ratios and pump stoichiometries at sites II and III in liver mitochondria. *J. Biol. Chem.* 266: 1034–1042.

Maison Peteri, B. and Malmström, B.G. (1989) Intrinsic uncoupling in proton-pumping cytochrome c oxidase: pH dependence of cytochrome c oxidation in coupled and uncoupled phospholipid vesicles. *Biochemistry* 28: 3156–3160.

Malmström, B.G. (1985) Cytochrome c oxidase as a proton pump. A transition state mechanism. *Biochim. Biophys. Acta* 811: 1–12.

Malström, B.G. (1989) The mechanism of proton translocation in respiration and photosynthesis. *FEBS Lett.* 250: 9–21.

Malmström, B.G. (1990) Cytochrome c oxidase: some unsolved problems and controversial issues. *Arch. Biochem. Biophys.* 280: 233–241.

Mitchell, P. (1966) Chemiosmotic Coupling in Oxidative and Photosynthetic Phosphorylation. Glynn Research Ltd., Bodmin, UK.

Mitchell, P. (1976) Possible molecular mechanism of the protonmotive function of cytochrome system. *J. Theor. Biol.* 62: 327–367.

Mitchell, P., Mitchell, R., Moody, A.J., West, I.C., Baum, H. and Wrigglesworth, J.M. (1985) Chemiosmotic coupling in cytochrome oxidase. Possible protonmotive O loop and O cycle mechanism. *FEBS Lett.* 188: 1–7.

Mitchell, P. (1987a) Respiratory chain systems in theory and practice. *In:* C.H. Kim, H. Tedeschi, J.J. Diwan and J.C. Salerno (eds): *Advances in Membrane Biochemistry and Bioenergetics.* Plenum, New York and London, pp. 25–52.

Mitchell, P. (1987b) A new redox loop formality involving metal-catalysed hydroxide-ion translocation. A hypothetical Cu loop mechanism for cytochrome oxidase. *FEBS Lett.* 222: 235–245.

Mitchell, P. (1988) Possible protonmotive osmochemistry in cytochrome oxidase. *In:* M. Brunori and B. Chance (eds): *Cytochrome oxidase. Structure Function and Physiopathology.* Annals New York Acad. Sci. 550, pp. 185–198.

Mitchell, R., Mitchell, P. and Rich, P.R. (1992) Protonation states of the reaction cycle intermediates of cytochrome c oxidase. *Biochim. Biophys. Acta* 1101: 188–191.

Mitchell, R. and Rich, P.R. (1994) Proton uptake by cytochrome c oxidase on reduction and on ligand binding. *Biochim. Biophys. Acta* 1186: 19–26.

Moody, A.J. and Rich, P.R. (1990) The effect of pH on redox titration of haem a in cyanide-liganded cytochrome-c oxidase: Experimental and modelling studies. *Biochim. Biophys. Acta* 1015: 205–215.

Murphy, M.P. and Brand, M.D. (1988) The stoichiometry of charge translocation by cytochrome oxidase and the cytochrome bc_1 complex of mitochrondria at high membrane potential. *Eur. J. Biochem.* 173: 645–651.

Murphy, M.P. (1989) Slip and leak in mitochondrial oxidative phosphorylation. *Biochim. Biophys. Acta* 977: 123–141.

Nicholls, D.G. (1974) The influence of respiration and ATP hydrolysis on the proton-electrochemical gradient across the inner membrane of rat-liver mitochondria as determined by ion distribution. *Eur. J. Biochem.* 50: 305–315.

Nicholls, P. and Petersen, L.C. (1974) Haem-haem interactions in cytochrome aa_3 during the anaerobic-aerobic transition. *Biochim. Biophys. Acta* 357: 462–467.

Oliveberg, M., Hallen, S. and Nilsson, T. (1991) Uptake and release of protons during the reaction between cytochrome c oxidase and molecular oxygen: A flow-flash investigation. *Biochemistry* 30: 436–440.

Ohnishi, T. and Trumpower, B.L. (1980) Differential effect of antimycin on ubisemiquinone bound in different environments in isolated succinate-cytochrome reductase complex. *J. Biol. Chem.* 255: 3278–3284.

Papa S., Guerrieri, F., Lorusso, M. and Simone, S. (1973) Proton translocation and energy transduction in mitochondria. *Biochimie* 55: 703–716.

Papa, S., Lorusso, M. and Guerrieri, F. (1975) Mechanism of respiration-driven proton translocation in the inner mitochondrial membrane. Analysis of proton translocation associated to oxidation of endogenous ubiquinol. *Biochim. Biophys. Acta* 387: 425–440.

Papa, S. (1976) Proton translocation reaction in the respiratory chain. *Biochim. Biophys. Acta* 456: 39–84.

Papa, S., Guerrieri, F. and Izzo, G. (1979) Redox-Bohr effects in the cytochrome system of mitochondria. *FEBS Lett.* 105: 213–216.

Papa, S., Capuano, F., Markert, M. and Altamura, N. (1980a) The H^+/O stoichiometry of mitochondrial respiration. *FEBS Lett.* 111: 243–248.

Papa, S., Guerrieri, F., Lorusso, M., Izzo, G., Boffoli, D., Capuano, F., Capitanio, N. and Altamura, N. (1980b) The H^+/e^- stoichiometry of respiration-linked proton translocation in the cytochrome system of mitochondria. *Biochem. J.* 192: 203–218.

Papa, S., Guerrieri, F., Izzo, G. and Boffoli, D. (1983a) Mechanism of proton translocation associated to oxidation of NNN′N′-tetramethyl-p-phenylendiamine in rat-liver mitochondria. *FEBS Lett.* 157: 15–20.

Papa, S., Lorusso, M., Boffoli, D. and Bellomo, E. (1983b) Redox-linked proton translocation in the bc_1 complex from beef heart mitochondria reconstituted into phospholipid vesicles. General characteristics and control of electron flow by $\Delta\mu H^+$. *Eur. J. Biochem.* 137: 405–412.

Papa, S. and Lorusso, M. (1984) The cytochrome chain of mitochondria: Electron transfer reactions and transmembrane proton translocation. *In:* R.M. Burton and F.C. Guerra (eds): *Biomembranes.* Plenum, London, pp. 257–290.

Papa, S., Guerrieri, F. and Izzo, G. (1986) Cooperative proton transfer reactions in the respiratory chain: Redox-Bohr effect. *Methods Enzymol.* 126: 331–343.

Papa, S., Capitanio, N. and De Nitto, E. (1987a) Characteristics of the redox-linked proton ejection in beef heart cytochrome oxidase reconstituted in liposomes. *Eur. J. Biochem.* 164: 507–516.

Papa, S., Capitanio, N., Izzo, G. and De Nitto, E. (1987b) Characteristics of the protonmotive activity of the cytochrome chain of mitochondria. *In:* C.H. Kim, H. Tedeschi, J.J. Diwan and J.C. Salerno (eds): *Advances in Membrane Biochemistry and Bioenergetics.* Plenum Press, New York, pp. 333–346.

Papa, S. (1988) Cytochrome c oxidase and its protonmotive activity, an overview. *In*: T.E. King, H.S. Mason and M. Morrison (eds): *Oxidase and Related Systems*. Alan R. Liss, Inc., New York, pp. 707–730.

Papa, S., Lorusso, M., Cocco, T., Boffoli, D. and Lombardo, M. (1990) Protonmotive ubiquinol-cytochrome c oxidoreductase of mitochondria. A possible example of cooperative anisotropy of protolytic redox catalysis. *In*: G. Lenaz, O. Barnabei, A. Rabbi, and M. Battino (eds): *Highlights in Ubiquinone Research*. Taylor and Francis, London, pp. 122–135.

Papa, S., Capitanio, N., Capitanio, G., De Nitto, E. and Minuto, M. (1991) The cytochrome chain of mitochondria exhibits variable H^+/e^- stoichiometry. *FEBS Lett.* 288: 183–186.

Planques, Y., Capitanio, N., Capitanio, G., De Nitto, E., Villani, G. and Papa, S. (1989) Role of supernumerary subunits in mitochondrial cytochrome c oxidase. *FEBS Lett.* 258: 285–288.

Pietrobon, D., Zoratti, M. and Azzone, G.F. (1983) Molecular slipping in redox and ATPase H^+ pumps. *Biochim. Biophys. Acta* 723: 317–321.

Rich, P.R. (1991) The osmochemistry of electron-transfer complexes. *Bioscience Reports* 11: 539–568.

Scarpa, A., Carafoli, E. and Papa, S. (1992) Ion-motive ATPases: Structure, Function and Regulation, *Ann. New York. Acad. Sci.* 671.

Thomas, J.W., Puustinen, A., Alben, J.O., Gennis, R.B. and Wikström, M. (1993) Substitution of asparagine for aspartate-135 in subunit I of the cytochrome *bo* ubiquinol oxidase of *Escherichia coli* eliminates proton pumping activity. *Biochemistry* 32: 10923–10928.

Urban, P.F. and Klingenberg, M. (1969) On the redox potentials of ubiquinone and cytochrome b in the respiratory chain. *Eur. J. Biochem.* 519–525.

Van Gelder, B.F., Van Rijin, J.L.M.L., Schilder, G.J.A. and Wilms, J. (1977) The effect of pH on the half-reduction potential of cytochrome c oxidase. *In*: K. Van Dam and B.F. Van Gelder (eds): *Structure and Function of Energy Transduction Membranes*. Elsevier North Holland Biomedical Press, Amsterdam, pp. 61–68.

Varotsis, C., Zhang, Y., Appelman, E.H. and Babcock, G.T. (1993) Resolution of the reaction sequence during the reduction of O_2 by cytochrome oxidase. *Proc. Natl. Acad. Sci. USA* 90: 237–241.

Verkhovskaya, M., Verkhovsky, M. and Wikström, M.K.F. (1992) pH dependence of proton translocation by *Escherichia Coli*. *J. Biol. Chem.* 267: 14559–14562.

Von Jagow, G. and Sebald, W. (1980) b-type cytochromes. *Annu. Rev. Biochem.* 49: 281–314.

Weiss, H., Friedrich, T., Hofhaus, G. and Preis, D. (1991) The respiratory-chain NADH dehydrogenase (complex I) of mitochondria. *Eur. J. Biochem.* 197: 563–576.

Wikström, M.K.F. (1977) Proton pump coupled to cytochrome c oxidase in mitochondria. *Nature* 266: 271–273.

Wikström, M.K.F. and Krab, K. (1979) Proton pumping cytochrome c oxidase. *Biochim. Biophys. Acta* 549: 177–222.

Wikström, M.K.F., Krab, K. and Saraste, M. (1981) Cytochrome oxidase. A synthesis. Academic Press, New York and London.

Wikström, M.K.F. (1989) Identification of the electron transfers in cytochrome oxidase that are coupled to proton-pumping. *Nature* 338: 776–778.

Wikström, M.K.F. and Saraste, M. (1984) The mitochondrial respiratory chain. *In*: L. Ernster (ed.): *Bioenergetics*. Elsevier, Amsterdam, pp. 49–94.

Wrigglesworth, J.M., Elsden, J., Chapman, A., Van der Water, N. and Grahm, M.F. (1988) Activation by reduction of the resting form of cytochrome c oxidase: Tests of different models and evidence for the involvement of Cu_B. *Biochim. Biophys. Acta* 936: 452–464.

Wyman, J. (1968) Regulation in macromolecules as illustrated by haemoglobin. *Quart. Rev. Biophys.* 1: 35–81.

Zundel, G. and Brezinsky, B. (1992) Proton polarizability of hydrogen bonded systems due to collective proton motion with a remark to the proton pathways in bacteriorhodopsin. *In*: T. Bountis (ed.): *Proton Transfer in Hydrogen-Bonded Systems*. Plenum Press, New York, pp. 154–166.

Biochemistry of Cell Membranes
ed. by S. Papa & J. M. Tager
© 1995 Birkhäuser Verlag Basel/Switzerland

Energetics and formation of ATP synthase: Allosteric, chemiosmotic and biosynthetic control of $\alpha\beta$ and $\alpha\beta\gamma$ core

S. Akiyama, C. Matsuda and Y. Kagawa

Department of Biochemistry, Jichi Medical School, Tochigi-ken 329-04, Japan

Summary. ATP synthase (F_0F_1) plays a central role in cellular energy metabolism, and is regulated to maintain ATP via an allosteric effect, membrane electrochemical potential and protein synthesis. To understand these, both the energetics and synthesis of F_0F_1 were studied.

Allosteric control. The $\alpha_3\beta_3$ oligomer and the $\alpha_1\beta_1$ protomer are the fundamental structures of F_1, which is the ATPase portion of F_0F_1. The crystallline $\alpha_3\beta_3$ oligomer is structurally symmetrical ($3 \times \alpha_1\beta_1$), but kinetically asymmetrical (one inhibitor/oligomer) and cooperative.

Chemiosmotic control. F_0F_1 incorporated into a lipid bilayer transported H^+ across the membrane on ATP hydrolysis. The V_{max}, but not the K_m, of H^+-current was controlled by the electrochemical potential of protons ($\Delta\mu H^+$) across the bilayer. Comparison of the phosphate potential ([ATP]/[ADP][Pi]) with the $\Delta\mu H^+$ revealed $3H^+/ATP$ stoichiometry.

Biosynthetic control. The gene structures of the α, β, and γ subunits of human F_0F_1 were determined, because the $\alpha_3\beta_3\gamma$ complex is the major component of F_1. Two tissue-specific isoforms of the γ subunit were produced by H^+-dependent alternative splicing. There are three common sequences in the 5′-upstream region of the genes for the α, β and γ subunits, which may regulate the formation of F_0F_1 depending on the energy demand of cells.

Introduction

ATP synthase (F_0F_1) is an ion-motive membrane ATPase that catalyzes oxidative phosphorylation (Mitchell, 1979; Racker, 1976; Kagawa, 1972). F_0F_1 is a multisubunit complex composed of two subcomplexes: an ATPase, F_1, and an ion channel, F_0 (Kagawa and Racker, 1965, 1966). F_1 is composed of five subunits with a stoichiometry of $\alpha_3\beta_3\gamma\delta\varepsilon$ (Kagawa, 1978; Futai et al., 1989), F_0 is composed of nine subunits, a, b, c, d, e, I, OSCP, F_6 and A6L (Lutter et al., 1993). F_0F_1 uses the energy of an electrochemical potential ($\Delta\mu H^+$) across the membrane generated by the redox chain to synthesize ATP (Mitchell, 1979; Skulachev, 1988). In fact, a proton current by F_0F_1 generating the $\Delta\mu H^+$ has been demonstrated in both liposomes (Kagawa, 1972) and planar lipid bilayers (Muneyuki et al., 1989).

Energy metabolism varies greatly depending on the activity of cells. To maintain the ATP level, the activity of F_0F_1 is regulated by three different mechanisms: allostery of F_1, chemiosmosis of F_0F_1 by $\Delta\mu H^+$, and biosynthetic control of F_0F_1. We analyzed the regulation of F_0F_1 at the molecular level.

Allosteric control: Structural symmetry with functional asymmetry of the $\alpha_3\beta_3$ oligomer composed of three $\alpha_1\beta_1$ protomers

Positive catalytic cooperativity and negative nucleotide-binding cooperativity are essential for both energy transduction and its regulation in F_0F_1 according to the binding change theory of Boyer (1993). Although his theory is based on the $\alpha_3\beta_3$ oligomeric structure, the $\alpha_3\beta_3$ oligomer and $\alpha_1\beta_1$ protomer of only thermophilic F_1 (TF$_1$) have been reconstituted (Kagawa et al., 1992). The isolated α and β subunits are homologous and both bind AT(D)P (Kagawa, 1978, 1984). Thus, there are six AT(D)P binding sites in F_1 and the $\alpha_3\beta_3$ oligomer. The catalytic site of ATP synthase may be located at the $\alpha\beta$ interface, because the α and β of $\alpha_1\beta_1$ protomer in F_1 (Schäfer et al., 1989) and the $\alpha_3\beta_3$ oligomer are cross-linked with diazido-ATP, and there are only three catalytic nucleotide binding sites. On electrophoresis, both the protomer and oligomer gave bands with ATPase activity, that of the oligomer being much higher owing to strong positive cooperativity (Kagawa et al., 1992). Miki and coworkers have succeeded in the crystallization of the mutated β (βY341L; orthorhombic, space group I2$_1$2$_1$2$_1$ or I222; a = 232 Å, b = 66 Å, c = 80 Å, V_M = 3.0 Å^3/Dalton, diffracted up to 2.5 Å resolution). Detailed X-ray crystallography of this mutant will help in analysis of the ATP binding sequence (-GGAGVGKT-). Both the $\alpha_3\beta_3$ oligomer and TF$_1$ were crystallized by a method similar to that reported by Shirakihara et al. (1991) (Fig. 1). X-ray crystallographic analysis of F_1 (Abrahams et al., 1993; Bianchet et al., 1991) revealed partial threefold and sixfold symmetry. There is a minor hollow cavity in the $\alpha_3\beta_3$ oligomer determined by small-angle X-ray scattering (Harada et al., 1991), and this corresponds to the pit of the F_1 molecule (Abrahams et al., 1993). On electron microscopy, F_0F_1 was seen as a sphere (F_1) connected by a stalk to a basal piece (F_0) (Kagawa and Racker, 1965, 1966; Kagawa, 1972). This structure was confirmed by Lutter et al. (1993) whose group determined that the stalk of F_1 was 40 Å by X-ray crystallography (Abrahams et al., 1993). The $\alpha_3\beta_3\gamma$ complex is connected to F_0 via the δ and ε subunits, which may form the stalk (Kagawa, 1978). The inherent asymmetry of the F_1 structure is caused by the $\gamma\delta\varepsilon$ subunits, because in the complete absence of nucleotide, X-ray crystallography of the $\alpha_3\beta_3$ oligomer (Fig. 1A) revealed a threefold symmetry (Shirakihara et al., 1993). However, the $\alpha_3\beta_3$ oligomer was functionally asymmetrical, because in the presence of AT(D)P, the three catalytic sites are tight, loose, and open owing to negative cooperativity as predicted by Boyer (1993). Moreover, the

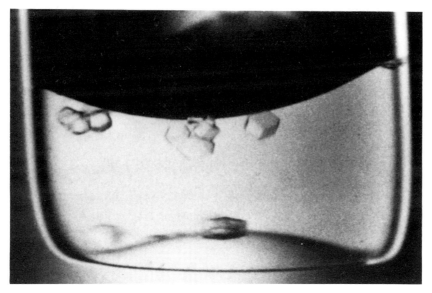

A   — 0.1 mm

B   — 0.1 mm

Figure 1. Crystals of the $\alpha_3\beta_3$ oligomer (A) and TF_1 (B) without nucleotide. (A) The $\alpha_3\beta_3$ oligomer was crystallized in a solution of 15% polyethylene glycol 6000, 0.2M Na_2SO_4, pH 8.0, at 15°C. Crystallography (cubic, a = b = c = 160 Å, 3.7 Å resolution, V_M = 3.2 Å^3/Dalton, 4 × $\alpha_3\beta_3$ oligomer per unit cell) showed threefold symmetry (Shirakihara et al., 1993). (B) The crystals of TF_1 were grown in 5% polyethylene glycol 20 000, pH 5.3, at 25°C as described previously (Shirakihara et al., 1991).

$\alpha_3\beta_3$ oligomer was inhibited by only one mole of [^{3}H] benzoyl-benzoyl-ADP per oligomer just like mitochondrial F_1 and TF_1 (Aloise et al., 1991; Kagawa et al., 1992). To inhibit the protomer activity completely, one mole of inhibitor was needed per mole of $\alpha_1\beta_1$ protomer. When structural asymmetry was introduced by adding the three minor sub-units, the resulting $\alpha_3\beta_3\gamma$, $\alpha_3\beta_3\gamma\delta$, $\alpha_3\beta_3\gamma\varepsilon$, $\alpha_3\beta_3\delta$, and $\alpha_3\beta_3\gamma\delta\varepsilon$ complexes showed characteristic inhibitor sensitivity (Paik et al., 1993). Similar modifications of the catalytic activity of a complex of major subunits (I, II and III) by minor subunits have been reported in cytochrome oxidase (Anthony et al., 1993).

The kinetic cooperativity of the $\alpha_3\beta_3$ oligomer (Kagawa et al., 1992) is essential in regulation of F_1 (Boyer, 1993). The rapid nucleotide-dependent conformational changes of the $\alpha_3\beta_3$ oligomer which is stable without nucleotide, have been analyzed by synchrotron radiation (Kagawa et al., 1992).

Chemiosmotic control: ATP-driven H^+-current through an F_0F_1-lipid bilayer

So-called respiratory control of oxidative phosphorylation is a unique regulatory mechanism via chemiosmotic equilibration between the membrane potential and phosphate potential (Mitchell, 1979). To date, the movements of ions through ATPases have been analyzed mainly using liposomes (Kagawa, 1972) or crude membrane vesicles. However, study of H^+-translocation requires strict control of the components of $\Delta\mu H^+$, which are the pH gradient and the membrane potential across the membrane (Mitchell, 1979; Skulachev, 1988). In this respect, a planar phospholipid bilayer plugged through by F_0F_1 is the most suitable model, since both sides of the membrane are accessible and controllable (Fig. 2) (Muneyuki et al., 1989). Passive transport of H^+ was demonstrated, when the F_0 portion alone was incorporated into lipid bilayers. Active transport of H^+ by F_0F_1 upon addition of MgATP resulted in a steady-state H^+-current that showed simple Michaelis-Menten-type kinetics, and a $K_m(ATP)$ of 140 μM under our conditions. This value was close to the $K_m(ATP)$'s for the ATPases of TF_1 and TF_0F_1 (thermophilic F_0F_1) in the steady-state catalytic cycle, indicating that proton translocation is coupled to the steady-state ATPase reaction. In contrast to the voltage-dependent V_{max}, the K_m showed no apparent dependence on the membrane voltage. An essentially similar ion translocating mitochondrial F_0F_1 bilayer has been obtained (Muneyuki et al., 1987). At -180 mV (negative on the F_1 side) the current was entirely suppressed. Assuming that the standard free energy change of ATP hydrolysis is -30.5 kJ and $[(ADP)(Pi)/(ATP)]$ is 10^{-4}, the free energy change of ATPase was calculated to be -552 mV.

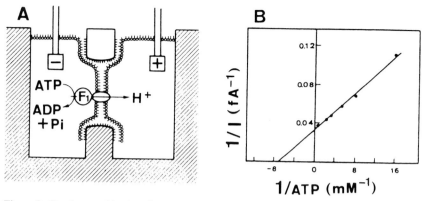

Figure 2. Steady-state kinetics of proton current through F_0F_1 reconstituted in planar bilayer membrane. (A) Teflon chambers (1.5 ml) separated with a septum. The hole (0.2 mm diameter) in the lipid bilayer membrane contains F_0F_1, which was incorporated from the left side chamber. (B) Lineweaver-Burk plot of ATP-driven proton current (Muneyuki et al., 1989).

Therefore, the H^+/ATP stoichiometry is 3. These results suggest that proton translocation is not directly coupled to the ATP binding step, but is coupled to some nucleotide dependent conformational change.

Biosynthetic control: Genes for the major subunits of human F_1

Because the V_{max} for ATP synthesis is limited by the amount of complexes of oxidative phosphorylation, their synthesis is increased in response to the continuous energy demand of muscle contraction (Williams et al., 1987), and neuronal activity (Hevner and Wong-Riley, 1993), etc. Of all the enzymes in mammals, only four enzyme complexes (F_0F_1, complexes I, III, and IV) are encoded by both mitochondrial and nuclear DNA. But this response must be caused by the nuclear genes encoding subunits of these complexes, because the expression of the mitochondrial DNA is dependent on the nuclear DNA, and mitochondrial DNA is identical in various tissues. In fact, age-dependent decrease of oxidative phosphorylation is caused by nuclear DNA, not by mitochondorial DNA, as shown by the cell fusion method (Hayashi et al., 1994).

For elucidation of this phenomenon, the human genes encoding the three major subunits of F_1, α, β, and γ, were sequenced (α, Akiyama et al., 1994; β, Tomura et al., 1990; γ, Matsuda et al., 1993a). These genes were shown to have the common features of housekeeping genes which have no TATA box. There are only single *bona fide* genes encoding the α, β, and γ subunits in the human genome, as determined by Southern

A

```
CAGGTTTGTTACACAGGTAAATGTGTGCCATGTGGTTTGCTGTAACTATCATCCCATCA  -721
                              SV40 enhancer core
CCTAGGTATTAAGCTCCGCATGTATTAGCTATTTATCTTGATGCTCCCCGGTTA        -661
TCAGGATTTTTAAACAGTTTGCTCTAATTCACTGACAGGATTCTTTTAGAATGAATCTAA  -601
AAAAGCTGGGAATCTTTAACCTGAGCGCAGATCGGATCACGTGGAACGGTGCTGGTGTTT  -541
                           GFII/MLTF          Mt4
GTTAAAATTCAGATTGCCCCGAATTTTCGACTCAGAGGGTCTGAAGTGGGGCTAAGATC   -481
     CS2                          OXBOX   Sp1
TGCAGATTAACGGGCTCCGTGGGAGGCTCCCTGCTGGTGGCCAGTCCCGGGGTTTAGCTG  -421
TTCACTGTTGCTACTCGTGCTAAACCACGTCTAACACTACCGACACTATTAATCGCCCT   -361
GGTCCTCAAGACAGCCGCGACCCCAGACGCGGTGAGGCCCTCATCCTCTAGCACCAGATGG -301
GGCCTGGAACTGCTGCTTAGCAGATTCAGCTCACCCGGGGCAACCACA              -241
AP-2          GFII    GCF    NRF-2
AGGTCGACTCAAAAGACCAAAATAAATGTATAACGAATGCCAGGACCAAAGCCGGAATAT  -181
                               Mt3          Mt1
ATCTAATCATTCATTTTGAGGTTTAAACTTAAAAATGTTACATAAAAAAATAAGTAACCT  -121
GCCAAAAAATGCGAGTGGGCGGCGGGGGGGCGGGGGAGAGGTGGTGAACGCGAGGGCAGT  -61
                                                  AP-2
ACTTCCGGGTCAGGTGGGCCGGCTGTCTTGACCTTCTTTGCGGCTCGGCCATTTGTCCC   -1
                                                Initiator
ETS-1  GFII    GCF
AGTCAGTCCGGAGGCTGCGGCTGCAGAAGTACCGGCCTGCGAGTAACTGCAAACATGCTG   60
       CS3       AP-2                                     MetLeu
```

B

Common Structure 1 (CS1)

Human	α	-936	ttggCAGAGGAAaaag	-921
	β	-261	taatCAGAGGAActac	-276
	γ	-2753	caacCAGAGGAActac	-2738
Bovine	α	102	gctgCgGAGGAAgcac	117

CS2

Human	α	-536	aaaaTTCAGATTgccc	-521
	β	-173	gagtTTCAGATTagca	-158
	γ	-2314	taaaTTCAGAaTatag	-2299
Bovine	α	-472	aacgTTCAGcTTgccc	-457

CS3

Human	α	9	cggaGGCTGCGGctgc	24
	β	-26	aggcGGCTGCGGttgc	-11
	γ	-541	cgggGGCTGCGGGggag	-526
Bovine	α	91	ctcgaGCTGCGGctgc	106

Figure 3. Putative regulatory cis-elements of the genes for the major subunits of F_1. These motifs were searched using database TFD release 6.0 (NLM, NIH, USA). (A) The 5′-upstream region of the human F_1 α subunit gene. (B) Common structures (CS1 to CS3) found in the 5′-flanking regions of the F_1 α, β and γ genes. The nucleotide numbers of the γ subunit gene are based on the nucleotide sequence of the cDNA encoding the human γ subunit (Akiyama et al., 1994).

blot hybridization, though Walker et al. (1989) reported a heart-type isoform of the bovine α subunit. Comparison of the 5'-upstream region of the human α gene with those of the genes for the bovine α, human β, and human γ subunits indicated three common sequences (CS1, CS2, and CS3; Fig. 3), suggesting that putative *cis*-elements coordinate the expressions of the genes for these three subunits of the ATP synthase. As shown in Figure 3B, CS1 contained the GGAA sequence recognized by the ETS domain protein. CS2 and CS3 were identified as nuclear factor binding sites by DNase I footprinting of the human β gene (Inohara et al., unpublished results). Multiple transcription initiation sites were found upstream of the initiation codon of the human α and β genes. In contrast to the α and γ genes, the β gene had no G-rich sequence. On the other hand, CAAT boxes, which are abundant in the β and γ genes, were not present in the α gene.

The gene for the α subunit was 14 kbp long and contained 12 exons interrupted by 11 introns. All the introns had GT and AG consensus dinucleotides at their 5'- and 3'-boundaries, respectively. The intron/exon structure of the α gene is similar to that of the bovine gene reported by Pierce et al. (1992). The intron/exon junctions of the α gene were homologous to those of the β gene. At least 13 Alu repeating sequences were found in the α gene. There were two pseudogenes for the human α subunit. The 5'-flanking region of the α gene contains several binding motifs for *cis*-elements that are considered to be important for the coordinated regulation of some nuclear genes encoding mitochondrial proteins (Fig. 3). For example, sequences similar to Mt1, Mt3 and Mt4 (Suzuki et al., 1991) are present in this region, although these sequences have not been found in the 5' upstream region of the mitochondrial transcription factor 1 (Tominaga et al., 1992).

Cross-talk of the controls: Proton-dependent expression of the genes for ATP synthase

The expressions of the genes for the proteins for ATP synthesis were shown to be coordinated under several conditions (Kagawa and Ohta, 1990). The expression of the bovine α gene is higher in heart and skeletal muscle than that in liver and kidney (Pierce et al., 1992). This is also true in the human β and γ genes (Matsuda et al., 1993a). The transcriptional activities derived from the 5'-deletion mutants of the human α promoter were different in cell lines from three different human tissues (Fig. 4). In particular, the promoter of the Alexander cell line showed marked activity even without the CS1 and CS2 shown in Figure 3. The results of gel retardation assays using a nucleotide probe corresponding to that region and nuclear extracts of Alexander and HeLa cells were

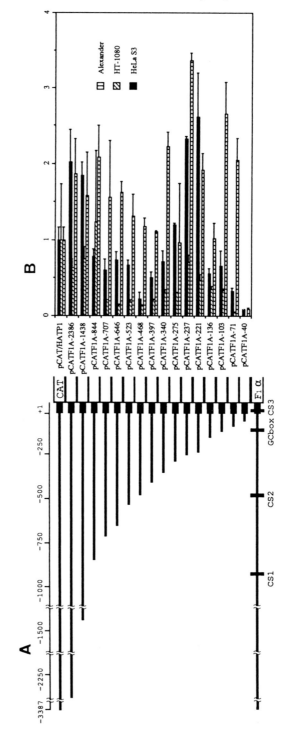

Figure 4. Chloramphenicol acetyl transferase (CAT) assays of sequential 5'-deletion mutation of the human F$_1$ α subunit gene. (A) Deletions of the α gene in the plasmids. The locations of CS1 to CS3, and GC-rich regions (see Fig. 3) are shown below. (B) Relative activity of expressed CAT. Each of the plasmids (5 μg) was co-transfected with 2 μg of pEF-BOS/βGAL into Alexander (hepatocarcinoma), HT-1080 (fibrosarcoma) and HeLa S3 (epithelioid carcinoma) cells. CAT activity was normalized with respect to the β-galactosidase activity in the same cell lysate. The CAT activity of pCAT/HATP1 was taken as 1 (Akiyama et al., 1994).

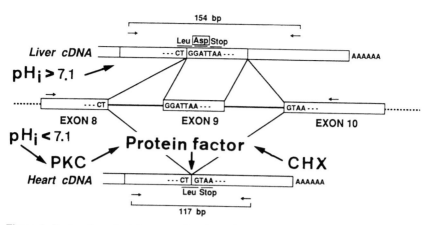

Figure 5. Proton-dependent tissue-specific alternative splicing of exon 9 of the human F_1 γ subunit. Exon 9 is spliced out in heart and muscle (Matsuda et al., 1993a). In cultured HT1080 cells, intracellular acidosis induced exon 9 exclusion. The cytosolic pH (pHi) was measured with 2′,7′-biscarboxyl-ethyl-5(6)-carboxyfluorescein-tetraacetoxymethyl ester. *De novo* protein synthesis of *trans*-acting protein factor was blocked by both cyclohexamide (CHX) and Calophostin, a protein kinase C (PKC) inhibitor (Endo et al., 1994).

significantly different, suggesting the presence of a cell-type specific mechanism in the region between −40 and −237 in expression of the α gene.

The catalytic major subunits I, II, and III of cytochrome oxidase in various tissues are identical, but the oxidase activity is controlled by the tissue-specific minor subunits (Anthony et al., 1993). Similarly, the α and β subunits of F_1 are common to various tissues, but the γ subunit is tissue-specific. However, the muscle-specific expressions of genes, including the minor subunits of cytochrome oxidase, are controlled by muscle-specific factors such as MyoD. In contrast to these, isoforms of the γ subunit are generated by alternate splicing: the heart/skeletal muscle-specific transcript lacks exon 9 in a cassette fashion (Matsuda et al., 1993a, b), while the liver/brain-specific transcript contains exon 9 (Fig. 5). Rapid muscle activity lowers the cytoplasmic pH. A low cytoplasmic pH (<7.1) induced exclusion of the exon 9, and this induction was inhibited by cycloheximide treatment, suggesting the presence of a *trans*-acting factor for the exclusion (Endo et al., 1994). In contrast, a high cytoplasmic pH resulted in mRNA splicing of the liver type and this reaction to the high pH was not inhibited by cycloheximide. The signal of low pH was blocked by the protein kinase C inhibitor Calphostin C (Endo et al., 1994). The expressions of the β gene (Kagawa et al., 1990), and cytochrome c1 gene (Suzuki et al., 1989) were sensitive to phorbol ester, because the genes contain a responsive *cis*-element.

Conclusion

The specific activity of isolated F_0F_1 is determined by the co-operativity of the $\alpha_3\beta_3$ oligomer. Detailed analyses of the crystals of the $\alpha_1\beta_1$ protomer, $\alpha_3\beta_3$ oligomer, and F_1 with and without nucleotide are needed to understand the allostery of F_0F_1.

F_0F_1 in the membrane is chemiosmotically regulated by $\Delta\mu H^+$. Owing to the fewer ions transported through F_0F_1, than those through a channel, detailed electrophysiological studies are needed to elucidate the interaction of the amino acid residues with protons.

The putative *cis*-elements of the human α, β, and γ genes were determined. There are some tissue-specific signal transduction pathways from the phosphate potential and pHi to the genes encoding F_0F_1 subunits, and to their splicing machinery. The *trans*-elements and molecular mechanisms of the pathway are yet to be elucidated.

Acknowledgments
We thank Dr. Y. Shirakihara of Hyogo University of Education for crystallographic data of TF_1 and the $\alpha_3\beta_3$ complex, Dr. E. Muneyuki of Tokyo Institute of Technology for the bilayer data, and Drs. S. Ohta, H. Endo, and N. Inohara for the data on DNA and its expression. This work was supported by a Grant-in-Aid for Scientific Research from the Ministry of Education, Science and Culture of Japan.

References

Abrahams, J.P., Lutter, R., Todd, R.J., van Raaij, M.J., Leslie, G.G.W. and Walker, J.E. (1993) Inherent asymmetry of the structure of F_1-ATPase from bovine heart mitochondria at 6.5 Å resolution. *EMBO J.* 12: 1775–1780.

Akiyama, S., Endo, H., Inohara, N., Ohta, S. and Kagawa, Y. (1994) Gene structure and cell type-specific expression of the human ATP synthase α subunit. *Biochim. Biophys. Acta* 1219: 129–140.

Aloise, P., Kagawa, Y. and Coleman, P.S. (1991) Comparative Mg^{2+}-dependent sequential covalent binding stoichiometries of 3'-O-(4-benzoyl)benzoyl adenosine 5'-diphosphate of MF_1, TF_1 and the $\alpha_3\beta_3$ core complex of TF_1. The binding change motif is independent of the $\gamma\delta\varepsilon$ subunits. *J. Biol. Chem.* 266: 10368–10376.

Anthony, G., Reimann, A. and Kadenbach, B. (1993) Tissue-specific regulation of bovine heart cytochrome-c oxidase activity by ADP via interaction with subunit VIa. *Proc. Natl. Acad. Sci. USA* 90: 1652–1656.

Bianchet, M., Ysern, X., Hullihen, J., Pedersen, P.L. and Amzel, L.M. (1991) Mitochondrial ATP synthase. Quaternary structure of the F_1 moiety at 3.6 Å determined by X-ray diffraction analysis. *J. Biol. Chem.* 266: 21197–21201.

Boyer, P.D. (1993) The binding change mechanism for ATP synthase – Some probabilities and possibilities. *Biochim. Biophys. Acta* 1140: 215–250.

Endo, H., Matsuda, C. and Kagawa, Y. (1994) Exclusion of an alternatively spliced exon in human ATP synthase γ-subunit pre-mRNA requires *de novo* protein synthesis. *J. Biol. Chem.* 269: 12488–12493.

Futai, M., Noumi, T. and Maeda, M. (1989) ATP synthase (H^+-ATPase): results by combined biochemical and molecular biological approaches. *Annu. Rev. Biochem.* 58: 111–136.

Harada, M., Ito, Y., Sato, M., Aono, O., Ohta, S. and Kagawa, Y. (1991) Small-angle X-ray scattering studies of Mg-AT(D)P-induced hexamer to dimer dissociation in the reconstituted $\alpha_3\beta_3$ complex of ATP synthase from thermophilic bacterium PS3. *J. Biol. Chem.* 266: 11455–11460.

Hayashi, J-I., Ohta, S., Kagawa, Y., Kondo, H., Kaneda, H., Yonekawa, H., Takai, D. and Miyabayashi, S. (1994) Nuclear but not mitochondrial genome involvement in human age-related mitochondrial dysfunction. Functional integrity of mitochondrial DNA from aged subjects. *J. Biol. Chem.* 269: 6878–6883.

Hevner, R.F. and Wong-Riley, M.T.T. (1993) Mitochondrial and nuclear gene expression for cytochrome oxidase subunits are diproportionately regulated by functional activity in neurons. *J. Neurosci.* 13: 1805–1819.

Kagawa, Y. (1972) Reconstitution of oxidative phosphorylation. *Biochim. Biophys. Acta.* 265: 297–338.

Kagawa, Y. (1978) Reconstitution of the energy tranformer, gate and channel. *Biochim. Biophys. Acta* 505: 45–93.

Kagawa, Y. (1984) Proton motive ATP synthesis. *In*: L. Ernster (ed.): *Bioenergetics, New Comprehensive Biochemistry*. Elsevier, Amsterdam, pp. 149–186.

Kagawa, Y. and Ohta, S. (1990) Regulation of mitochondrial ATP synthesis in mammalian cells by transcriptional control. *Internatl. J. Biochem.* 22: 219–229.

Kagawa, Y., Ohta, S., Harada, M., Sato, M. and Ito, Y. (1992) The $\alpha_3\beta_3$ and $\alpha_1\beta_1$ complexes of ATP synthase. *Annals N.Y. Acad. Sci.* 671: 366–376.

Kagawa, Y. and Racker, E. (1965) A factor conferring oligomycin-sensitivity to mitochondrial ATPase. *Fed. Proc.* 24: 363.

Kagawa, Y. and Racker, E. (1966) Partial resolution of the enzymes catalyzing oxidative phosphorylation X. Correlation of morphology and function in submitochondrial particles. *J. Biol. Chem.* 241: 2475–2482.

Lutter, R., Saraste, M., van Walraven, H.S., Runswick, M.J., Finel, M., Deatherage, J.F. and Walker, J.E. (1993) F_0F_1-ATP synthase from bovine heart mitochondria: development of the purification of a monodisperse oligomycin-sensitive ATPase. *Biochem. J.* 295: 799–806.

Matsuda, C., Endo, H., Ohta, S. and Kagawa, Y. (1993a) Gene structure of human mitochondrial ATP synthase γ-subunit: Tissue specificity produced by alternative RNA splicing. *J. Biol. Chem.* 268: 24950–24958.

Matsuda, C., Endo, H., Hirata, H., Morosawa, H., Nakanishi, M. and Kagawa, Y. (1993b) Tissue specific isoforms of the bovine mitochondrial ATP synthase γ-subunit. *FEBS Lett.* 325: 281–284.

Mitchell, P. (1979) Keilin's respiratory chain concept and its chemiosmotic consequences. *Science* 206: 1148–1159.

Muneyuki, E., Kagawa, Y. and Hirata, H. (1989) Steady state kinetics of proton translocation catalyzed by thermophilic F_0F_1 ATPase reconstituted in planar bilayer membranes. *J. Biol. Chem.* 264: 6092–6096.

Muneyuki, E., Ohno, K., Kagawa, Y. and Hirata, H. (1987) Reconstitution of the proton translocating ATPase from bovine heart mitochondria into planar phospholipid bilayer membrane. *J. Biochem.* 102: 1433–1440.

Paik, S.R., Yokoyama, K., Yoshida, M., Ohta, T., Kagawa, Y. and Allison, W.S. (1993) The ATPase activities of assembled $\alpha_3\beta_3\gamma$, $\alpha_3\beta_3\gamma\delta$ and $\alpha_3\beta_3\gamma\varepsilon$ complexes are stimulated by low and inhibited by high concentrations of rhodamine 6G whereas the dye only inhibits the $\alpha_3\beta_3$ and $\alpha_3\beta_3\delta$ complexes. *J. Bioenerg. Biomembrane* 25: 679–684.

Pierce, D.J., Jordan, E.M. and Breen, G.A.M. (1992) Structural organization of a nuclear gene for the α-subunit of the bovine mitochondrial ATP synthase complex. *Biochim. Biophys. Acta* 1132: 265–275.

Racker, E. (1976) *A New Look at Mechanisms in Bioenergetics*, Academic Press, New York.

Schäfer, H-J., Rathgeber, G., Dose, K. and Kagawa, Y. (1989) Photoaffinity cross-linking of F_1 ATPase from the thermophilic bacterium PS3 by 3′-arylazideo-β-alanyl-2 azido ATP. *FEBS Lett.* 253: 264–268.

Shirakihara, Y., Sekimoto, Y., Ueda, T., Yoshida, M., Saiga, K., Okada, M. and Kagawa, Y. (1993) Crystallographic study of TF1 and its $\alpha_3\beta_3$ complex. *Proc. Japan. Bioenerg.* 19: 22–23.

Shirakihara, Y., Yohda, M., Kagawa, Y., Yokoyama, K. and Yoshida, M. (1991) Purification by dye-ligand chromatography and a crystallization study of the F_1-ATPase and its major subunits, β and α, from a thermophilic bacterium, PS3. *J. Biochem.* 109: 466–471.

Skulachev, V.P. (1988) *Membrane Bioenergetics*. Springer-Verlag, Berlin, Heidelberg, pp. 19–25.

Suzuki, H., Hosokawa, Y., Nishikimi, N. and Ozawa, T. (1989) Structural organization of the human mitochondrial cytochrome c1 gene. *J. Biol. Chem.* 264: 1368–1374.

Suzuki, H., Hosokawa, Y., Nishikimi, N. and Ozawa, T. (1991) Structural organization of the human mitochondrial cytochrome c1 gene. *J. Biol. Chem.* 266: 2333–2338.

Tomura, H., Endo, H., Kagawa, Y. and Ohta, S. (1990) Novel regulatory enhancer in the nuclear gene of the human mitochondrial ATP synthase β subunit. *J. Biol. Chem.* 265: 6525–6527.

Tominaga, K., Akiyama, S., Kagawa, Y. and Ohta, S. (19920 Upstream region of a genomic gene for human mitochondrial transcription factor 1. *Biochim. Biophys. Acta* 1131: 217–219.

Walker, J.E., Powell, S.J., Vinas. O and Runswick, M.J. (1989). ATP synthase from bovine mitochondria: Complementary DNA sequence of the import precursor of a heart isoform of the α subunit. *Biochemistry* 28: 4702–4708.

Williams, R.S., Garcia-Moll, M., Mellor, J., Salmons, S. and Harlan, W. (1987) Adaptation of skeletal muscle to increased contractile activity. Expression of nuclear genes encoding mitochondrial proteins. *J. Biol. Chem.* 262: 2764–2767.

Biochemistry of Cell Membranes
ed. by S. Papa & J. M. Tager
© 1995 Birkhäuser Verlag Basel/Switzerland

Mitochondrial ATP synthase: Progress toward understanding the relationship between its unique structure and its biological function

P.L. Pedersen, M. Bianchet, L.M. Amzel, J. Hullihen, D.N. Garboczi and P.J. Thomas

Department of Biological Chemistry and Department of Biophysics and Physical Chemistry, Johns Hopkins University School of Medicine, 725 North Wolfe Street, Baltimore, Maryland 21205-2185, USA

Summary. One of the most challenging problems in biology today is to understand the molecular mechanism by which mitochondria synthesize ATP. The process is catalyzed by an unusually complex molecule called the ATP synthase, the substructure of which is comprised of a number of different polypeptide chains. Here, we briefly review progress made to date in obtaining nucleotide binding and structural information about the F_1 moiety of the rat liver ATP synthase complex. Also, implications of this information for the function of the enzyme in synthesizing ATP is discussed.

Introduction

Mitochondrial ATP synthases are comprised of two major units, one called F_0 and the other F_1 (reviewed in 1–6). The F_0 moiety spans the mitochondrial inner membrane and directs protons to the F_1 moiety which binds ADP and P_i and synthesizes ATP. The view as to how the process of ATP synthesis and release from F_1 takes place is depicted in its simplest form in Figure 1. It is known that ADP and P_i bind to the surface of F_1 and carry out the synthesis of bound ATP with an equilibrium constant near 1 (7). Thus, little or no energy input is believed to be necessary to make bound ATP. Rather, the coupling event (energy input step) is believed to involve the release ("de-binding") of bound ATP at the expense of the electrochemical proton gradient (8). Whether this release is direct or effected via a conformational change in the F_1 molecule is unclear and remains the subject of debate.

In order to understand how F_1 works at a molecular/chemical level, three fundamental questions must be answered: (1) What constitutes an ATP domain?; (2) Where is it located?; (3) How does an electrochemical proton gradient interact with this domain to effect the release of bound ATP? The problem is complicated because F_1 is not a simple molecule. In fact, it is quite large (~ 370 kDa) and consists of five different

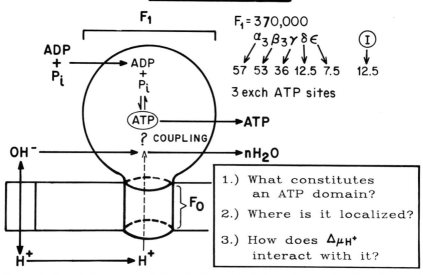

Figure 1. Simplified diagram of the mitochondrial ATP synthase and a list of several questions that must be answered to understand how the enzyme complex functions. "I" refers to an inhibitor peptide which is not present in the F_1 preparations used in studies reported here.

subunit types in the stoichiometric ratio $\alpha_3\beta_3\delta\gamma\varepsilon$ (9, 10). The asymmetric nature of its substructure is believed to have functional consequences as depicted in most models depicting how the enzyme functions (1–6). Thus, in one such model one or more of the small subunits are believed to undergo some movement or repositioning so that each of the $3\alpha\beta$ pairs are specified "in turn" for ATP synthesis (1, 11).

In this article we will review briefly what we have learned to date about the *rat liver* enzyme as it concerns questions 1 and 2 in Figure 1, i.e., "What constitutes an ATP domain?" and "Where is it located?" We will indicate also how we believe the subunit asymmetry within F_1 affects nucleotide (ATP and ADP) binding. Finally, the possible relationship of this information to the function of F_1 during ATP synthesis will be discussed.

Results and discussion

To address the above two questions (1) "What constitutes an ATP domain?" and (2) "Where is it located?", we have subjected purified rat liver F_1 to four different types of studies as indicated below.

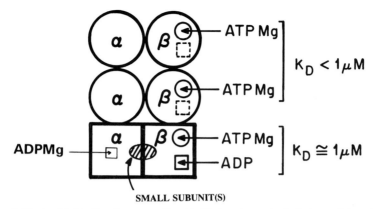

SMALL SUBUNIT(S)

Figure 2. Nucleotide binding sites on rat liver F_1 filled under nonhydrolytic conditions. Note, there are five sites that can be readily loaded, four of which are exchangeable and all located preferentially on β subunits. Only the "asymmetric center" has a readily detectable, exchangeable ADP site under these conditions, which makes it the obvious candidate for the substrate site during ATP synthesis. As all three $\alpha\beta$ pairs are believed to participate in ATP synthesis, the small subunits would be predicted to reposition from the $\alpha\beta$ pair shown in the figure and specify "in turn" the other two $\alpha\beta$ pairs for binding ADP and therefore for ATP synthesis. (*It should be noted that the three small subunits in intact F_1 may be "touching" the $3\alpha\beta$ pairs simultaneously, but that only specified amino acid sequences can induce asymmetry. Hence, only one $\alpha\beta$ pair at a time may experience the structural change that induces it to bind the substrate for ATP synthesis, i.e., ADP.*) Note also that some sites may be at subunit interfaces.

Nucleotide binding studies on intact F_1 and subfractions thereof

Prior to discussing these results, four points should be made. First, rat liver F_1 has been shown previously to restore ATP synthesis to inner membrane vesicles depleted of this enzyme (12). Secondly, rat liver F_1 can be readily separated into an $\alpha_3\gamma$ fraction and a $\beta_3\delta\varepsilon$ fraction by incubating overnight in $MgCl_2$. Third in the studies described below we have avoided the use of covalent nucleotide binding analogs which may preferentially bind to reactive groups not at nucleotide binding sites. Finally, AMP-PNP, a nonhydrolyzable ATP analog which competitively inhibits ATP hydrolysis of F_1, has been used to detect ATP sites.

Results depicted in Figure 2 show that rat liver F_1 can load maximally 5 moles nucleotide/mole enzyme. Significantly, when F_1 is first analyzed prior to the addition of nucleotide (following chromatography in Sephadex G-25 in KP_i and two ammonium sulfate precipitation steps), it is shown to contain about two moles of bound nucleotide and 1 mole Mg^{++} per mole. One mole of nucleotide is recovered as ADP in the $\alpha_3\gamma$ fraction where the bound Mg^{++} site is located (11), while the other mole of nucleotide, is recovered as ATP in the $\beta_3\delta\varepsilon$ fraction. The ADP site is nonexchangeable whereas the ATP site is exchangeable. These results alone emphasize the inherent asymmetry within F_1 and indicate that both the nonexchangeable ADP site and the ATP site are located on an $\alpha\beta$ pair in which one or more of the small subunits, γ, δ, and ε, have induced a structural change which favors tight nucleotide binding.

In addition to the two sites noted above three additional sites are filled on rat liver F_1 when nucleotide is added. These three sites all readily exchange bound nucleotide. One is an ADP site also located at the asymmetric center and two others are ATP sites located elsewhere. It is important to note that all exchangeable nucleotide binding sites on F_1, of which there are four, are located within the $\beta_3\gamma\delta\varepsilon$ fraction. Only the single nonexchangeable ADP site is located within the $\alpha_3\gamma$ fraction. It is important to note also that the single exchangeable ADP site located in the $\beta_3\delta\varepsilon$ fraction is separate and distinct from the ATP sites located in the same fraction as shown by double labeling experiments.

The above results, which will be reported in detail elsewhere, have been collected over a 5-year period on over 50 different F_1 preparations from rat liver. They verify our previous work on the ligand binding properties of the rat liver F_1 (11) and extend these studies to the subunit location of the nucleotide binding sites. These studies are in agreement with reports that both the bovine heart and E. coli enzymes have three exchangeable ATP sites located preferentially on β subunits (reviewed in 2, 4) and in agreement with reports that β subunits have a second distinct nucleotide binding site (2, 13). (The number of nonexchangeable sites reported for F_1 preparations from some species is three (also reviewed in 2, 4), whereas for the rat liver F_1 it is only one. This may reflect species differences or the presence of the inhibitor peptide contaminating other F_1 preparations. The inhibitor peptide is known to induce enhanced nucleotide binding (14)).

In other studies still under investigation we find that following ATP hydrolysis F_1 retains 1 or less mole ATP/mole enzyme while increasing its ADP load from 2 to near 4. This finding suggests that upon ATP hydrolysis F_1 switches to a different form in that ADP sites normally silent on two of the three β subunits are now able to load ADP, whereas previously only the asymmetric β was able to do so. Thus, during ATP hydrolysis it might be argued that F_1 switches to a conformation distinctly different from that operative during ATP synthesis.

In summary, and as it relates to the original question addressed, these nucleotide binding studies tell us that rat liver F_1 contains ATP domains preferentially located on β subunits, and that β subunits, when complexed with one or more of the small subunits contain also a distinct domain for binding ADP, the substrate for ATP synthesis.

Nucleotide binding to a synthetic peptide fragment of the F_1-β subunit

The above studies indicate that β subunits, or more specifically the $\beta_3\delta\varepsilon$ complex, contains distinct sites for ATP and ADP, but they do not tell

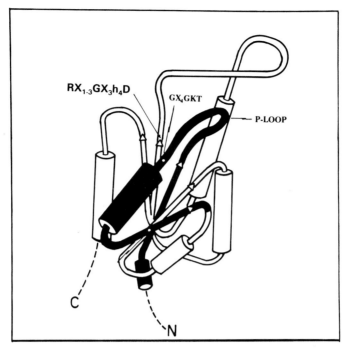

Figure 3. Region of the F_1-β subunit predicted to exhibit a three-dimensional fold similar to that characteristic of adenylate kinase. The region depicted in black consisting of amino acid residues form aspartic acid 141 through threonine 190 has been chemically synthesized and shown to bind ATP. The NMR structure of this 50 amino acid peptide in the presence of ATP is now being completed in collaborative studies with Mildvan and colleagues (see text).

us where within this complex these sites are. One approach to this problem is to model the F_1-β subunit after a homologous enzyme also known to bind ATP and ADP. One such enzyme is adenylate kinase which exhibits significant sequence homology to the F_1-β subunit (15) and, in addition, exhibits a known high-resolution x-ray structure (16). Like F_1, adenylate kinase makes ATP, in this case via the reaction 2 ADP $\rightleftharpoons$ ATP + AMP. Moreover, both F_1-β and adenylate kinase contain the Walker A (GX_4GKT) and B (RX_7h_4D) consensus motifs which in adenylate kinase are known to reside within that region of the enzyme responsible for nucleotide binding (16, 17).

Figure 3 depicts that region of the F_1-β subunit that is predicted to fold in a manner similar to adenylate kinase. Specifically, the amino acid sequence depicted in "black" which consists of amino acid residues from aspartic acid 141 through threonine 190 was chosen as a likely candidate for the major ATP binding region. This decision was based both on the fact that this sequence contains the Walker A consensus and on previous studies demonstrating that a similar peptide fragment from adenylate kinase binds ATP (17).

For the above reasons, we chemically synthesized the F_1-β fragment called "PP-50" consisting of aspartic acid 141 through threonine 190. After purification, PP-50 was subjected to direct nucleotide binding analysis using the fluorescent nucleotide analog TNP-ATP. These studies which have been reported in detail elsewhere (18) showed clearly that this region of the β subunit binds TNP-ATP and that TNP-ATP can be displaced with authentic ATP. More recently, NMR studies carried out in collaboration with Mildvan and his colleagues, which are currently being processed for publication, have confirmed this work and identified a number of residues within PP-50 which are predicted to reside sufficiently near the ATP molecule to participate in direct bonding interactions. Significantly, these bonding interactions between PP-50 and ATP are those that must be broken when the electrochemical proton gradient interacts with F_1 to induce release of bound ATP.

In summary, and again as it relates to the original question addressed, the ATP domain within the F_1-β subunit is predicted to lie within an adenylate kinase fold. Within this putative fold a 50 amino acid stretch containing the Walker A consensus appears to play a major role in binding. The location of the second nucleotide binding site, i.e., the site for binding ADP has not been located, and would be predicted to properly form only when the β subunit interacts with one or two of the small subunits δ and ε (see previous section). The precise location of the ADP site will be a challenging problem for future investigation. The possibility that the ADP site corresponds to the AMP site in adenylate kinase, but with somewhat different properties, remains an attractive possibility.

Nucleotide binding to overexpressed wild type and mutant F_1-β subunit fragments

The above studies localize the ATP binding domain within F_1 to β subunits and then to a 50 amino acid sequence within a predicted adenylate kinase fold. However, this 50 amino acid region does not include the Walker B consensus, nor another region homologous to adenylate kinase which contains the motif $VXADX_3DX_8HLDA$ which we have referred to as the "C" consensus. In order to establish whether the N-terminal region of the β subunit preceding the A, B, and C consensus regions contribute to ATP binding, and to what extent the A, B, and C regions contribute, we resorted to the tools of molecular biology.

Five β subunit proteins were overexpressed in *E. coli*, purified to homogeneity and then subjected to nucleotide binding studies in which TNP-ATP was first bound, and then competed off with ATP. From these studies Kd values for ATP were calculated and then used to

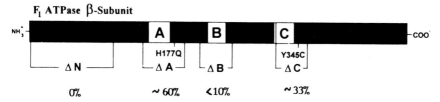

Figure 4. Schematic representation of the F_1-β subunit depicting those regions where deletions were made prior to testing ATP binding. Numbers below the deleted regions indicate the percentage contribution those regions make to the total energy for ATP binding to the β subunit.

obtain the free energy of binding. The five proteins subjected to this analysis were the "wild type" β subunit, a 2/3 β subunit fragment called C_4 in which 121 amino acids had been excised from the N-terminal region, and three different C_4 mutants containing deletions respectively in the A, B, and C consensus regions. The results of these studies summarized in Figure 4 and presented in detail elsewhere (19) show that the A consensus region makes the largest contribution to the ATP binding energy ($\sim 60\%$), the C consensus region contributes about 33%, while the B consensus region and the N-terminal 121 amino acid region make little or no contribution.

These studies confirm work on the 50 amino acid peptide called PP-50 which contained the A consensus and was shown to bind ATP (see above). They further indicate that the C consensus region may play a partial role in binding ATP, and suggest that the B consensus may have a role other than in ATP binding. For example, the B consensus may contribute a catalytic base, aid in stabilizing the transition state, or both.

Progress toward localizing ATP domains within the three-dimensional structure of the intact F_1 moiety of rat liver ATP synthase

The ultimate test of the view developed here that ATP domains within β subunits of the ATP synthase are very similar to those found in adenylate kinase is to establish whether these subunits within intact F_1 contain adenylate kinase folds which bind ATP. Earlier, our groups were successful in obtaining crystals of the F_1 suitable for x-ray diffraction studies (20). Recently, this work has led to a 3.6 Å map of F_1 (21) (Figs 5A and 5B). At this resolution the overall dimensions of the F_1 molecule are 120 Å × 120 Å × 74 Å, and the overall structure exhibits a threefold axis of symmetry. The fact that α subunits of the rat liver enzyme contain cysteine residues, whereas the β subunits do not, made it possible to identify both α and β subunits. The α and β subunits each

A

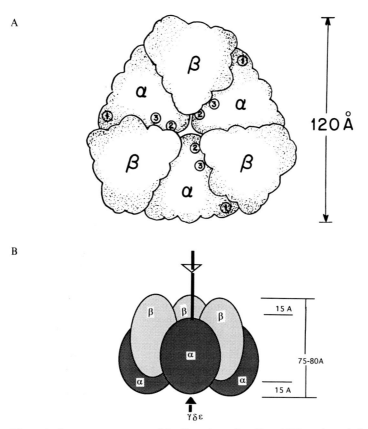

B

Figure 5. Quaternary structure of the F_1-moiety of rat liver ATP synthase derived from x-ray crystallographic data at 3.6 Å (21). (A) Top view, simplified representation. Numbers represent positions of heavy atom labels. Heavy atom position 1 corresponds to a cysteine residue in the α subunit which is located between the A and B consensus regions. (B) Diagram of the arrangement of the α and β subunits of F_1. The three-fold axis of symmetry relating the three copies each of α and β subunits is represented by a vertical line. The minor subunits (not shown) are thought to be located in the center. Located within β subunits is a structural fold very similar *but not identical* to the nucleotide binding fold found in adenylate kinase.

exist as trimeric arrays which are organized in two slightly offset, interdigitated layers along the three-fold axis. Higher resolution data near 3.0 Å has been obtained with retention of the three-fold symmetry and at this resolution we have been able to map most of the secondary structure of the F_1 molecule contributed by α and β subunits.

The above information together with significant chain tracing has permitted us to address the question of whether β subunits do in fact contain structural folds very similar to those found within the ATP binding region of adenylate kinase. In data to be described in detail elsewhere, it is quite clear that the F_1-β subunit does contain a structural fold very similar, *but not identical* to that found within the ATP binding region of adenylate kinase. Experiments are currently underway to obtain crystals in which ATP or nucleotide analogs thereof are present.

Implication for ATP synthase function during ATP synthesis and ATP hydrolysis

Studies summarized here emphasize that both the substrate (ADP) and the product (ATP) of ATP synthesis have distinct binding sites, on F_1. They indicate also that the ADP site readily binds ADP only when a β subunit interacts with either the δ subunit, the ε subunit, or both together. Thus, the induction of asymmetry to only one of the three β subunits may specify that subunit or the $\beta_3\delta\varepsilon$ complex to participate in ATP synthesis. If so, this ADP site becomes the key catalytic site of interest. It will be of importance in future studies to establish whether ADP bound at this site is converted to ATP, and then moved to an ATP site on the asymmetric β subunit before its release. It will be of importance also to establish where within the β subunit the ADP site resides.

Three types of evidence, i.e., studies with synthetic peptides, mutational analysis, and x-ray crystallography are all consistent in identifying the product site for ATP synthesis (i.e., the ATP site) within an adenylate kinase-like nucleotide binding fold in the β subunit. This fold contains the Walker A consensus sequence which makes a significant contribution to the free energy of binding. It is this site which is predicted to interact directly or indirectly with the electrochemical proton gradient during ATP synthesis resulting in the release of bound ATP.

Finally, perhaps one of the most interesting aspects of these studies still under investigation is the finding that the nucleotide binding properties of the F_1 moiety are quite different under nonhydrolytic, ATP synthetic conditions relative to hydrolytic conditions. Thus, under nonhydrolytic conditions one molecule of rat liver F_1 readily binds only one exchangeable ADP molecule and three exchangeable ATP (AMP-PNP) molecules, whereas under hydrolytic conditions almost the exact reverse is observed with ADP binding predominating. This strongly implicates two distinct conformational states of F_1, an ATP synthetic state in which asymmetry in ADP binding is very clear, and an ATP hydrolytic state in which asymmetry in ADP binding is less obvious.

Along these lines it is interesting to note that the x-ray structure of rat liver F_1 exhibits three-fold symmetry (21), whereas that recently reported for bovine heart F_1 (22), although exhibiting some three-fold symmetrical features, appears to reflect some very clear asymmetrical features. Significantly, rat liver F_1 was crystallized in the presence of ATP, which is almost completely hydrolyzed. It is possible, therefore, that the asymmetry induced in a β subunit (or $\alpha\beta$ pair) by one or more of the small subunits during ATP synthesis (in order to facilitate ADP binding) is switched off when the enzyme enters its ATP hydrolytic mode. In this case, it is important under physiological conditions to load the enzyme with ADP to facilitate product inhibition and therefore preserve the ATP pool. During this switch from an asymmetrical to a more symmetrical functional behavior the small subunits may dissociate in part from the larger subunits and form a more compact mass within the center of the F_1 molecule which either conforms to the three-fold symmetry and/or is not possible to clearly delineate by x-ray crystallography.

It is unfortunate that investigators working on the x-ray structure of the bovine heart F_1 (22) have failed to recognize that F_1 may exist in two functionally important conformations.

References

1. Pedersen, P.L. and Amzel, L.M. (1992) Structure, reaction center, mechanism, and regulation of one of natures most unique machines. *J. Biol. Chem.* 268: 9937–9940.
2. Penefsky, H.S. and Cross, R.L. (1991) Structures and mechanism of F_0F_1-type ATP synthases and ATPases. *Adv. Enzymol. and Related Areas* 64: 173–214.
3. Futai, M., Noumi, T. and Maeda, M. (1989) ATP synthese (H^+-ATPase): Results by combined biochemical and molecular biological approaches. *Ann. Rev. Biochem.* 58: 111–136.
4. Senior, A.E. (1990) The proton-translocation ATPase of *Escherichia coli*. *Annu. Rev. Biophys. Chem* 19: 7–41.
5. Nagley, P. (1988) Eukaryotic membrane genetics: The F_0 sector of mitochondrial ATP synthase. *Trends in Genetics* 4: 46–52.
6. Hatefi, Y. (1985) The mitochondrial electron transport chain and oxidative phosphorylation system. *Ann. Rev. Biochem.* 54: 1015–1069.
7. Grubmeyer, C., Cross, R.L. and Penefsky, H.S. (1982) Mechanism of ATP hydrolysis by beef heart mitochondrial ATPase: Rate constants for elementary steps in catalysis at a single site. *J. Biol. Chem.* 257: 12092–12100.
8. Penefsky, H.S. (1985) Energy dependent dissociation of ATP from high affinity catalytic sites of beef heart mitochondrial adenosine triphosphatase. *J. Biol. Chem.* 260: 13735–13741.
9. Catterall, W.A. and Pedersen, P.L. (1971) Adenosine triphosphatase from rat liver mitochondria I. Purification, homogeneity, and physical properties. *J. Biol. Chem.* 246: 4987–4994.
10. Catterall, W.A., Coty, W.A. and Pedersen, P.L. (1973) Adenosine triphosphatase from rat liver mitochondria III. Subunit composition. *J. Biol. Chem.* 248: 7427–7431.
11. Williams, N., Hullihen, J. and Pedersen, P.L. (1987) Ligand binding studies of the F_1 moiety of rat liver ATP synthase: Implications about the enzymes structure and mechanism. *Biochem.* 26: 162–169.
12. Pedersen, P.L. and Hullihen, J. (1978) Adenosine triphosphatase of rat liver mitochondria: Capacity of the homogeneous F_1 component of the enzyme to restore ATP synthesis in urea-treated membranes. *J. Biol. Chem.* 253: 2176–2183.
13. Gromet-Elhanan, Z. and Khananshvili, D. (1984) Characterization of two nucleotide binding sites on the isolated reconstitutively active β subunit of the F_0F_1 ATP synthase. *Biochem.* 23: 1022–1028.
14. DiPietro, A., Penin, F., Julliard, J.H., Godinot, C. and Gautheron, D.C. (1988) IF_1 inhibition of mitochondrial F_1-ATPase is correlated to entrapment of four adenine- or

guanine-nucleotides including at least one triphosphate. *Biochem. Biophys. Res. Commun.* 152: 1319–1325.

15. Dry, D.C., Kuby, S.A. and Mildvan, A.S. (1986) ATP-binding site of adenylate kinase: Mechanistic implication of its homology with ras-encoded p21, F_1-ATPase, and other nucleotide binding proteins. *Proc. Natl. Acad. Sci. (USA)* 83: 907–811.

16. Pai, E.F., Sachsenheimer, W., Schirmer, R.H. and Schulz, G.E. (1977) Substrate positions and induced-fit in crystalline adenylate kinase. *J. Mol. Biol.* 114: 37–45.

17. Dry, D.C., Kuby, S.A. and Mildvan, A.S. (1985) NMR studies of the MgATP binding site of adenylate kinase and of a 45-residue peptide fragment. *Biochemistry* 24: 4680–4694.

18. Garboczi, D.N., Shenbagamurthi, P., Kirk, W., Hullihen, J. and Pedersen, P.L. (1988) Interaction of a synthetic 50-amino acid, β subunit peptide with ATP. *J. Biol. Chem.* 263: 812–816.

19. Thomas, P.J., Garboczi, D.N. and Pedersen, P.L. (1992) Mutational analysis of the consensus nucleotide binding sequences in the rat liver mitochondrial ATP synthase β-subunit. *J. Biol. Chem.* 267: 20331–20338.

20. Amzel, L.M. and Pedersen, P.L. (1978) Adenosine triphosphatase from rat liver mitochondria. Crystallization and x-ray diffraction studies of the F_1-component of the enzyme. *J. Biol. Chem.* 253: 2067–2069.

21. Bianchet, M., Ysern, S., Hullihen, J., Pedersen, P.L. and Amzel, L.M. (1991) Mitochondrial ATP synthase: Quaternary structure of the F_1 moiety at 3.6 Å determined by x-ray diffraction analysis. *J. Biol. Chem.* 266: 21197–21201.

22. Abrahams, J.P., Lutter, R., Todd, R.J., van Raaij, M.J., Leslie, A.G.W. and Walker, J.E. (1993) Inherent asymmetry of the structure of F_1-ATPase from bovine heart mitochondria at 6.5 Å resolution. *EMBO J.* 12: 1775–1780.

Biochemistry of Cell Membranes
ed. by S. Papa & J. M. Tager
© 1995 Birkhäuser Verlag Basel/Switzerland

Structure-function relationships in the mitochondrial carrier family

M. Klingenberg[1] and D. Nelson[2]

[1]*Institute for Physical Biochemistry, University of Munich, Goethestrasse 33, D-80336 Munich, Germany*
[2]*Department of Biochemistry, The University of Tennessee, Memphis, TN 38163, USA*

Introduction

The intensive exchange of metabolites between the cytosol and matrix space of mitochondria requires a number of carriers which catalyze this transport. These carriers must have emerged with the symbiosis of the predecessor prokaryotes in the eukaryotic cell. Therefore, the evolutionary position is distinct from bacterial carriers or from plasma membrane carriers of the eukaryotes. In fact, the structures of these carriers show no similarity to any of the other known carriers. Since most other components of the mitochondrial oxidative phosphorylation system originate from the ancestor prokaryote only the acquired metabolite carriers are uniquely characteristic for mitochondria. During the last 12 years it has become clear that these mitochondrial transport systems are evolutionarily related and belong to a protein family. Addition of further structural genes of still unknown proteins even in non-mitochondrial organelles justified elevating the mitochondrial carrier family to a super family. In the present brief survey we will describe some features which are characteristic for the carrier family and discuss some structure-function relationships elucidated on these components in our laboratories.

Survey of carriers by function

A survey of the mitochondrial carriers is given in Table 1. They are divided into carriers involved in energy transfer, carbon translocation, and nitrogen transfer. The first group comprises those carriers which are directly involved in oxidative phosphorylation, for the ATP export and phosphate uptake and a carrier for H^+ or OH^- transport which is involved in the opposite function, i.e., in generating heat and thus

Table 1. Mitochondrial solute carriers

Carrier name	Substrates	Transport type	Control	Metabolic role
Energy transfer				
ADP/ATP	ADP^{3-} ATP^{4-}	exchange	$\Delta\psi$	Oxidative phosphorylation
Phosphate	$H^+ + P_i^-$	unidir.	ΔpH	Oxidative phosphorylation
Uncoupling	H^+ or OH^-	unidir.	ΔpH	Heat generation
Carbon transfer				
Dicarboxylate	Mal^{2-}, $Succ^{2-}$ P_i^{2-}	exchange	—	TCC-cycle Gluconeogenesis
Ketoglutarate	$Ketogl^{2-}/Mal^{2-}$	exchange	—	Hydrogen import
Citrate	$H^+ + Cit^{3-}/Mal^{2-}$	exchange	ΔpH	Hydrogen export Fatty acid synthesis
Pyruvate	$Pyr^- + H^+$	unidir.	ΔpH	Glucose oxidation
Carnitine	Car/Acylcar	exchange	—	Fatty acid oxidation
Nitrogen transfer				
Aspartate/ Glutamate	$Asp^-/Glu^- + H^+$	exchange	$\Delta\psi$	Urea cycle
Glutamate	$Glu^- + H^+$	unidir.	$\Delta\psi$	—
Ornithine	Ornit, Citrul	exchange	—	Urea cycle

bypassing oxidative phosphorylation. The group of carbon transloca-
tors includes five carriers involved in the tricarboxylic acid cycle,
in carbohydrate oxidation, and fatty acid oxidation. The group involved in nitrogen transfer is more heterogeneous in its function since
the asparatate/glutamate carrier is also involved in hydrogen transport
and in the urea cycle. A previous survey on the functional properties of
these carriers is given by Krämer and Palmieri (1992). A most recent
series of minireviews on mitochondrial carriers can be found in the
October 1993, Vol. 25, No. 5 issue of the *Journal of Bioenergetics and
Biomembranes.*

Most of these carriers are of the exchange type, i.e., the uptake is
associated with the concomitant release of a cooperating substrate. This
has the advantage for an intracellular organelle, such as the mitochondria, to remain osmotically neutral. In this context the question of
electroneutrality of transport or exchange is most important. The large
membrane potential across the inner mitochondrial membrane should
drastically affect an electrically active transport, whereas the low ΔpH
imposes only a rather small gradient whenever H^+ are cotransported.
So far, only three carrier systems are known to be electrically active,
that for the ADP/ATP exchange (Wulf et al., 1978; La Noue et al.,
1978), the uncoupling protein (UCP) (Klingenberg and Winkler, 1985),
and the aspartate-glutamate exchanger (La Noue and Schoolwerth,
1979). Either the charge difference, such as between ATP^{4-} and ADP^{3-}
or the cotransport of an H^+ neutralizing the glutamate branch of the

Table 2. Control by $\Delta\psi$ or Δ pH

Carrier	Substrate	Control
ADP/ATP	ADP^{3-}/ATP^{4-}	$\Delta\psi$
Phosphate	$P_i^- + H^+$	Δ pH
Uncoupling protein	H^+ or OH^-	$\Delta\psi$
Dicarboxylate	Mal^{2-}/P_i^{2-}	—
Ketoglutarate	KG^{2-}/Mal^{2-}	—
Citrate	$Citr^{3-}/Mal^{2-} + H^+$	—
Pyruvate	$Pyr^- + H^+$	—
Carnitine	Carn/Acyl Carn	—
Aspartate/Glutamate	$Asp^-/Glu^- + H^+$	$\Delta\psi$
Glutamate	$Glu^- + H^+$	Δ pH
Ornithine	Orn/Citr	—

aspartate-glutamate transport, makes this exchange system electrical (Tab. 2). On the other hand, electroneutrality can be achieved by H^+-cotransport when a charge difference exists between citrate^{3-} in exchange for malate^{2-} (McGivan and Klingenberg, 1971; Palmieri et al., 1972) or to neutralize Pi^- (McGivan and Klingenberg, 1971) in a unidirectional carrier and also pyruvate$^-$ (Halestrap and Denton, 1974) and glutamate$^-$. The control by $\Delta\psi$ is of utmost importance for the large thermodynamic differences in the cytosol and matrix space of mitochondria, which originate from the symbiotic import of the mitochondrial ancestors into the eukaryotic host. Thus, a 30% higher phosphorylation potential of free energy of ATP in the cytosol as compared to the mitochondria can be bridged in oxidative phosphorylation only when the ATP export is driven by the high membrane potential (Klingenberg, 1976). The same is true for the high redox potential difference of the NADH system in both compartments which is coordinated or bridged by the $\Delta\psi$-driven aspartate-glutamate exchanger (LaNoue and Schoolwerth, 1979).

One can also argue that some of the carriers are of more basic importance for all mitochondria and others are more specialized. For example, the carriers for oxidative phosphorylation, i.e., the AAC and the phosphate carrier, should be present in all mitochondria from all cells since they share the common function of supplying ATP to the surrounding cytosol. They are thus the most characteristic components of mitochondria. The other factors in oxidative phosphorylation, the respiratory chain and the ATP synthase are carried over from the prokaryotes. The distribution of the di-carboxylate and tri-carboxylate carriers is only known in mammalian organs. Thus, it seems that the di- and tri-carboxylate carriers are less involved in the actual energy transfer, but rather play a role in biosynthetic processes, such as hydrogen export into the cytosol, glyconeogenesis, etc. On the other hand, the aspartate/glutamate carriers as well as the associated ketoglutarate/ malate carrier will be found where hydrogen from glucolysis is generated and has to be imported into the mitochondria. Thus, it functions in coordination with oxidative phosphorylation and will be present in most cells. The same is true for the acyl carnitine/carnitine exchanger which will be more prominent wherever fatty acids are oxidized. The ornithine/citrulline exchanger mostly associated with the urea cycle will be confined primarily to the urea producing organs, such as liver.

A particular case is uncoupling protein (UCP). It is confined only to mammals and there only to one organ, the brown fat adipose tissue. Although UCP in general serves an unspecific purpose, i.e., heat generation, it is highly specialized to one small group of phyla and there only to one small organ. This can be explained only by assuming that the way the heat is generated by UCP and in particular the way this heat production is regulated must be different from other ways of heat generation. The transport function of UCP is a most primitive one, i.e., facilitating the transport of the simplest ion, H^+, through the membrane. On the other hand, this H^+-transport activity in UCP is highly regulated by a number of cofactors, in particular fatty acids and nucleotides. These are indirectly controlled by c-AMP mediated signaling pathways. Obviously, on the basis of its regulatory properties UCP meets the particular requirements in its host for heat generation where it is utilized in highly special situations, such as birth or hibernation.

It therefore seems that UCP is the latest addition to the mitochondrial carrier family. We have proposed UCP to originate from a H^+-substrate cotransporter by deleting the transport of the substrate (LaNoue et al., 1978). For example, the fundamental phosphate H^+ cotransporter could be an ancestor of UCP. However, as we will see below there is no special phylogenetic relationship of UCP to the phosphate carrier.

Identification of carriers by sequences

Many of the mitochondrial carriers have in the meantime been identified and isolated. There are 65 carrier sequences known, but only five functions have been assigned. This leaves many sequences without a known function. The molecular weight of the isolated proteins, as seen on SDS gels, reveals that the size of these proteins is quite similar, around 32 kD (Tab. 3) (Krämer and Palmieri, 1992). That these molecular weights give a fair estimate is seen from the molecular weight calculated from those carriers of which the sequences are known. These are slightly higher and cluster around 33 kD. The similar molecular weights alone suggest that these carriers are similar proteins and might therefore form a protein family.

Table 3. Molecular weights of isolated carriers

Carrier	Source	M_r (kD) from SPAGE	from Sequence
ADP/ATP	bovine heart (-1)	30	32.2
ADP/ATP	Sacch. cerv. (-2)	29.5	34.5
ADP/ATP	neurosp. crassa	33	33.9
Phosphate	bovine heart	33	35.0
Oxoglut./Mal.	bovine heart	31.5	34.2
Dicarboxylate	rat liver	28	—
Citrate	rat liver	30	—
Pyruvate	bovine heart	34	—
Carnitine	rat liver	32.5	—
Aspartate/Glutamate	bovine heart	31.5	—
Ornithine	rat liver	33.5	—
Uncoupling protein	hamster, rat brown adipose cells	32	33.2

The protein sequencing of the AAC provided the first primary sequence of any biomembrane carrier and, of course, of the first member of the mitochondrial carrier family (Aquila et al., 1982). This was followed by sequencing the uncoupling protein (UCP) where for the first time the similarity between both the proteins was noticed and the possibility of a carrier family postulated (Aquila et al., 1985). With the sequence of the third carrier, the phosphate carrier, again a similarity to the two previously known proteins AAC and UCP was found and the existence of a mitochondrial carrier family was substantiated (Aquila et al., 1987). Recently, the sequences of the ketoglutarate/ malate (Runswick et al., 1990) and the citrate/malate (Kaplan et al., 1993) carriers extend primary structure similarity to five different carrier types. Some characteristic features of amino acid composition of the mitochondrial carriers are given in Table 4. The high content of polar residues in these carriers is striking. There is a high excess of basic residues in all these proteins.

Sequences of these carriers, in particular of the AAC, have been obtained from many different organisms. Also isoforms from the AAC in several organisms or cells have been found. Thus, in humans three different isoforms exist which have been associated with different organs. In yeast (Saccharomyces) also three different genes for the AAC are found. Only one gene is normally expressed, and it seems that one other gene is activated only under anaerobic conditions. In addition to the known mitochondrial carriers, further structural genes have been found. These gene products are not necessarily even associated with mitochondria. Their function is still unknown. With these findings it appears that there exists an even larger family, a "super family" of carriers, of which the mitochondrial carrier family is only a subgroup.

Table 4. Molecular characteristics of several mitochondrial transport proteins

Feature	AAC	UCP	P$_i$C	KMC	CMC
Molecular mass (kDA)	32.8	33.2	34.7	34.2	32.6
Polarity (%)	37.7	41.8	35.9	38.2	40.3
Net charge	+ 23	+ 10	+ 11	+ 16	+ 17
Number of amino acids					
Total	297	306	312	314	298
Acidic (D + E)	21	19	25	20	22
Basic (K + R)	45	29	36	36	39
Aromatic (F + W + Y)	38	26	41	32	25
Cysteine (C)	4	7	8	3	6
Histidine (H)	3	3	2	3	6

The mitochondrial carrier superfamily, phylogenetic tree

Figure 1 gives an alignment of sequences which can be attributed to the mitochondrial superfamily. A phylogenetic tree constructed from the carriers is given in Figure 2. Some of these sequences are fragments from the yeast, *Arabidopsis*, *C. elegans* and human genome projects. They are included because they are generally long enough to span a whole carrier domain, and this is sufficient to place them on the tree. Of course, when full sequences become available, some revision will probably be necessary. For example, the *C. elegans* fragment T00593 is only 65 amino acids long. Though it has the sequence typical of the first transmembrane segment of an ADP/ATP carrier, it is too short to cluster with its nearest cousin the ADP/ATP carrier of *C. elegans*.

The tree shown is a UPGMA tree that has been computed using a unit matrix that scores identities. Other trees have been made with the PAM250 matrix, and trees have been made with both matrices using only complete sequences. There are some changes that occur in these trees when different alignments of different scoring matrices are used. Note that there are five major branches on the tree. The branching pattern inside these main clusters stays pretty much the same no matter what method is used to make the tree. However, the order of the major branches does vary, with the ADP/ATP carriers always on the top.

One striking feature of the tree is the large number of ADP/ATP carriers represented. Almost 40% of the carrier sequences are in this family. Multiple isoforms exist in yeast, plants, *C. elegans* and mammals, and there are seven pseudogenes reported in humans (Chen et al., 1990). The presence of isoforms outside of the ADP/ATP carriers is rare. Only the MRS3 and MRS4 genes in yeast are probable isoforms of an as yet unidentified function. The similar fragments from *Arabidopsis*, T12697 and T14791 may actually be the N- and C-terminal fragments of the same gene. There are additional accession numbers for *Arabidop-*

sis thaliana AAC sequences that are very similar to the sequences in the alignment. These are Z18781 (75 amino acid fragment, only three differences with X65549), Z17427 (96 amino acid fragment, only two differences with X65549).

An estimate of the number of unique carrier functions represented in this tree can be made by finding the most distant sequences with the same function. These would be the phosphate carriers. The yeast sequence is only about 39% identical to the mammalian sequences. The next sequence in this cluster is a *C. elegans* fragment that probably has a different function than the *C. elegans* phosphate carrier. Notice that the citrate carrier cluster also as a *C. elegans* sequence and a new yeast gene called YBR2039. The yeast citrate carrier is about 36% identical to the rat sequence. Below the yeast citrate carrier is another yeast sequence S36407. This is probably a different kind of carrier. Branches that occur further to the left than the phosphate or citrate carrier families will probably have unique functions. If one draws a line to the left of these groups, 24 branches are intersected. These probably transport 21 different substrates or substrate pairs. Only five of these functions are assigned, leaving 16 to be determined. Thirteen yeast sequences with ten probable funtions are present in this tree. Nine are assigned to chromosomes, and four (AAC2, AAC3, RIM2, and the citrate carrier) are on chromosome II. This may mean that chromosome II in yeast is unusually rich in mitochondrial carriers.

How many more carriers are there? Today, the yeast genome project is about 25% complete, and 13 carriers have been identified so far. Therefore, one could predict that there will be about 50 carriers in yeast. The project is scheduled for a 1996 completion date, so we should know if this is a reasonable estimate by then. In *C. elegans*, however, eight carriers have been identified so far and only 3–4% of the *C. elegans* genome has been sequenced. That suggests that in animals, the number of carriers may be in the range of 100–200. Determining what they do will be a very difficult task. Recently, a *C. elegans* sequence has been tentatively identified as an oxoglutarate/malate carrier (Walker and Runswick, 1993). This sequence has 38.6% identity to the rat oxoglutarate/malate carrier. In contrast, the *C. elegans* citrate carrier and phosphate carrier have 62.8% and 74.7% identity to their rat homologues. Therefore, the new *C. elegans* sequence is probably not an oxoglutarate/malate carrier, but a closely related sequence with another function.

Mitochondrial carrier family is functionally and structurally different

It should be stressed that no similarity of the mitochondrial carrier family with any other solute carrier has been found even with the well

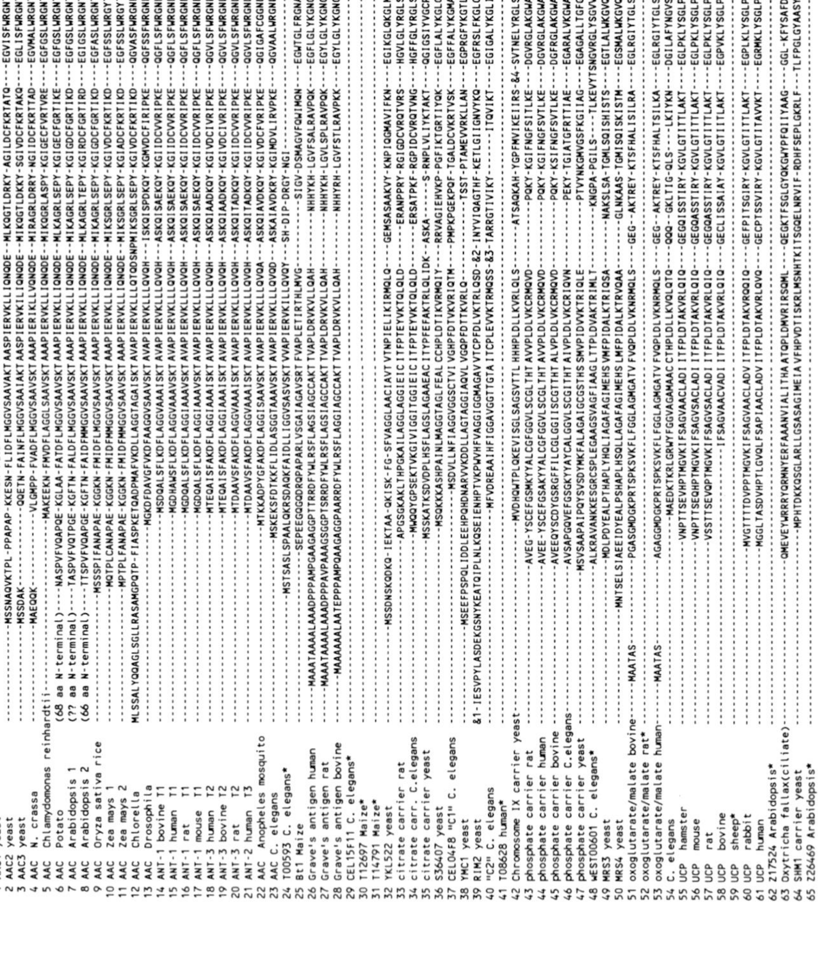

1 AAC1 yeast
2 AAC2 yeast
3 AAC3 yeast
4 AAC N. crassa
5 AAC Chlamydomonas reinhardtii
6 AAC Potato
7 AAC Arabidopsis 1
8 AAC Arabidopsis 2
9 AAC Oryza sativa rice
10 AAC Zea mays 1
11 AAC Zea mays 2
12 AAC Chlorella
13 ANT-1 Drosophila
14 ANT-1 bovine T1
15 ANT-1 human T1
16 ANT-1 rat T1
17 ANT-1 mouse T1
18 ANT-3 human T2
19 ANT-1 bovine T2
20 ANT-3 rat T2
21 ANT-2 human T3
22 AAC Anopheles mosquito
23 AAC C. elegans
24 T00593 C. elegans
25 B11 Maize
26 Grave's antigen human
27 Grave's antigen rat
28 Grave's antigen bovine
29 CEL15F11 C. elegans*
30 T11269? Maize*
31 T14791 Maize*
32 YKL522 yeast
33 citrate carrier rat
34 citrate carrier C.elegans
35 citrate carrier yeast
36 S36407 yeast
37 CEL04F8 "C1" C. elegans
38 phosphate carrier human
39 YMC1 yeast
40 R1M2 yeast
41 T00628 human
42 Chromosome IX carrier yeast
43 phosphate carrier rat
44 phosphate carrier human
45 phosphate carrier bovine
46 phosphate carrier C.elegans
47 phosphate carrier yeast
48 WES100601 C. elegans*
49 MRS3 yeast
50 MRS4 yeast
51 oxoglutarate/malate bovine
52 oxoglutarate/malate rat*
53 oxoglutarate/malate rat*
54 oxoglutarate/malate human*
55 UCP hamster
56 UCP mouse
57 UCP rat
58 UCP bovine
59 UCP sheep*
60 UCP rabbit
61 UCP human
62 217524 Arabidopsis*
63 Oxytricha fallax(ciliate)*
64 SHM1 carrier yeast
65 226469 Arabidopsis*

Figure 1.

Figure 1. (Continued)

Figure 1. Alignment of 65 mitochondrial carriers. The symbol "&" followed by a number indicates a segment of sequence has been deleted from the alignment to save space. &1 = MPKKSIEEWEEDA, &2 = IFLKAYKSQAVNISKGSTRPKS, &3 = RGLDAQSGPSTSSG, &4 = SANSGR, &5 = KPGETQLKGVGND, &6 = TPPGFLLKAVIGMTA, &7 is identical to &6, &8 = LPFYQKALLAGFA, &9 = KERSIEKFGYQAEGTK. Sequence 46 has an extra Leu between amino acids 130 and 131 that is not shown. Sequence numbering on the right assumes these missing amino acids are counted. An * indicates a sequence fragment. Thirty-seven of the 62 sequences were included in Nelson et al. (1993) and will not be referenced again here. The 27 new sequences are given with their alignment number followed by the reference and a Genbank accession number unless otherwise stated. (7) AAC *Arabidopsis* 1, Saint-Guily, A. et al., unpublished (1993) GenEMBL: X65549; (8) AAC *Arabidopsis* 2, Schuster, W. et al. (1993) GenEMBL: X68592; (13) AAC *Drosophila*, Louvi, A. et al. (1992) GenEMBL: S43651; (16) rat T1, Shinohara, Y. et al. (1993) GenEMBL: X61667; (17) mouse T1, Laplace, C. and Costet, P. unpublished (1993) GenEMBL: X74510; (20) rat T2, Shinoara, Y. et al. (1993) GenEMBL: D12771; (22) AAC *Anopheles* mosquito, Beard, C.B., Crews-Oyen, A.E. and collins, F.H., unpublished (1993) GenEMBL: L11617, L11618, Z21814, Z21815; (24) T00593 *C. elegans*, Kerlavage, A.R., unpublished (1992) GenEMBL: T00593; (25) Btl maize, Sullivan, T.D. et al. (1991) GenEMBL: M79333; (28) Grave's antigen bovine, Fiermonte, G. et al. (1992) GenEMBL: S51482; (30) T12697 maize, Baysdorfer, C. unpublished (1993) GenEMBL: T12697; (31) T14791 maize, Helentjaris, T. unpublished (1994) GenEMBL: T14791; (33) rat citrate carrier, Kaplan, R.S. et al. (1993) GenEMBL:L12016: (34) citrate carrier *C. elegans*, Sulston, J. et al. (1992) GenEMBL: Z22180; (35) citrate carrier yeast, Holmstroem, K., unpublished (1993) GenEMBL: X76053; (36) S36407 yeast, Fernandez, M., Fernandez, E. and

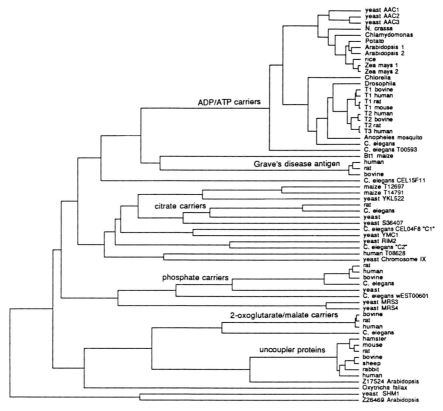

Figure 2. A phylogenetic tree of 65 mitochondrial carriers. A simple UPGMA (unweighted pair group method of averaging) tree was computed using a unit scoring.

Rodicio, R., unpublished (1992) PIR2: S36407, GenEMBL: Z25485; (37) C1 *C. elegans*, Walker, J.E. and Runswick, M.J. (1993) no accession number for whole sequence, N-terminal is CEL04F8; (38) YMC1 yeast, Graf, R. et al. (1993) GenEMBL: X67122; (39) RIM2 yeast, Demolis, N. et al. (1993) GenEMBL: Z21487; (40) C2 *C. elegans*, Walker, J.E. and Runswick, M.J. (1993) no accession number; (41) T08628 human, Adams, M.D. et al. (1993) GenEMBL: T08628; (42) yeast chromosome IX carrier, Walker, J.E. and Runswick, M.J. (1993) no accession number; (54) *C. elegans* similar to oxoglutarate/malate carrier, but probably different, Walker, J.E. and Runswick, M.J. (1993) no accession number; (56) UCP mouse, Kozak, L.P. et al. (1988) GenEMBL: M21222, M21244–M21247; (59) UCP sheep, Casteilla, L. et al. (1991) PIR3: S16082; (60) UCP rabbit, Balogh, A.G. et al. (1989) GenEMBL: X14696; (62) *Arabidopsis* Z17524, Bardet, C., Axelos, M., Tremousaygue, D., Lebas, M., Lagravere, T. and Lescure, B. unpublished (1992) GenEMBL: Z17524; (64SHM1 yeast, Kao, L.-R., unpublished (1994) GenEMBL: U08352, (65) Z26469 *Arabidopsis*, Desprez, T., Amselem, J., Chiapello, H., Caboche, M. and Andhofte, H., unpublished (1993) GenEMBL: Z26469.

characterized 12 helical transporter family of the prokaryotes (Maloney and Wilson, 1992). This is not surprising if one considers the entirely different evolutionary position of mitochondrial carriers which might have emerged only with the development of the eukaryotic cell. Although mitochondrial cariers may catalyze transport of similar substrates as prokaryote carriers, for example transport of citrate or malate is also found in prokaryotes, the environmental conditions under which these carriers work are entirely different. Thus, bacteria can accumulate substrate from extremely dilute solutions and therefore have developed a complex machinery for the concentration of the substrates including an accumulation in the periplasmic space and ATP-driven transport mechanisms. Often a host of proteins is involved in the transport of one solute into the bacteria cell. The molecular weight of mitochondrial carriers is smaller than that of the bacterial carriers and they consist only of a single protein component.

The mitochondrial carriers deal with intracellular conditions in which the concentration of the transported solutes is relatively high. So the carriers are probably saturated with substrate. They have no concentrative but rather a catalytic function in transport. As stated above, in some carriers the $\Delta\psi$ can create a stronger gradient of substrate ratios but the more frequent influence of the ΔpH on the mitochondrial carriers can only weakly shift the concentration across the mitochondrial membrane. There are no phosphorylation, ATP hydrolysis sites or other molecular sites which are linked to energy delivering mechanisms incorporated into these carriers. They are essentially stripped-down proteins competent only for the translocation process.

Secondary structure prediction

It can be expected that in a molecular family the three-dimensional structure of their members will be similar. Most of the evidence on the carrier structure is obtained from the ADP/ATP carrier (AAC) since it was the first carrier to be isolated and is available in large amounts as an intact protein. Except for the uncoupling protein, the other carriers are only available in minute quantities, mainly only for reconstitutional purposes.

The isolated AAC is packaged in a large Triton X 100-micelle which, according to hydrodynamic studies, has been described as an oblate ellipsoid with the shortest axis corresponding to the two-fold symmetry axis of the AAC dimer (Klingenberg et al., 1979). Besides the AAC, also the UCP has been found to be a dimer in the isolated state and also by crosslinking studies in the mitochondrial membrane (Klingenberg, 1990; Lin et al., 1980). Therefore, all these carriers can be assumed to form homodimers. In the isolated micelle the AAC dimer binds a large excess of detergent molecules amounting to about 170 mol/dimer. This extensive detergent envelope explains the peculiar properties on isolation of these proteins by a passthrough of hydroxyapatite.

Although this might suggest that the AAC is very hydrophobic, the primary structure of AAC showed a low average degree of hydrophobicity. A hydrophobicity algorithm search for transmembrane helices identified only rather weakly hydrophobic sections. The tri-partition of the whole sequence into similar domains of about 100 residues each strongly facilitated the possible assignment of the hydrophobic sections (Saraste and Walker, 1982). This prediction was reinforced when further sequences became known, first of UCP (Aquila et al., 1985) and then of the phosphate carrier (Aquila et al., 1987; Runswick et al., 1987). By this comparison and by introducing algorithms for amphiphatic α-helices in each of the three repeats, two transmembrane helices could be assigned. Particularly the vertical alignment of the repeat segments, as shown in Figure 3 showed the conservation of acidic residues with large non-polar stretches which we assume are delimiting the transmembrane helices in this carrier family. Based on this repeat structure we have arranged five different carrier structures for maximum similarity and conservation of critical residues by allowing insertions or additions of residues. In each of the repeats two hydrophobic regions can be discerned corresponding to two transmembrane helices. This would amount to a total of six transmembrane helices. Thus, these carriers form a six transmembrane α-helix family.

With this model it must be assumed that both the N- and C-terminal face the same side of the membrane (Fig. 4). Evidence from various laboratories is in line with the generally accepted conclusion that both the C- and N-terminals face the cytosolic side (Brandolin et al., 1989; Eckerskorn and Klingenberg, 1987). This is based on immunodetection with peptide antibodies (Brandolin et al., 1989), limited digestion with trypsin (Marty et al., 1992), and crosslinking studies. More ambivalent are the results on the transmembrane folding of these proteins.

In each domain, the transmembrane helices are separated by about 40 residue long hydrophilic sections. Often the polar groups come in charge pairs. These sections start with an acidic group at the end of helix one and terminate with a basic group at the start of helix two in each domain. Also at the start of each section a common motif is seen according to the general motif (D, E) XX (R, K) X (R, K) in which X are neutral groups. Later, we will discuss how these charges may participate in charge pair interaction between adjacent domains. There are further well conserved positions of basic groups, some in pairs. So it seems that these hydrophilic sections are more conserved than the transmembrane helices. These large concentrations of conserved polar groups indicate that we deal here with functionally important parts of the structure.

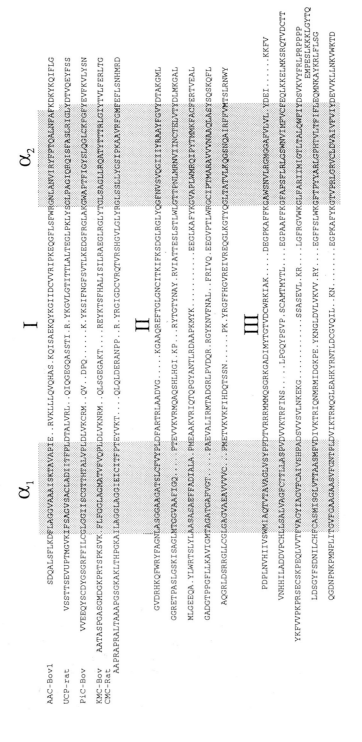

Figure 3. Three repeat alignments of five functionally different mitochondrial carriers. Deletions are introduced to obtain in the three repeat domains residues in maximum positional conservation. The transmembrane helices are shaded. References are given in the text.

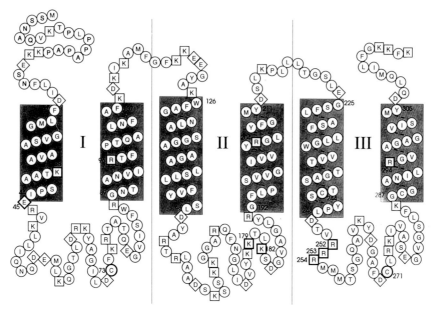

Figure 4. The putative transmembrane arrangement of the AAC2 from *Saccharomyces cerevisiae*. Both the N- and C-terminals face the cytosol. The three repeat domains are numerated.

Non-helical segments within the membrane

Several types of evidence show that these hydrophilic central sections are located at the matrix side. However, such a simple model is contradicted by some other data in which these sections are probed with covalent reagents from the cytosolic ·site. One type of evidence which was difficult to reconcile with the exclusively matrix-sidedness of the central section was obtained early with the membrane impermeant lysine reagent, pyridoxalphosphate (Bogner et al., 1986). Thus, in the bovine heart AAC, lysines equivalent to positions K179 and K182 in AAC2 from yeast were labeled from the cytosolic side in mitochondria (Mayinger et al., 1989). Furthermore, the incorporation of 8-azido or 2-azido ATP was found in bovine heart and yeast mitochondria to be confined to the region 172 to 210 in yeast or the equivalent in bovine heart (Dalbon et al., 1988) which again corresponds to the C-terminal portion of the central hydrophilic region and helix 4 in the second domain. Particularly 8-azido ATP is quite impermeant and therefore transport to the inside and labeling from the matrix side is improbable. More striking is the evidence obtained by covalent incorporation of nucleotide analogues in UCP (Winkler and Klingenberg, 1992; Mayinger and Klingenberg, 1992). Here, the nucleotide binding site exclusively faces the cytosolic side. Yet, 2-azido ATP and also FDNP ATP both are incorporated in the central hydrophilic section of the third domain. Again, the incorporation is confined to about the last 12 residues in this section.

For these reasons we have proposed that this section accessible from the cytosolic side protrudes as a loop into the membrane from the matrix side (Bogner et al., 1986; Klingenberg, 1989). Although it is rather hydrophilic, it is conceived to be surrounded by the helices and thus not in contact with the hydrophobic region of the membrane. In terms of the threefold pseudo symmetry, the carrier contains three of these loops, or as a dimer, six loops. The localization of the binding site for ATP in the AAC suggests that these loops are associated with the binding region and therefore line the translocation channel of the carrier. Also, Marty et al. (1992) propose for the bovine AAC that in addition to the transmembrane helices further sections protrude into the membrane surrounding the translocation channel, however as shortened helices. Also their model has not a threefold symmetry but prefers only six transmembrane helices.

Similar models with intramembranous loops have been suggested recently for various cation carriers and also possibly for the acetylcholine receptor (for a review see Durell and Guy, 1992). It remains to be seen whether these loops are rather short, as in our earlier model, or protrude as a β-hairpin through the membrane as suggested in Figure 5. At any rate these models are instrumental in selecting sites for mutagenesis of the AAC protein.

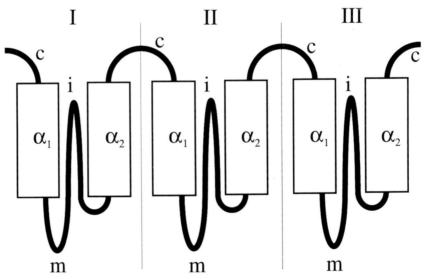

Figure 5. Scheme of the putative transmembrane arrangement of a typical mitochorial carrier, including three loops protruding from the matrix side into the membrane space.

Site-directed mutagenesis

For elucidating structure-function relationships we have employed site-directed mutagenesis of the AAC2 in *S. cerevisiae*. For this purpose first the two genes AAC1 and AAC2, which can be expressed under aerobic conditions, had to be deleted (Lawson et al., 1990). The elimination of the AAC3 was not necessary since AAC3 does not interfere with the expression of the AAC2 in plasmids because it is expressed only under strictly anaerobic conditions. In the first work related to this project AAC1 and AAC2 were reintroduced in the deleted yeast strains on a plasmid in order to compare the properties of these carrier isoforms (Gawaz et al., 1990). The AAC1 was expressed on a multicopy plasmid and thus sufficient amounts could be isolated for reconstitution. The AAC2 was available in equal amounts from the plasmid strain or the wild type. The reconstitution of these pure isoforms revealed no difference in the affinity for either ADP or ATP. However, the maximum rates for both the ADP and ATP were about 40% of the AAC2-rate. It was further shown that the cytochrome content, i.e., respiratory-chain capacity, was adjusted in the two plasmid strains to the exchange capacity rather than to the number of AAC1 or AAC2 expressed. In this work, small 5′ and 3′ sections of the coding region for AAC2 remained after gene disruption in the chromosomal DNA. A number of site-directed mutagenesis experiments were performed with these strains, in particular those relating to the cysteine mutations (Klingenberg et al., 1992; Hoffmann et al., 1994). In the further work of site-directed mutagenesis, strains were constructed in which the genes both for AAC2 and AAC1 were completly deleted in order to avoid recombination between the plasmid and nuclear DNA (Nelson et al., 1993).

The site-directed mutagenesis focused first on possibly "essential" cysteines in the AAC, based on the inhibition of the transport by SH reagents. Another residue detected by functional studies is the intrahelical lysine K38, the reaction of which with pyridoxalphosphate in bovine heart AAC (at K22) was strongly inhibited by atractylate (Bogner et al., 1986). Furthermore, in the folding model, conserved residues which are located at conspicuous positions were selected. These are first of all the three intrahelical arginines positioned in the second helix of each repeat R96, R204 and R294. A further target was the striking arginine triplet R252 to R254. One intrahelical proline was also mutated to assay its possible role in the functionally essential mobility of helices. The striking delimitation of the first helices in each domain by an acidic group led us also to mutate E45 and D149. K179, found in the analysis of bovine AAC (K162) to be accessible from the cytosolic site by pyridoxalphosphate, was also an interesting target for mutation. In Table 5 the effects of these mutations on the growth properties of these mutants on non-fermentable sources, such as glycerol, are summarized. All three

Table 5. Mutations in the ATP/ADP translocator

	Growth on glycerol
The three repeat intrahelical (α2) arginines R96, R204, R294	– – –
The two repeat and third (essential?) cysteines C73, C244, C271	+ + +
An intrahelical proline P247	+
One of the three repeat positive charge clusters in the matrix loop R252, R253, R254	– – –
Two of the three repeat negative charge terminals on the matrix side of α1 E45, D149	+ –
Lysines at the intrahelical loop tip K179, K182	+

(Nelson et al., 1993)

cysteine mutations are glycerol positive, all three intrahelical arginine mutations to neutral residues are glycerol negative. Also, the mutation of any arginine in the arginine triplet R252 to R254 disabled the cell to grow on glycerol. The mutation of intrahelical proline P247G did not affect growth. The replacement of the only intrahelical lysine K38 made the growth on glycerol negative, whereas the neutralization of the lysine at the central hydrophilic sections K179 and K182 permitted growth on glycerol.

In order to understand the changes in the AAC function due to these mutations, we have investigated the AAC on the level of mitochondria and after isolation in the reconstituted system. For this purpose the mutants were grown on a large scale for isolation of mitochondria. Respiratory activity and cytochrome content were measured on the isolated mitochondria as related functions to the AAC activity. The content of AAC was determined either by ^{3}H-CAT binding or by immuno quantitation. Most important was the ATP-synthesis activity of these mitochondria as one measure of the translocation activity. In all cases the ATP-synthesis activity turned out to be inhibited by carboxyatractylate indicating the involvement of AAC in measuring this activity. These data are summarized in Table 6, where the rate of ATP synthesis is related to the protein content, but also to the ^{3}H-CAT binding capacity of the various mitochondria as a measure for the content of AAC molecules. Only this specific activity (v') reflects the influence of mutation on the AAC transport capability.

All the mutations of the basic groups and also of the two acidic groups decrease the capability of the mitochondria to synthesize ATP to about 12% as a maximum with R294A mutant or to zero with the R204L and R96L mutants. Also the acidic D149S mutant was completely inactive. The content of AAC as measured by the ^{3}H-CAT

Table 6. ATP synthesis and carrier content in mitochondria of yeast mutants with polar residue mutations

Strain	V(ATP) $\frac{\mu\text{mol}}{\text{min} \cdot \text{g}}$	³H-CAT $\frac{\mu\text{mol}}{\text{g}}$	$V' = \frac{V(ATP)}{^3H\text{-}CAT}$	$\frac{V'}{V'_{WT}}$ %
WT	124	0.66	188	$\equiv 100$
R96H	3.8	0.25	15.2	8.0
R204L	0	0.028	0	0
R294A	2.0	0.53	3.8	2.0
R252I	0.48	0.07	6.6	3.4
R253I	0.20	0.06	3.3	1.7
R254I	0.86	0.015	7.2	3.8
K38A	0.80	0.28	2.9	1.5
K179M	75	0.45	167	88
K179I + K182I	2.0	0.31	6.5	3.4
E45G	46	0.28	163	86
D149S	0.5	0.11	4.7	2.4

(V. Müller, D. Nelson, M. Klingenberg, manuscript in preparation)

binding is only slightly decreased. In the R294A mutant it is even as high as in the wild type, and in the R96H mutant it still attains 40% of the wild type. Therefore, the efficiency of oxidative phosphorylation as compared to the amount of carrier molecules is also in general strongly decreased in these mutants. It can be assumed that the ATP synthesis rate, particularly in the mutant, is to a large extent limited by the ATP-translocation rate. Therefore, these data should reflect the changes of transport activity of the AAC in the various mutants.

The isolation and reconstitution of the AAC was a major effort to understand the mutational effects on the transport function of the AAC. There are several problems in the reconstitution, such as the lability of the mutant AAC and the dependence on certain lipid compositions. For this purpose for the reconstitution the mitochondria have been preloaded with atractylate in order to protect the AAC against proteolysis. Furthermore, as will be shown below, a number of mutant carriers have a strong dependence on cardiolipin and therefore cardiolipin has been routinely incorporated into the proteoliposomes. The basic exchange system which has, in general, the highest activity is that of ADP/ADP. In this "homo mode" of exchange no charge difference is generated across the membrane and the transport is independent of any membrane potential.

The exchange activity (Fig. 6) is generally much lower as expected, but it differs strongly between the various glycerol⁻ mutants. Most strikingly, the intrahelical mutants R294A and also the R96H still have a considerable exchange activity in the reconstituted system. In the

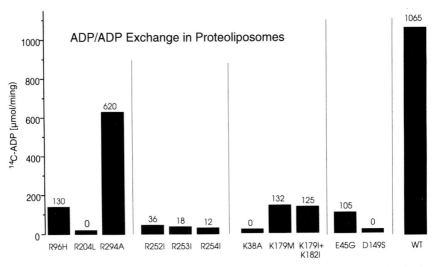

Figure 6. The exchange activity of isolated mutant AAC on reconstitution. The basic exchange activity ADP/ATP is determined at 25° with the inhibitor stop method. (V. Müller, M. Klingenberg, D. Nelson, manuscript in preparation).

R96H it amounts even to two-thirds of the wild-type activity. Paradoxically, the mutants at K179 and K179 + K182, which are glycerol$^+$, have still a significant, yet much lower activity. Among the three intrahelical arginines the R204L in the central domain is more strongly affecting the AAC activity. The neutralization of each of the arginines in the R252-254 triplet also results in a strong, however not complete inactivation.

The paradox is the glycerol$^-$ state of the R294A mutant and the high exchange activity for ADP. The heterologous exchange ADP/ATP or vice versa (D/T, T/D) are most active when a K$^+$-diffusion potential is placed across the membrane with a K$^+$ gradient plus valinomycin. In the wild-type AAC this activity is equal or even higher than the D/D exchange. In contrast, in the R294A mutant the D/T and T/D activity reaches only about 10–15% of the D/D activity. A similar suppression of the ATP employing transport activities versus the ADP exchange is also observed for the R96H AAC mutant, although it is less strongly expressed. With the absolute D/T exchange activity in the R294A AAC being about equal to the exchange activity of the glycerol$^+$ K179M mutant, one might expect that this activity in the R294A cell should sustain growth on glycerol. However, if one considers that the total capacity of the exchange activity is distributed between various exchange modes, such as between the unproductive D/D and the productive D/T exchange mode, the unproductive mode may suppress the D/T mode to such an extent that the ATP export rate does not sustain enough oxidative phosphorylation. This also agrees with the low ATP synthesis rate in the R294A mutant as compared to the high D/D exchange activity. This low ATP-synthesis activity is coupled to the low D/T exchange. In the cell, the competition between the D/D and D/T exchange may be even stronger than the in vitro measurements suggested. Although an assessment of this competition is difficult to assess because of the lack of knowledge on the free ADP and ATP concentratons both intra and extra mitochondrially, in a first approximation, one could visualize that the competitive productive D/T exchange would follow the formula.

Superficially it would seeem at first that the R294A residue is important for the extra charge of ATP^{4-} versus ADP^{3-}. However, the intrahelical arginines are conserved also in several other carriers at homologous positions e.g. in UCP and in the ketoglutarate-malate carrier (Fig. 3). Moreover, we will show below that a revertant of the R294A mutant exists in which a nearby negatively charged residue is also neutralized. All three intrahelical arginine mutants are important for optimal activity in transport. The importance must be considered in the efficiency to maintain transport activity. As mentioned above, ADP exchanges more readily than ATP in single substrate exchange experiments. Mutations that reduce the efficiency of ATP transport relative to ADP transport could result in high ADP/ADP exchange and a glycerol$^-$ phenotype. In other words these mutations may reflect a more generalized effect of decreased transport efficiency which affects a weaker component, such as the ATP transport, more strongly.

The role of the R252-254 triplet is also only partially specific for the AAC. It must be conceded that the triplet only occurs in AAC, however in four out of five different carriers a doublet according to the formula (R, K) X (R, K) is seen at the same position. Since this doublet is also quite well conserved in the other two domains it suggests that it forms a kind of pseudo ring of positive charges at the matrix site. It might be suggested that this ring of positive charges surrounds the vestibule of the carrier in order to attract the negatively charged substrate into the carrier channel. The necessity for such a positive charge ring can be argued to be particularly important for mitochondrial carriers where an extremely strong negative surface charge of cardiolipin on the inner phase of the matrix membrane would repel the substrates and thus only allow a very low effective concentration. Evidence is accumulating that these charges are not free, but interact with specific negative charges (see below) and other polar groups in AAC2 (such as N53 and possibly the 95% conserved Q52). Fluctuations in these interactions may be critical to the function of the carrier.

Another set of charged residues repeated in each domain are the acid terminals of the odd numbered helices. These always end in the sequence (I, L, F, W) (D or E). Two of the three negative charges have been neutralized by mutations of D149 and E45. The D149S mutant produces a completely inactive AAC, whereas E45G still permits the mutant to grow on glycerol and also produce a high activity in the ATP synthesis rate and in the reconstituted exchange rates. Selection for revertants of D149S identified a mutant R252T that suggested a charge pair between those two amino acids. First, we find again that the mutations in domain I seem to have more serious consequences on the AAC function than those in domain I and III. This is reminiscent of the finding that incorporation of azido-ATP occurs also in the second domain. Two residues K179 and K179 + K182 were neutralized by mutations in the second domain. However, these mutations did not produce a glycerol⁻ strain although the ATP synthesis in mitochondria and the reconstituted exchange rates were reduced to only 15% of wild type.

So far, all the mutations occurred at repeat residues, i.e., at positions which approximately are repeated in the three domains. One striking exception is the intrahelical K38. Neutralization of this residue nearly completely inactivated ATP synthesis in mitochondria and also the exchange activity in the proteoliposomes. It is this residue which is more or less directly involved in the binding of carboxyatractylate and probably therefore also the substrates ADP or ATP (Bogner et al., 1986).

Charge pairs in the yeast AAC2

The yeast genetic system permits selection for regain of function mutations in non-functional AAC2 mutants. To date, every non-functional mutant (eight different residues) in AAC2 has involved a charged amino acid. To learn more about the structure-function relationships in the molecule, we used all eight of these mutants in a selection for second site revertants. Six of the eight have given second site revertants, and three of these identified charge pair interactions in the protein. We were not anticipating this result, which illustrates the power of yeast genetic methods in studying protein structure. The charge pairs detected were R294 plus E45, K38 plus E45 and R252 plus D149.

R294A was glycerol minus, but became glycerol plus three independent times by mutation of E45 to Gly(twice) or Gln(once). R252I never gave second site revertants, but selection for revertants of D149S identified a mutation R252T that restored function. Thus, D149 and R252 appear to be in a charge pair. Both of these residues are 100% conserved in all ADP/ATP carriers and the negative and positive charges are conserved in nearly all mitochondrial carriers, no matter what they transport. The negative charge is not conserved in four of 58 carriers, chromosome IX carrier, YKL522 and RIM2, all from yeast, and C2 from *C. elegans*. The C2 protein the RIM2 appear to be homologues. The positive charge is not conserved in 6 of 59 carriers, five are phosphate carriers and one is YKL522. It is remarkable that YKL522 has lost both charges and not just one. This suggests that retention of one of the two charges is detrimental to the structure of the protein. The phosphate carriers, the chromosome IX carrier, C2 and the RIM2 protein may form an alternative charge pair. The chromosome IX carriers, C2 and the RIM2 protein all have NPIW at this site, so they are probably closely related.

The mitochondrial carriers are noted for their internal triplicate sequence homology. The implications of finding a charge pair between two resides are that there may be other charge pairs between homologous positions. We tested this idea with R96 and D149 double mutants. These residues are homologous to R294 and E45. The double mutant was still glycerol minus, so the argument from homology did not hold up for this pair. There is a better chance that this will be true for the homologues of D149 and R252. D249 and K48 are 100% conserved in ADP/ATP carriers. The positive charge at K48 is conserved in 53/53 carriers, with no exceptions. The negative charge at D249 is conserved in 54/55 sequences, with the only exception being in a fragmentary sequence of the rat oxoglutarate malate carrier. This sequence has not appeared in the databases yet, and it could have a sequencing error at this point. Both the human and bovine oxoglutarate/malate carrier sequences have a negative charge. The other possible charge pair predicted by homology is E45 and R152. E45 is already involved in two charge pairs (see below), so a third may be unlikely, but this could also reflect the dynamic nature of the transport cycle. The negative charge at E45 is 100% conserved in 57 sequences that cover this position. The positive charge at R152 is conserved at 57/58 sequences. The only exception is *Oxytricha fallax*, and there is some sequence ambiguity at this point in the database entry GenEMBL:M63174.

The mutant K38A was also glycerol minus, but became glycerol plus when E45 mutated to Gly. This is the same mutation that restored R294A. It is unlikely that E45 is in a permanent interaction with K38 and R294. The result suggests that there are dynamic interactions between E45 and the two positive side chains. Both of the amino acid alpha carbons are in the middle portion of the membrane according to the topology model. Therefore, the rather long side chains of the K38 and 294 residues must be pointing down toward the matrix at least part of the time, and they probably fluctuate during transport. One could imagine that the c state and the m state may involve different charge pairs.

It is worth noting that the particular mutation used as a starting mutant is important for success. R252I never gave any revertants at D149 or anywhere else, even though R253I and R254I gave 15 different revertant mutations between them (Nelson et al., 1993). However, starting with D149S, a revertant was possible. This reflects the genetic code, and what is possible by a single point mutation. The Asp codon GAT could not mutate to Ser when R252I was selected, because it requires two nucleotide changes. Even if it could, it is not clear that D149S plus R252I would be glycerol plus. The combination of Ser at D149 and Thr at R252 may be critical for function. Because of this restriction, it is necessary to start selections for revertants from more than one mutant at a given site. Six mutations have been made at R96 and only two have resulted in second site revertants. R96L gave the revertant S33I. R96A gave three revertants D26V, C288Y, and Y305H. Another revertant arose starting from a double mutant R96T plus D149G. This mutant was constructed because we thought there might be a charge pair between R96 and D149 (homologous residues to R294 and E45). The construct was glycerol minus, but reverted by a third site mutation G30V. There is a pattern developing here that needs to be mentioned. D26, G30, S33, and Y305 are all sites of second site revertants from R253I and R254I (Nelson and Douglas, 1993). These same residues are being affected again by mutations in R96 and D149. It seems there is a mild change in the protein structure that is making the yeast glycerol minus. Similar effects are caused by mutations at R96, D149, R253 or R254. The side of the protein near the cytoplasmic surface of the inner membrane is the site of the alteration, and second site revertant mutations in this region compensate for the alteration. As was discussed earlier, the relative binding of ADP and ATP may be affected, causing the protein to exchange ADP for itself rather than for ATP. The region where most of the second site mutations are occurring may be involved in discrimination between these two substrates. Even though the D26V mutant caused the loss of a charge, we don't think this mutation is indicative of a charge pair between R96 and D26. There are multiple mutations that restore R96 mutants and none of the others involve charged residues. In addition, D26E is a revertant of R254I, suggesting that the geometry around the residue is important. There is probably a more subtle alteration than disruption of a charge pair involved here. If the D26V mutation is glycerol minus, then selection for revertants will be possible. If R96 is affected, then a charge pair between these two residues might be more likely.

Dependence on cardiolipin

One important spin-off in these mutations has been the discovery of a complete dependence on cardiolipin of the AAC function. Previously it had been shown that the isolated AAC of bovine heart contains six moles/mole AAC dimer tightly bound cardiolipin (Beyer and Klingenberg, 1985). Removal of the cardiolipin however caused irreversible

Table 7. Cardiolipin dependence of ADP/ATP exchange activity

Mutant	− CL	+ CL
μmol/min/g prot		
Wild	1050	1100
C73S	5	290
C244S	900	1050
C271S	420	600
K179I	10	350
K179I + K182I	10	200
R96H	120	130
R294A	600	630

inactivation and it was therefore not possible to conclude whether cardiolipin was only necessary to maintain the proper structure or whether it also is important for function. This could only be proven by a reversible removal and addition of cardiolipin in the reconstituted system. Such a case was first observed in the reconstitution of the C73S mutant of AAC2 (Hoffman et al., 1994). An absolute dependence on cardiolipin was found which could not be replaced by any other acidic phospholipid or cardiolipin analogues. This mutant is glycerol$^+$ and nearly as effective in oxidative phosphorylation as wild type. It was shown by NMR measurement that the isolated C73S AAC is deficient in cardiolipin as compared to the wild AAC, which does not require cardiolipin for reconstitution. A similar strong cardiolipin dependence was observed for the K179I + K182I mutants (Tab. 7). Also here, at least 5% of cardiolipin addition produces maximum activity in the proteoliposomes. This case is more interesting since basic groups in the AAC have to be postulated to be involved in the binding of the high amount of cardiolipin. K179 and K182 might be candidates for binding the acidic cardiolipin head groups at the membrane interface and therefore a mutation of these groups also leads consequently to a loss of cardiolipin.

It is clear that only AAC from glycerol$^+$ mutants shows sufficient exchange activity in order to demonstrate cardiolipin dependence. No other mitochondrial protein is known to have this absolute dependence on cardiolipin and also to bind cardiolipin in such excessive amounts. It seems reasonable to suggest on the basis of these results that most other mitochondrial carriers also require cardiolipin, although this has not yet been clearly demonstrated in reconstitution (Krämer and Palmieri, 1989; Stappen and Krämer, 1993). Possibly also these carrier preparations carry variable amounts of cardiolipin into the reconstitution, similar to the AAC preparations, and thus mask the dependence on cardiolipin. The high abundance of positive charges in these carriers and the high abundance of their occurrence, in particular of the AAC, may result in binding of the major portion of the mitochondrial cardiolipin to the mitochondrial carriers

Conclusions

The mitochondrial carriers seem to be structurally the simplest carriers known. Their low molecular weight and their pseudo threefold symmetry would provide for molecules with no or few accessory functions. The extremely wide variety of substrates handled by these similar structures from very large solutes, such as ATP, to the smallest H^+, illustrates the great flexibility in adapting to widely different substrates in these carriers. This is even true for the narrow mitochondrial carrier family.

The finding of similar sequence motifs in non mitochondrial proteins, i.e., within the superfamily of "mitochondrial" carriers may indicate that the precursor genes existed already before the arrival of mitochondria.

The three-dimensional structure of these proteins remains unknown because of their failure to crystallize in two or three dimensions. This unfortunate situation is common for nearly all typical membrane proteins. It seems that probably other methods, such as NMR, will have to intervene in this structural analysis although the barriers due to detergent requirements etc. are formidable. Therefore, until now the fundamental structure-function analysis had to be based on models. However, these models are becoming more and more limited by constraints from experimental results, in particular from site-directed mutagenesis. At any rate, the mitochondrial carriers have a good chance to remain on the frontier in the analysis of carrier mechanisms.

Acknowledgements
This work was supported by grants from Deutsche Forschungsgemeinschaft (Kl 134/24), from Fonds der Chemischen Industrie and from NATO.

References

Adams, M.D., Soares, M.B., Kerlavage, A.R., Fields, C. and Venter, J.C. (1993) Rapid cDNA sequencing (expressed sequence tags) from a directionally cloned human infant brain cDNA library. *Nature Genetics* 4: 373–380.

Aquila, H., Misra, D., Eulitz, M. and Klingenberg, M. (1982) Complete amino acid sequence of the ADP/ATP carrier from beef heat mitochondria. *Hoppe-Seyler's Z. Physiol. Chem.* 363: 345–349.

Aquila, H., Link, T.A. and Klingenberg, M. (1985) The uncoupling protein from brown fat mitochondria is related to the mitochondrial ADP/ATP carrier. Analysis of sequence homologies and of folding of the protein in the membrane. *EMBO J.* 4: 2369–2376.

Aquila, H., Link, T.A. and Klingenberg, M. (1987) Solute carriers involved in energy transfer of mitochondria from a homologous protein family. *FEBS Lett.* 212: 1–9.

Balogh, A.G., Ridley, R.G., Patel, H.V. and Freeman, K.B. (1989) Rabbit brown adipose tissue uncoupling protein mRNA: Use of only one of two polyadenylation signals in its processing. *Biochem. Biophys. Res. Commun.* 161: 156–161.

Beyer, K. and Klingenberg, M. (1985) ADP/ATP carrier protein from beef heart mitochondria has high amounts of tightly bound cardiolipin, as revealed by ^{31}P nuclear magnetic resonance. *Biochemistry* 24: 3821–3826.

Bogner, W., Aquila, H. and Klingenberg, M. (1989) The transmembrane arrangement of the ADP/ATP carrier as elucidated by the lysine reagent pyridoxal 5-phosphate. *Eur. J. Biochem.* 161: 611–620.

Brandolin, G., Boulay, F., Dalbon, P. and Vignais, P.V. (1989) Orientation of the N-terminal region of the membrane-bound ADP/ATP carrier protein explored by antibodies and an arginine-specific endoprotease: evidence that the accessibility of the N-terminal residues depends on the conformational state of the carrier. *Biochemistry* 28: 1093–1100.

Casteilla, L., Nougues, J., Reyne, Y. and Ricquier, D. (1991) Differentiation of bovine brown adipocyte precursor cells in a chemically defined serum-free medium. Importance of glucorticoids and age of animals. *Eur. J. Biochem.* 198: 195–199.

Chen, S.-T., Chang, C.-D., Huebner, K., Ku, D.-H., McFarland, M., DeRiel, J.D., Baserga, R. and Wurzel, J. (1990) A human ADP/ATP translocase gene has seven pseudogenes and localizes to chromosome X. *Somat. Cell Molec. Genet.* 16: 143–149.

Dalbon, P., Brandolin, G., Boulay, F., Hoppe, J. and Vignais, P.V. (1988) Mapping of the nucleotide-binding sites in the ADP/ATP carrier of beef heart mitochondria by photolabeling with 2-azido adenosine diphosphate. *Biochemistry* 27: 5141–5149.

Demolis, N., Mallet, L., Bussereau, F. and Jacquet, M. (1993) RIM2, MSl1 and PGl1 are located within and 8 kb segment of *S. cerevisiae* chromosome II, which also contains the putative ribosomal gene L21 and a new putative essential gene with a leucine zipper motif. *Yeast* 9: 797–806.

Durell, S.R. and Guy, H.R. (1992) Atomic scale structure and functional models of voltage-gated potassium channels. *Biophys. J.* 62: 238–250.

Eckerskorn, C. and Klingenberg, M. (1987) In the uncoupling protein from brown adipose tissue the C-terminus protrudes to the c-side of the membrane, as shown by tryptic cleavage. *FEBS Lett.* 226: 166–170.

Fiermonte, G., Runswick, M.J., Walker, J.E. and Palmieri, F. (1992) Sequence and pattern of expression of a bovine homologue of a human mitochondrial transport protein associated with Grave's disease. *DNA Sequence* 3: 71–78.

Gawaz, M., Douglas, M.G. and Klingenberg, M. (1990) Structure-function studies of adenine nucleotide transport in mitochondria. II. *J. Biol. Chem.* 265: 14202–14208.

Graf, R., Baum, B. and Braus, G.H. (1993) YMC1, a yeast gene encoding a new putative mitochondrial carrier protein. *Yeast* 9: 301–305.

Halestrap, A.P. and Denton, R.M. (1974) Specific inhibition of pyruvate transport in rat liver mitochondria and human erythrocytes by a-cyano-4-hydroxycinnamate. *Biochem. J.* 138: 313–316.

Hoffmann, B., Stöckl, A., Schlame, M., Beyer, K. and Klingenberg, M. (1994) The reconstituted ADP/ATP carrier activity has an absolute requirement for cardiolipin as shown in cysteine mutants. *J. Biol. Chem.* 269: 1940–1944.

Kaplan, R.S., Mayor, J.A. and Wood, D.D. (1993) The mitochondrial tricarboxylate transport protein. cDNA cloning, primary structure and comparison with other mitochondrial transport proteins. *J. Biol. Chem.* 268: 13682–13690.

Kaplan, R.S., Mayor, J.A. and Wood, D.O. (1993) The mitochondrial tricarboxylate transport protein. *J. Biol. Chem.* 269: 13682–13690.

Klingenberg, M. (1976) The ADP/ATP Carrier in Mitochondrial Membranes. *In:* A.N. Martonosi (ed): *The Enzymes of Biological Membranes*, Vol. 3, Plenum Publishing Corp., New York/London, pp. 383–438.

Klingenberg, M., Aquila, H. and Riccio, P. (1979) Isolation of Functional Membrane Proteins Related to or Identical with the ADP/ATP Carrier of Mitochondria. *In:* S. Fleischer and L. Packer (eds): *Methods in Enzymology*, Vol. VI, Academic Press, New York, pp. 407–414.

Klingenberg, M. and Winkler E. (1985) The reconstituted isolated uncoupling protein is a membrane potential driven proton translocator. *EMBO J.* 4: 3087–3092.

Klingenberg, M. (1989) Molecular aspects of the adenine nucleotide carrier from mitochondria. *Arch. Biochem. Biophys.* 270: 1–14.

Klingenberg, M. (1990) Mechanisms and evolution of the uncoupling protein of brown adipose tissue *TIBS* 15: 108–112.

Klingenberg, M., Gawaz, M., Douglas, M.G. and Lawson, J.E. (1992) Mutagenized ADP/ ATP Carrier from Saccharomyces. *In:* E. Quagliariello and F. Palmieri (eds): *Molecular Mechanisms of Transport*, Elsevier Science, Amsterdam, pp. 187–195.

Kozak, L.P., Britton, J.H., Kozak, U.C. and Wells, J.M. (1988) The mitochondrial uncoupling protein gene. Correlation of exon structure to transmembrane domains. *J. Biol. Chem.* 263: 12274–12277.

Krämer, R. and Palmieri, F. (1989) Molecular aspects of isolated and reconstituted carrier proteins from animal mitochondria. *Biochim. Biophys. Acta* 974: 1–23.

Krämer, R. and Palmieri, F. (1992) Metabolite carriers in mitochondria. *In:* L. Ernster (ed): *Molecular Mechanisms in bioenergetics*, Elsevier Science Pulishers, Amsterdam, pp. 359–384.

LaNoue, K., Mizani, M. and Klingenberg, M. (1978) Electrical imbalance of adenine nucleotide transport across the mitochondrial membrane. *J. Biol. Chem.* 253: 191–198.

LaNoue, K.F. and Schoolwerth, A.C. (1979) Metabolite transport in mitochondria. *Ann. Rev. Biochem.* 48: 871–922.

Lawson, J.E., Gawaz, M., Klingenberg, M. and Douglas, M.G. (1990) Structure-function studies of adenine nucleotide transport in mitochondria. I. *J. Biol. Chem.* 265: 14195–14202.

Lin, C.S., Hackenberg, H. and Klingenberg, M. (1980) The uncoupling protein from brown adipose tissue mitochondria is a dimer. A hydrodynamic study. *FEBS Lett.* 113: 304–306.

Louvi, A. and Tsitilou, S.G. (1992) A cDNA clone encoding the ADP/ATP translocase of Drosophila melanogaster shows a high degree of similarity with the mammalian ADP/ATP translocase. *J. Molec. Evol.* 35: 44–50.

Maloney, P.C. and Wilson, T.H. (1992) The evolution of membrane carriers. *Soc. Gen. Physiologists* 46: 148–160.

Marty, I., Brandolin, G., Gagnon, J., Brasseur, R. and Vignais, P.V. (1992) Topography of the membrane-bound ADP/ATP carrier assessed by enzymatic proteolysis. *Biochemistry* 31: 4058–4065.

McGivan, J.D. and Klingenberg, M. (1971) Correlation between proton and anion movement in mitochondria and the key role of the phosphate carrier. *Eur. J. Biochem.* 20: 392–399.

Mayinger, P. and Klingenberg, M. (1992) Labeling of two different regions of the nucleotide binding site of the uncoupling protein from brown adipose tissue mitochondria with two ATP analogs. *Biochemistry* 31: 10536–10543.

Mayinger, P., Winkler, E. and Klingenberg, M. (1989) The ADP/ATP carrier from yeast (AAC2) is uniquely suited for the assignment of the binding center by photoaffinity labeling. *FEBS Lett.* 244: 421–426.

Nelson, D.R., Lawson, J.E., Klingenberg, M. and Douglas, M.G. (1993) Site-directed mutagenesis of the yeast mitochondrial ADP/ATP translocator. Six arginines and one lysine are essential. *J. Mol. Biol.* 230: 1159–1170.

Nelson, D.R. and Douglas, M.G. (1993) Function based mapping of the yeast mitochondrial ADP/ATP translocator by selection for second site revertants. *J. Molec. Biol.* 230: 1171–1182.

Palmieri, F., Stipani, I., Quagliariello, E. and Klingenberg, M. (1972) Kinetic study of the tricarboxylate carrier in rat liver mitochondria. *Eur. J. Biochem.* 26: 587–594.

Runswick, M.J., Walker, J.E., Bisaccia, F., Iacobazzi, V. and Palmieri, F. (1990) Sequence of the bovine 2-oxoglutarate/malate carrier protein: Structural relationship to other mitochondrial transport proteins. *Biochemistry* 29: 11033–11040.

Runswick, M.J., Powell, S.J., Nyren, P. and Walker, J.E. (1987) Sequence of the bovine mitochondrial phosphate carrier protein: structural relationship to ADP/ATP translocase and the brown fat mitochondria uncoupling protein. *EMBO J.* 6: 1367–1373.

Saraste, M. and Walker, J.E. (1982) Internal sequence repeats and the path of polypeptide in mitochodrial ADP/ATP translocase. *FEBS Lett.* 144: 250–254.

Schuster, W., Kloska, S. and Brennicke, A. (1993) An adenine nucleotide translocator gene from *Arabidopsis thaliana*. *Biochim. Biophys Acta* 1172: 205–208.

Shinohara, Y., Kamida, M., Yamazaki, N. and Terada, H. (1993) Isolation and characterization of cDNA clones and a genomic clone encoding rat mitochondrial adenine nucleotide translocator. *Biochim. Biophys. Acta* 1152: 192–196.

Stappen, R. and Krämer, R. (1992) Functional properties of the reconstituted phosphate carrier from bovine heart mitochondria: Evidence for asymmetric orientation and characterization of three different transport modes. *Biochem. Biophys. Acta* 1149: 40–48.

Sullivan, T.D., Strelow, L.I., Illingworth, C.A., Phillips, R.L. and Nelson, O.E., Jr. (1991) Analysis of maize brittle 1 alleles and a defective suppressor-mutator-induced mutable allele. *Plant Cell* 3: 1337–1348.

Sulston, J., Du, Z., Thomas, K., Wilson, R., Hillier, L., Staden, R., Halloran, N., Green, P., Thierry-Mieg, J., Qiu, L., Dear, S., Coulson, A., Craxton, M., Durbin, R., Berks, M., Metzstein, M., Hawkins, T., Ainscough, R. and Waterson, R. (1992) The C. elegans sequencing project: a beginning. *Nature* 356: 37–41.

Walker, J.E. and Runswick, M.J. (1993) The mitochondrial transport superfamily. *J. Bioenerg. Biomembranes* 25: 435–445.

Winkler, E. and Klingenberg, M. (1992) Photoaffinity labeling of the nucleotide-binding site of the uncoupling protein from hamster brown adipose tissue. *Eur. J. Biochem.* 203: 295–304.

Wulf, R., Kaltstein, A. and Klingenberg, M. (1978) Proton and cation movements associated with ADP/ATP transport in mitochondria. *Eur. J. Biochem.* 82: 585–592.

Biochemistry of Cell Membranes
ed. by S. Papa & J. M. Tager
© 1995 Birkhäuser Verlag Basel/Switzerland

Metabolic conversions of NAD+ and cyclic ADP ribose at the outer surface of human red blood cells

E. Zocchi, L. Guida, L. Franco, U. Benatti, F. Malavasi[1]
and A. De Flora

*Institute of Biochemistry, University of Genoa and Advanced Biotechnology Center,
I-16132 Genoa*
[1]*Department of Genetics, Biology and Medical Chemistry, University of Turin and CNR,
I-10126 Turin, Italy*

Summary. In this review the metabolism of some conventional and unconventional adenine dinucleotides in human red blood cells (RBC) is discussed. Adenosine diphosphate ribose (ADPR) and Adenosine diphosphate ribulose (ADPRu) have been shown to be normal metabolites in RBC and both can bind specific cytoskeletal membrane proteins covalently. In the RBC cytosol, ADPR and ADPRu, besides being interconvertible, are degraded to AMP. Human RBC were demonstrated to display at their outer surface three enzyme activities involved in the turnover of NAD+, ADPR and a cyclic nucleotide, cyclic ADP ribose (cADPR), previously shown to be a new second messenger involved in calcium mobilization in several cell types. Immunopurification of these ectoenzymes showed that they are present at the extracellular portion of a single membrane-spanning 46 KDa glycoprotein (CD38), long known as a surface antigen of some lymphocyte subpopulations. These data raise a number of questions: (a) about the possible extracellular functions of cADPR in RBC and in other CD38+ cells and (b) about the possible mechanisms mediating these still undefined functions.

Introduction

Recently, considerable attention has been paid to ADP-ribose as an NAD+-derived covalent tag of several proteins in various cell types (Moss and Vaughan, 1990). Mono-ADPribosylation and de-ADPribosylation mechanisms proved to be quite a widespread system of protein modification, whose potential in cellular regulation is still largely undefined. Besides NAD+-dependent transferase mechanisms of ADP ribosylation, free ADP-ribose (ADPR) by itself is able to bind covalently to some target proteins. Non-enzymatic ADPribosylation has been putatively implicated in calcium release processes from mitochondrial pools (Richter et al., 1983). In addition, conclusive evidence for the occurrence of free ADPR as an intracellular metabolite was provided in human red blood cells (RBC) where hemoglobin prevents any artefactual formation of ADPR from NAD+ during procedures of acid extraction (Guida et al., 1992).

The unequivocal demonstration that free ADPR occurs at 0.4–0.5 µM concentrations in RBC stimulated further investigations aimed at defining its intraerythrocytic metabolism. It soon became apparent that the major pathway of ADPR conversion in the RBC cytosol is its hydrolysis to AMP and ribose 5-P, catalyzed by a dinucleotide pyrophosphatase which was partially purified and characterized (Zocchi et al., 1993a). Although this enzyme displays some activity on NAD(H), it shows a clear substrate preference toward ADPR and other adenine dinucleotides bearing multiple connecting pyrophosphate bonds and generally classified as "alarmones" (Lee et al., 1983a,b).

The pyrophosphatase-catalyzed breakdown of ADPR is inhibited by ATP. In the presence of ATP, it was possible to demonstrate, both in the RBC cytosol and in ADPR-loaded intact RBC, that metabolism of ADPR is diverted to yield ADPribulose (ADPRu) (Zocchi et al., 1993a). This new metabolite that is apparently formed by the pentose phosphate isomerase operating in the pentose phosphate pathway, was identified by means of combined techniques of mass spectrometry and NMR and of enzymatic analyses (Zocchi et al., 1993a; Franco et al., 1993). In addition, besides being formed in *in vitro* experiments, ADPRu was shown to be, like its isomer ADPR, a physiological metabolite in human RBC, where it is found at approximately 0.1 µM concentrations (Franco et al., 1993).

The biological role(s) of ADPR and of ADPRu and, more specifically, their functions in RBC are completely unknown. It is however of interest that both metabolites are able to covalently label some selected cytoskeletal membrane proteins and specifically bands 3, 4.1 and 4.2, and Glyceraldehyde 3-phosphate dehydrogenase (Ga3PDH) as well. The patterns of protein binding are identical for ADPR and ADPRu, both quantitatively and qualitatively. Their interaction with Ga3PDH is particularly intriguing, despite the apparently low stoichiometry of binding (Zocchi et al., 1993a), because this RBC protein has been recently reported to be the target of nitric oxide-stimulated NAD^+-dependent auto ADPribosylation (Kots et al., 1992). In fact, recent experiments demonstrated that this covalent modification of Ga3PDH is produced by NAD^+ *per se*, rather than by ADPR (Zocchi et al., 1993a), and that NO enhances NAD^+ binding through nitrosylation of cysteine at the active site (McDonald and Moss, 1993).

The functional implications of the linkage of ADPR and ADPRu to RBC membrane proteins are so far unknown. Binding of ADPR to purified Ga3PDH produces inhibition of this enzyme activity, yet with K_i values as high as 0.2 mM. The selectivity of interaction of both ADPR and ADPRu with membrane target proteins, together with the occurrence of their metabolism in the RBC cytosol only, may be related to control of RBC shape and deformability. This, however, has not yet been proved experimentally.

This review deals with a discussion on the metabolism of ADPR and ADPRu. Specifically, emphasis will be given to a structurally related metabolite, cyclic ADP-ribose (cADPR). This NAD$^+$ metabolite, discovered by Lee, has been identified in sea urchin eggs (Lee et al., 1989) and in several other cell types including mammalian ones (Rusinko and Lee, 1989). Growing interest toward cADPR is justified by its remarkable activity in mobilizing intracellular Ca^{2+} in both invertebrate and mammalian cell systems (Dargie et al., 1990; Koshiyama et al., 1991; Takasawa et al., 1993), by its role as agonist for the Ca^{2+}-induced Ca^{2+} release (CICR) mechanism (Lee, 1993), and by its possible role as a second messenger in response to extracellular stimuli (Takasawa et al., 1993). The data reported herein point to unexpected patterns of cADPR turnover in human RBC that may be useful to elucidate a number of still undefined processes taking place in other cells as well.

Discussion

Elucidation of the pathways of transformation of ADPR in RBC contrasts with the lack of knowledge about its mechanisms of synthesis. Availability of extracellular NAD$^+$ to the ectoenzyme NAD-glyco-hydrolase could produce ADPR on the outer surface of the erythrocyte membrane. Free ADPR, once formed, can permeate the RBC membrane to reach the cytosol (Kim et al., 1993). Indeed, NAD$^+$ is present in freshly drawn and immediately extracted plasma at 30–50 µM concentrations (L. Guida, unpublished data).

Another possible source of free ADPR in RBC may be de-ADPribosylation of membrane G proteins. This is suggested by the presence in human RBC of an enzyme catalyzing transfer of ADPR from NAD$^+$ to the α-subunit of G$_i$ and of the corresponding hydrolase to generate ADPR (Tanuma et al., 1988; Tanuma and Endo, 1990). The low amounts of G$_i$ and in general of G proteins in the RBC membrane (Tanuma and Endo, 1990) seem however to disfavor this pathway of ADPR synthesis, unless postulating a futile cycle of ADP-ribosylation and de-ADPribosylation of G$_{i\alpha}$ in certain conditions.

A further reasonable candidate for generation of ADPR in RBC is cyclic ADPR. Recently, its crystal structure has been worked out, thus resolving the controversial issue of the cyclization site (Lee et al., 1994). This has been unequivocally established to occur between N1 of the adenine ring and C1″ of the terminal ribose.

Despite the wide occurrence of cADPR in several cells, it has never been sought in RBC from any species. In an attempt to establish whether cADPR is present in RBC and whether it is a metabolic precursor of ADPR, we first assayed the two enzyme activities known to be responsible for its turnover in other cellular systems. These are (Fig. 1),

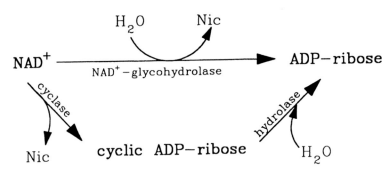

Figure 1. Enzyme activities involved in the turnover of cyclic ADP-ribose.

(i) ADP-ribosyl cyclase, catalyzing the conversion of NAD^+ to cADPR and nicotinamide and (ii) cADPR hydrolase, hydrolyzing cADPR to ADPR. As shown in Figure 1, the sum of the cyclase and cADPR hydrolase activities results in the hydrolysis of NAD^+ to ADPR and nicotinamide, thus reproducing what is generally considered an NAD^+-glycohydrolase reaction. Whether in RBC, as well as in other cell types, NAD^+-glycohydrolase really exists as a separate enzymatic entity or is rather the sum of the cyclase and of the cADPR hydrolase activities, is still an open question (see below).

Hemoglobin-free membranes from human erythrocytes display both ADP-ribosyl cyclase and cADPR hydrolase activities (Lee et al., 1993). The co-existence of the two activities precludes a precise estimate of the cyclase activity that is actually measured as rate of appearance of cADPR. This cyclic nucleotide is routinely measured by means of a specific Ca^{2+}-releasing assay on sea urchin egg homogenates (Lee et al., 1993) or by HPLC analysis. NAD^+-glycohydrolase activity is also present, as expected (see above). The three enzyme activities are located at the outer surface of the RBC membrane, as indicated by experiments on right-side-out resealed RBC ghosts and on intact erythrocytes as well (Lee et al., 1993).

Following these findings, in an attempt to separate the various enzyme activities occurring at the outer surface of RBC, we undertook their purification from Triton X-100-solubilized RBC membranes. For this purpose, we exploited the finding that ADP-ribosyl cyclase from *Aplysia californica* has considerable sequence homology with CD38, a lymphocyte surface antigen (Funaro et al., 1990). In addition, the amino acid sequence of both proteins showed a high content of histidine and cysteine residues, suggesting metal ion affinity chromatography as a potentially useful step of purification. Indeed, combination of this step with immunoaffinity chromatography on an immobilized anti-human CD38 MoAb resulted in the approximately 30 000-fold purification of a

46 KDa glycoprotein that was obtained in an apparently homogeneous form (Zocchi et al., 1993b). This M_r value and the selectivity of binding to anti-CD38 MoAb identify the purified protein as CD38. It is very interesting that during the various steps of purification the ratios between ADP-ribosyl cyclase, cADPR hydrolase, and NAD⁺-glyco-hydrolase activities keep fairly constant at $1:10:100$, respectively.

Conclusions

These data raise a number of questions beyond the biochemical characterization of CD38 in human RBC (Tab. 1). Indeed, CD38, previously held as a multilineage surface antigen of early and late stages of differentiation in lymphocyte subpopulations (Malavasi et al., 1992) and of the erythroid and myeloid progenitors in bone marrow, has been found also in other cell types. These include, besides human RBC, retinoic acid-differentiated HL-60 promyelocytic cells (Kontani et al., 1993) and pancreatic β-cells (Takasawa et al., 1993).

In RBC, CD38 displays at least two ectoenzyme activities, i.e., ADP-ribosyl cyclase and cADPR hydrolase. Whether CD38 also has intrinsic NAD⁺-glycohydrolase activity should be addressed by characterization of the catalytic mechanism through mapping of the active site and elucidation of kinetic properties of this multienzyme glycoprotein. It is significant, in this respect, that an NAD⁺-glycohydrolase purified to homogeneity from canine spleen microsomes and retrospectively identifiable as CD38, also displays ADP-ribosyl cyclase and cADPR activities (Kim et al., 1993).

Localization of CD38 at the outer surface of RBC and mapping of its enzyme activities at its ectoerythrocytic region suggest some extracellular function(s) for cADPR (Tab. 1). This possibility is rather unexpected, in view of the established role of cADPR as an intracellular

Table 1. Open questions on CD38 and cADPR

1. Is the NAD⁺-glycohydrolase reaction a true enzymatic activity or the combination of ADP-ribosyl cyclase and cADPR hydrolase?
2. Are the ectoenzyme activities of CD38 involved in intracellular metabolism upon CD38 internalization?
3. Is the extracellularly formed cADPR permeating through the membrane of RBC and other CD38⁺ cells?
4. Does the extracellularly formed cADPR act as a signal to be transduced across the membrane through specific receptors?
5. Are there extracellular functions of cADPR in RBC and other CD38⁺ cells (e.g., cell to cell interactions)?
6. Do the catalytic ectoenzyme activities of CD38 take place physiologically because of availability of NAD⁺?
7. Is the extracellular formation of cADPR regulated?

Ca^{2+} mobilizer. However, it cannot be excluded that catalytic activity of CD38 occurs intracellularly following internalization of CD38. This process might be triggered by physiological stimuli. Indeed, preliminary evidence seems to support disappearance of CD38 from the surface of cells upon interaction with agonistic Mabs (F. Malavasi, unpublished data). Search for physiological agonists present in extracellular fluids and mimicking these Mabs with respect to internalization can be fruitful for elucidating catalytic roles of CD38 inside the cells. Moreover, topological studies exploring active site(s) of CD38 during pathways of internalization in endocytic vesicles would also be required. Availability of NAD^+ as substrate of ADP-ribosyl cyclase is another related question that deserves investigation since endocytosis *per se* is expected to shield the catalytic site(s) from cytosolic NAD^+.

As mentioned, extracellular formation and metabolism of cADPR might suggest still unidentified roles of this nucleotide outside the cells (Tab. 1). This holds, irrespective of whether such extracellular functions occur in the same cADPR-producing cells or in adjacent cells by virtue of autocrine or paracrine mechanisms, respectively. In either case, occurrence of cADPR receptors may be a pre-requisite for such functions to occur. Binding of cADPR to putative ectocellular receptors could trigger either of two events: modulation of cells other than those displaying bound cADPR at their surface, or initiation of signal transduction across the plasma membrane for intracellular responses, taking place in the same cells bearing receptor-bound cADPR. In human RBC, the search for cADPR receptors at the outer surface has been unsuccessful so far (L. Guida, unpublished data), but this negative result might reflect a very low density of such receptors. The recent synthesis of photoactive analogs of cADPR that act as antagonists of cADPR-induced Ca^{2+} release in the sea urchin egg homogenates (Walseth et al., 1993) is certainly a valuable tool for exploring the presence of cADPR-binding proteins at the outer surface of RBC or of other $CD38^+$ cells.

Another, although not necessarily alternative hypothesis, would be influx of this nucleotide into cells as soon as it is produced at the cell surface (Tab. 1). Demonstration of cADPR uptake is not an easy task, however, since cADPR itself is rapidly converted by the cADPR hydrolase activity of CD38 to free ADPR at the outer surface and the latter nucleotide is then internalized by human RBC where it is eventually degraded to AMP (Kim et al., 1993).

Assuming that cADPR displays any functions, either extracellularly (cell to cell interactions?) or intracellularly (Ca^{2+} mobilization?), occurrence of its turnover at the external surface of $CD38^+$ cells seems to suggest possible mechanisms of regulation of ADP-ribosyl cyclase *versus* cADPR hydrolase (Tab. 1). Unrestrained operation of the two enzyme activities would in fact yield minute steady-state levels of cADPR at the outer side of cell membranes, and, in addition, the comparatively high

NAD$^+$-glycohydrolase activity would remove NAD$^+$ rapidly. We have two lines of evidence supporting a possible shift of the ratio of cyclase to cADPR hydrolase activities in favor of the former enzyme (Zocchi et al., 1993b); (i) Zn^{2+} selectively activates ADPribosyl cyclase; and (ii) activity of CD38 on 1,N^6-etheno-NAD$^+$ rather than on its natural substrate NAD$^+$, displaces the cyclase to cADPR hydrolase ratio from 0.01 to 2.0. Accordingly, formation of cADPR could prevail over its degradation to ADPR, although physiological mechanisms of such putative regulation are not yet known. On the other hand, cGMP has been reported to mobilize intracellular Ca^{2+} in sea urchin eggs by stimulating the ADP ribosyl cyclase in a still undefined way (protein phosphorylation?) (Galione et al., 1993). This finding has disclosed new regulatory pathways of cADPR metabolism because of the known activation of guanylate cyclase by NO. Whether this or related mechanisms may regulate extracellular ADP-ribosyl cyclase activity of CD38 selectively, leaving the cADPR activity unmodified, is not known so far.

CD38 ectoenzyme activities in physiological conditions require the presence of its substrate, i.e., NAD$^+$. As mentioned, NAD$^+$ is present at low but detectable levels in human plasma (L. Guida, unpublished data). These levels might be increased under certain physiological conditions, or at specific districts of the organism either through secretion of the dinucleotide from cells or as a result of cell death. While these fluctuations of plasmatic NAD$^+$ and the relevant changes in cADPR metabolism at the outer surface of RBC are purely speculative so far, similar alterations of NAD$^+$ levels in interstitial fluids (e.g., following apoptosis and consequent release of cellular metabolites) seem to be more likely to occur. Accordingly, cells characterized by higher CD38 expression (e.g., lymphocytes) could respond by enhancing the formation of cADPR.

All these considerations point to the need for intensive studies on structural, functional, metabolic, and regulatory properties of CD38 and its enzymatic activities before any clear conclusion on the several still unresolved aspects listed in Table 1 can be reached. Some facts now deserve attention: (1) For the first time the unknown roles of CD38 can be approached in terms of mechanisms, and enzymology is an important tool for this goal to be reached. (2) Although the ectocellular nature of its multiple enzyme activities is a puzzle complicating these studies further, it nevertheless represents an additional example of a steadily growing area of biochemistry ("ectobiochemistry"). (3) Once more, RBC provide invaluable help in elucidating functional aspects of other cells because of their metabolic simplicity. (4) cADPR might represent a trace metabolite involved in mediating processes of cell to cell communication much in the same way as more easily diffusable gaseous metabolites like NO and CO.

Acknowledgments
This study was supported in part by the Italian C.N.R. (Target Projects "Genetic Engineering" (ADF) and "Biotechnology and Bioinstrumentation" and "ACRO" (FM), by the Ministry of University and Scientific Research (ADF), and by EEC (Biotechnology #B102/CT92/0269 (FM)).

References

Dargie, P.J., Agre, M.C. and Lee, H.C. (1990) Comparison of Ca^{2+} mobilizing activities of cyclic ADP-ribose and inositol triphosphate. *Cell Regulation* 1: 279–290.

Franco, L., Guida, L., Zocchi, E., Silvestro, L., Benatti, U. and De Flora, A. (1993) Adenosine diphosphate ribulose in human erythrocytes: a new metabolite with membrane binding properties. *Biochem. Biophys. Res. Commun.* 190: 1143–1148.

Funaro, A., Spagnoli, G.C., Ausiello, C.M., Alessio, M., Roggero, S., Delia, D., Zaccolo, M. and Malavasi, F. (1990) Involvement of the multilineage CD38 molecule in a unique pathway of cell activation and proliferation. *J. Immunol.* 145: 2390–2396.

Galione, A., White, A., Willmott, N., Turner, M., Potter, B.V.L. and Watson, S.P. (1993) cGMP mobilizes intracellular Ca^{2+} in sea urchin eggs by stimulating cyclic ADP-ribose synthesis. *Nature* 365: 456–459.

Guida, L., Zocchi, E., Franco, L., Benatti, U. and De Flora, A. (1992) Presence and turnover of Adenosine diphosphate ribose in human erythrocytes. *Biochem. Biophys. Res. Commun.* 188: 402–408.

Kim, H., Jacobson, E.L. and Jacobson, M.K. (1993) Synthesis and degradation of cyclic ADP-ribose by NAD glycohydrolase. *Science* 261: 1330–1333.

Kim, U.H., Han, M.K., Park, B.H., Kim, H.L. and An, N.H. (1993) Function of NAD^+ glycohydrolase in ADP-ribose uptake from NAD by human erythrocytes. *Biochim. Biophys. Acta* 1178: 121–126.

Kontani, K., Nishina, H., Ohoka, Y., Takahashi, K. and Katada, T. (1993) NAD glycohydrolase specifically induced by retinoic acid in human leukemic HL-60 cells. Identification of the NAD glycohydrolase as leukocyte cell surface antigen CD38. *J. Biol. Chem.* 268: 16895–16898.

Koshiyama, H., Lee, H.C. and Tashjian, A.H., Jr. (1991) Novel mechanism of intracellular calcium release in pituitary cells. *J. Biol. Chem.* 266: 16985–16988.

Kots, A.Y., Skurat, A.V., Sergienko, E.A., Bulargina, T.V. and Severin, E.S. (1992) Nitroprusside stimulates the cysteine-specific mono (ADP-ribosylation) of glyceraldehyde-3-phosphate dehydrogenase from human erythrocytes. *FEBS Lett.* 300: 9–12.

Lee, H.C., Walseth, T.F., Bratt, G.T., Hayes, R.N. and Clapper, D.L. (1989) Structural determination of a cyclic metabolite of NAD^+ with intracellular Ca^{2+}-mobilizing activity. *J. Biol. Chem.* 264: 1608–1615.

Lee, H.C. (1993) Potentiation of calcium- and caffeine-induced calcium release by cyclic ADP-ribose. *J. Biol. Chem.* 268: 293–299.

Lee, H.C., Zocchi, E., Guida, L., Franco, L., Benatti, U. and De Flora, A. (1993) Production and hydrolysis of cyclic ADP-ribose at the outer surface of human erythrocytes. *Biochem. Biophys. Res. Commun.* 191: 639–645.

Lee, H.C., Aarhus, R. and Levitt, D. (1994) The crystal structure of cyclic ADP-ribose. *Nature Structural Biology* 1: 143–144.

Lee, P.C., Bochner, B.R. and Ames, B.N. (1983a) Diadenosine 5′,5‴-P1,P4-tetraphosphate and related adenylated nucleotides in *Salmonella typhimurium. J. Biol. Chem.* 258: 6827–6834.

Lee, P.C., Bochner, B.R. and Ames, B.N. (1983b) ApppppA, heat-shock stress, and cell oxidation. *Proc. Natl. Acad. Sci. USA* 80: 7496–7500.

Malavasi, F., Funaro, A., Alessio, M., De Monte, L.B., Ausiello, C.M., Dianzani, U., Lanza, F., Magrini, E., Momo, M. and Roggero, S. (1992) CD38: a multi-lineage cell activation molecule with a split personality. *Int. J. Clin. Lab. Res.* 22: 73-80.

Mc Donald, L.J. and Moss, J. (1993) Stimulation by nitric oxide of an NAD^+ linkage to glyceraldehyde-3-phosphate dehydrogenase. *Proc. Natl. Acad. Sci. USA* 90: 6238–6241.

Moss, J. and Vaughan, M. (1990) *ADP-ribosylating toxins and G proteins*, American Society for Microbiology, Washington, D.C.

Richter, C., Winterhalter, K.H., Baumhüter, S., Lötscher, H.R. and Moser, B. (1983) ADP ribosylation in inner membrane of rat liver mitochondria. *Proc. Natl. Acad. Sci. USA* 80: 3188–3192.

Rusinko, N. and Lee, H.C. (1989) Widespread occurrence in animal tissues of an enzyme catalyzing the conversion of NAD$^+$ into a cyclic metabolite with intracellular Ca^{2+}-mobilizing activity. *J. Biol. Chem.* 264:11725–11731.

Takasawa, S., Nata, K., Yonekura, H. and Okamoto, H. (1993) Cyclic ADP-ribose in insulin secretion from pancreatic β cells. *Science* 259: 370–373.

Tanuma, S., Kawashima, K. and Endo H. (1988) Eukaryotic mono (ADP-ribosyl) transferase that ADP-ribosylates GTP-binding regulatory G$_i$ protein. *J. Biol. Chem.* 263: 5485–5489.

Tanuma, S. and Endo, H. (1990) Identification in human erythrocytes of mono(ADP-ribosyl) protein hydrolase that cleaves a mono(ADP-ribosyl) G$_i$ linkage. *FEBS Lett.* 261: 381–384.

Walseth, T.F., Aarhus, R., Kerr, J. and Lee, H.C. (1993) Identification of cyclic ADP-ribose-binding proteins by photoaffinity labeling. *J. Biol. Chem.* 268: 26686–26691.

Zocchi, E., Guida, L., Franco, L., Silvestro, L., Guerrini, M., Benatti, U. and De Flora, A. (1993a) Free ADP-ribose in human erythrocytes: pathways of intra-erythrocytic conversion and non-enzymic binding to membrane proteins. *Biochem. J.* 295: 121–123.

Zocchi, E., Franco, L., Guida, L., Benatti, U., Bargellesi, A., Malavasi, F., Lee, H.C. and De Flora, A. (1993b) A single protein immunologically identified as CD38 displays NAD$^+$ glycohydrolase, ADP-ribosyl cyclase and cyclic ADP-ribose hydrolase activities at the outer surface of human erythrocytes. *Biochem. Biophys. Res Commun.* 196: 1459–1465.

Biochemistry of Cell Membranes
ed. by S. Papa & J. M. Tager
© 1995 Birkhäuser Verlag Basel/Switzerland

Mg^{2+} Homeostasis in cardiac ventricular myocytes

A. Romani, C. Marfella and A. Scarpa

Department of Physiology and Biophysics, Case Western Reserve University, School of Medicine, Cleveland, OH 44106-4970, USA

Introduction

Magnesium ion (Mg^{2+}) is abundantly represented in mammalian and non mammalian cells. The concentration of total Mg^{2+} in cardiac ventricular myocytes ranges between 17 and 11 mM (Polimeni and Page, 1973; Paradise et al., 1978; Gupta et al., 1984; Wu et al., 1981; Hess et al., 1982; Blatter and McGuigan, 1986; Fry, 1986; Masuda et al., 1990; Romani and Scarpa, 1992; Romani et al., 1993; Murphy et al., 1989), depending on the technique used and the animal species investigated. The majority of cellular Mg^{2+} is present as a complex with ATP (Polimeni and Page, 1974), bound to cytosolic proteins (Page and McCallister, 1973) or metabolites (Garfinkel et al., 1986), or internalized within intracellular structures (Polimeni and Page, 1974; Garfinkel et al., 1986), so that only a small fraction of total Mg^{2+} is *free* within the cytosol. Yet, because of the indirect techniques used (Paradise et al., 1978; Hess et al., 1982; Blatter and McGuigan, 1986; Fry, 1986) and/or the difficulty in proper quantitation (Polimeni and Page, 1973; Gupta et al., 1984; Hess et al., 1982; Fry, 1986; Romani et al., 1993; Murphy et al., 1989), the values of cytosolic free Mg^{2+} reported to be present in cardiac ventricular myocytes under physiological conditions vary considerably. Table 1 summarizes data from various laboratories which measured the absolute value for cytosolic Mg^{2+} by using the equilibria of Mg^{2+}-sensitive cellular enzymes (Masuda et al., 1990), or the spectral shift of ATP by ^{31}P NMR (Gupta et al., 1984), or fluorescent indicators (Murphy et al., 1989), or selective Mg^{2+} electrodes and intracellular recordings (Hess et al., 1982; Blatter and McGuigan, 1986; Fry, 1986), or $^{28}Mg^{2+}$ distribution (Polimeni and Page, 1973; Romani et al., 1993). Although the basal value of Mg^{2+} broadly ranges between 0.4 and 3.5 mM, it is unanimously accepted that the cytosolic free Mg^{2+} concentration does not change at all, or changes only minimally, even under drastic hormonal stimulation (Murphy et al., 1989; Murphy et al., 1991).

Table 1. Total and free magnesium in mammalian cardiac cells

Total (Mg^{2+})	Free (Mg^{2+})	Method	Reference
17 mM	1.0 mM	28 Mg	(Polimeni and Page, 1973)
	1.5–3.6 mM	Mg-flux	(Paradise et al., 1978)
	0.8 mM	31P–NMR	(Gupta et al., 1984)
	2.5 mM	31P–NMR*	(Wu et al., 1981)
	3.0–3.5 mM	Mg^{2+}-electrode	(Hess et al., 1982)
	0.4 mM	Mg^{2+}-electrode	(Blatter and McGuigan, 1986)
	2.4 mM	Mg^{2+}-electrode	(Fry, 1986)
17 mM	0.44 mM	13C–NMR citrate/ isocitrate ratios	(Masuda et al., 1990)
11–12 mM		28 Mg redistribution	(Romani and Scarpa, 1990)
11–12 mM	0.5–0.7 mM	atomic absorbance	(Romani et al., 1993)
	0.5 mM	Mag-Fura	(Murphy et al., 1989)

*It has been demonstrated by Gupta et al., 1984, and also by other groups, that a wrong K_d has been used to estimate these data.

Recently, a large redistribution of total Mg^{2+} to or from cellular compartments has been identified and characterized in perfused rat heart and in isolated cardiac ventricular myocytes (Romani et al., 1993; Vormann and Gunther, 1987; Romani and Scarpa, 1990). Under conditions that may be physiologically occurring *in vivo*, such as hormonal stimulation, it was observed that approx. 5–10% of total Mg^{2+} is extruded from or accumulated in the cell within a few min (Romani et al., 1993; Vormann and Gunther, 1987; Romani and Scarpa, 1990). The observation that the amount of Mg^{2+} transported across the membrane is equal to or larger than the whole content of Mg^{2+} free within the cytosol and that negligible changes in cytosolic free Mg^{2+} content are detectable, lead to the conclusion that Mg^{2+} redistributes among intracellular organelles, and that the observed fluxes of Mg^{2+} across the plasma membrane change the total but not the free cellular Mg^{2+}.

Mg^{2+} Efflux from rat hearts perfused in a Langendorff system

Figure 1A shows a typical experiment of Mg^{2+} efflux from a perfused rat heart stimulated by an adrenergic agonist. Mg^{2+} content in the perfusate was measured by atomic absorbance spectrophotometry (AAS). To gain sufficient sensitivity for AAS measurements, 10 min before 0 time of the trace, the perfusion medium was switched to one containing no added Mg^{2+}. Each bar of the trace represents the amount of Mg^{2+} (released plus contaminant) present in the perfusate, collected every 20 s and measured by AAS. Mg^{2+} content in the perfusate was approx. 7.5 µM at the beginning of the experiment and decreased to 3–4 µM during the 30-min wash-out period. By contrast, as shown in

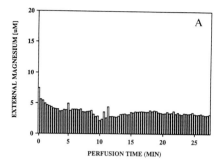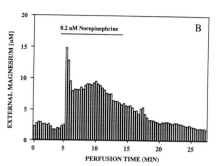

Figure 1. Mg²⁺ efflux in perfused control (Fig. 1A) and stimulated (Fig. 1B) rat hearts. A typical experiment out of seven for the control and out of three for the stimulated rat hearts is reported. Male Sprague-Dawley rats (220–250 g) were used as heart donors. The hearts were removed and perfused in a Langendorff open system (no recirculation), at a rate of 7 ml/min, with the following medium: NaCl 120 mM, KCl 3 mM, Na-HEPES 10 mM, glucose 10 mM, KH_2PO_4 1.2 mM, $NaHCO_3$ 12 mM, $CaCl_2$ 1.2 mM, $MgCl_2$ 1.2 mM (pH 7.2), equilibrated with $O_2:CO_2$, 95:5 v/v, at 37°C. As reported in the text, 10 min before the 0 time indicated in the figure, the perfusion medium was substituted with one having the same composition, but no added Mg²⁺ (*Mg²⁺-free medium*). The perfusate was collected every 20 s and the Mg²⁺ content measured by atomic absorbance spectrophotometry (AAS). Mg²⁺ content in the *Mg²⁺-free medium* (contaminant plus wash-out) was measured by AAS and found to range between 5 and 10 μM. Where indicated in Figure 1B, 0.2 μM norepinephrine was directly added to the *Mg²⁺-free medium*.

Figure 1B, the perfusion with 0.2 μM norepinephrine (NE) results in an extrusion of Mg²⁺ from cardiac cells. This efflux (three- to fourfold more Mg²⁺ in the perfusate) continued for approx. 5–8 min after NE addition. NE appears to induce Mg²⁺ efflux in a dose-dependent manner, the smallest effective dose being 0.04 μM and the largest (approx. twofold the amount extruded in Figure 1B, data not shown) at 1 μM NE. Qualitatively similar results were also observed in hearts perfused with epinephrine or isoproterenol.

The sequential stimulation of a perfused heart by small additions of NE (0.2 μM each) resulted in progressively reduced Mg²⁺ effluxes (Romani and Scarpa, 1990). Originally, we hypothesized that these results could depend either on the down regulation of the adrenoceptors or on the depletion of an intracellular Mg²⁺ pool(s) (Romani and Scarpa, 1990). More recent evidence would indicate that the latter explanation is likely to be the more accurate.

Cardiac cells possess both α and β adrenergic receptors (Bode and Brunton, 1989). Therefore, α and β blocking agents (Weiner, 1985) were used to discriminate the class of receptor involved.

Figure 2A shows that the $α_2$-antagonist yohimbine has little effect in preventing the NE-induced Mg²⁺ efflux from perfused hearts. Identical results were also obtained by perfusing the heart with the $α_1$-antagonist prazosin (not shown). When the concentration of yohimbine or prazosin or phentolamine was increased by 10 fold over the dose of NE, the

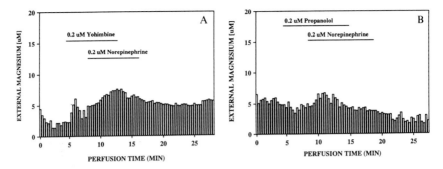

Figure 2. Mg^{2+} efflux in perfused rat hearts pre-treated with 0.2 μM yohimbine (Fig. 2A) or 0.2 μM propanolol (Fig. 2B) and stimulated by 0.2 μM norepinephrine. One experiment typical of two for both inhibitors is reported. The perfusion of the hearts and the measurements of Mg^{2+} efflux were as reported in the legend of Figure 1. The α$_2$-antagonist yohimbine or the non selective β-antagonist propanolol were directly added to the *Mg^{2+}-free medium* 5 min before the addition of norepinephrine.

Mg^{2+} efflux was inhibited less than 50% (Romani et al., 1990). On the contrary, when the hearts were perfused with 0.2 μM or 10 μM propanolol (a non-selective β-blocker), the Mg^{2+} efflux induced by NE was largely reduced (Fig. 2B) or completely abolished (Romani and Scarpa, 1990), respectively. These results were confirmed in isolated ventricular myocytes (Romani and Scarpa, 1990), supporting the conclusion that the activation of β-adrenergic receptors prompts a major extrusion of Mg^{2+} from cardiac cells.

Since the β-adrenergic stimulation affects the heart inotropism or chronotropism, it could be possible that the observed Mg^{2+} efflux is the consequence of a change in cardiac cell contractility. However, when the hearts were paced at variable frequencies through A–V node stimulation, no significant extrusion of Mg^{2+} was observed (unpublished results). Moreover, when these hearts were perfused with NE, a Mg^{2+} efflux qualitatively similar to that reported in Figure 1B was observed, independent of the rate at which the hearts were paced (unpublished results). The fact that Mg^{2+} efflux occurs independently of cardiac cell contraction is strengthened by the experiments illustrated in Figure 3, showing that the increase in heart contraction induced by 2 mM caffeine (Fig. 3A) or 20 mM KCl (Fig. 3B) does not elicit Mg^{2+} efflux.

To exclude the possibility that Mg^{2+} mobilization was due to a NE-induced change in plasma membrane permeability or to the death of a relevant percentage of cell population, we measured both the release of cytosolic enzyme lactate dehydrogenase and cellular K$^+$ in the extracellular compartment. Following the addition of NE, no detectable increase in lactic dehydrogenase or K$^+$ content in the perfusate was observed (data not shown).

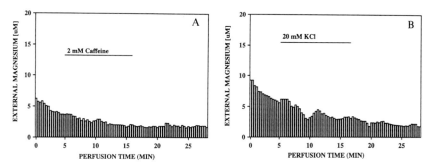

Figure 3. Mg²⁺ efflux in rat hearts perfused with caffeine (Fig. 3A) or KCl (Fig. 3B). A typical experiment out of two for both caffeine and KCl is reported. Heart perfusion and Mg²⁺ measurements were performed as previously described (see legend of Fig. 1). Caffeine (2 mM, Fig. 3A) or KCl (20 mM, Fig. 3B) were directly added to the *Mg²⁺-free medium*.

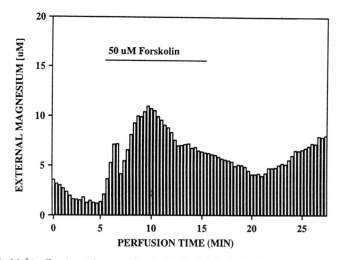

Figure 4. Mg²⁺ efflux in rat heart perfused with 50 μM forskolin. One typical experiment out of three is reported. Experimental procedures and Mg²⁺ determinations were as reported in the legend to Figure 1.

The stimulation of β-adrenergic receptors by NE or isoproterenol increases the cytosolic level of cAMP (Jakobs et al., 1979; Nishizuka, 1984). The direct or indirect involvement of this intracellular second messenger in Mg²⁺ extrusion was investigated by perfusing rat hearts or by stimulating isolated cardiac ventricular myocytes with forskolin or permeant cAMP analogs. As Figure 4 shows, the perfusion of isolated hearts with forskolin prompts a release of Mg²⁺ comparable to that obtained with NE. Similar results were also obtained when the hearts were perfused with the permeant cAMP analogs dibutyril-cAMP or 8-Cl-cAMP (data not shown).

The experiments described permit us to draw the following conclusions:

1. The addition of norepinephrine or isoproterenol elicits a large Mg^{2+} efflux from rat perfused hearts.
2. This efflux is specific and is not secondary to an increase in heart contractility, heart rate or heart depolarization.
3. It is the consequence of the rise in cytosolic cAMP level which follows the stimulation of β-adrenergic receptors.

Mg^{2+} Fluxes in isolated cardiac ventricular myocytes

Because of the heart cellular heterogeneity and because of possible variations in perfusion rate and cell metabolism, the amount of Mg^{2+} extruded from perfused hearts following the β-adrenergic stimulation by NE or isoproterenol is difficult to quantitate. These limitations can be overcome by the use of a preparation of collagenase-dispersed rat ventricular myocytes. Isolated myocytes respond to the stimulation by NE, isoproterenol, forskolin or permeant cAMP analogs by extruding an amount of Mg^{2+} which is comparable to that previously observed in perfused hearts (approx. 5–8% of the total cellular Mg^{2+} content) (Romani and Scarpa, 1992; Romani et al., 1993; Romani and Scarpa, 1990).

The Mg^{2+} efflux from isolated cardiac ventricular myocytes occurs within 5 min stimulation by NE, adrenergic agonists, forskolin or permeant cAMP analogs (i.e., dibutyryl-cAMP, 8-Cl-cAMP or 8-B-cAMP) (Romani and Scarpa, 1990). As in the case of perfused hearts, this efflux is almost completely prevented by the β-blocking agents propanolol or sotalol, but it is only slightly affected by the α-blocking agents yohimbine or prazosin (Romani and Scarpa, 1990; Romani et al., 1990).

It has to be noted that in all the experiments described so far (i.e., perfused hearts or isolated myocytes) Mg^{2+} was measured in the perfusate or in the supernatant by AAS. Although accurate and selective, this technique cannot be used in the presence of a physiological concentration of Mg^{2+} in the perfusate or in the extracellular fluid. Hence, to gain the necessary sensitivity to measure Mg^{2+} in the supernatants, the hearts or the isolated cells were perfused or incubated, respectively, in the presence of unphysiologically low concentrations of extracellular Mg^{2+} (5–20 μM).

Therefore, it is in principle possible that a low concentration of external Mg^{2+} can lead to an unnatural gradient for the cation across the cell plasma membrane, which may favor or amplify the observed Mg^{2+} efflux.

The use of ^{28}Mg, which is rarely available, allowed us to investigate the observed Mg^{2+} efflux in the presence of a physiological concentra-

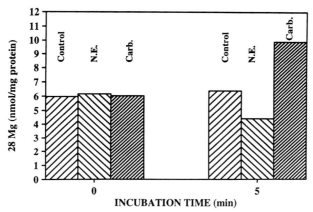

Figure 5. ²⁸Mg content in cardiac ventricular myocytes following the stimulation by nor-epinephrine or carbachol. Isolated cardiac ventricular myocytes were prepared according to the procedure of De Young et al., 1989, and resuspended in the medium reported in the legend to Figure 1, in the presence of 1.2 mM $MgCl_2$ labeled with 1 µCi/ml ²⁸Mg. After the radioisotopic equilibrium was achieved (approx. 3 h), excess radioisotope was removed by washing the cells three times with a medium having the same composition, but containing only "cold" $MgCl_2$. Cells were incubated therein at 37°C in the presence of a slow flow of $O_2:CO_2$ (95:5 v/v), at the final protein concentration of approx. 350 µg/ml. One µCi/ml ²⁸Mg was added to the incubation medium to detect Mg²⁺ accumulation. At selected times after the addition of 10 µM norepinephrine (NE) or 100 µM carbachol, aliquots of the incubation mixture were withdrawn in quadruplicate. The sedimentation of the cells was performed in microfuge tubes through an oil layer, as described by Grubbs et al., 1989. Both supernatant and oil layer were removed by vacuum suction and the radioactivity in the pellet was measured by liquid β-scintillation counting. The data are means ± S.E. of two different preparations, each one performed in duplicate.

tion of extracellular Mg²⁺ (approx. 1.2 mM). Myocytes were pre-loaded with ²⁸Mg²⁺ for 3 h, a period of time sufficient to achieve isotopic equilibrium between ²⁸Mg and cellular Mg²⁺. As Figure 5 shows, before stimulation, the samples contain similar amounts of Mg²⁺. Whereas control myocytes did not present significant changes in Mg²⁺ content within the period of incubation (5 min at 37°C), the samples stimulated by NE (Fig. 5) or forskolin (not shown) lost a considerable amount of ²⁸Mg (approx. 2–3 nmol ²⁸Mg/mg protein).

Interestingly, the treatment of cardiac myocytes with carbachol, which has been reported to decrease *in vitro* the activity level of adenylate cyclase (Jakobs et al., 1979), favors an accumulation of ²⁸Mg within the cells. The activation of a similar Mg²⁺ influx by carbachol was also observed in myocytes incubated in the presence of contaminant external Mg²⁺ (5–10 µM, Romani et al., 1993; Romani and Scarpa, 1990) and in rat hearts perfused with 50 µM Mg²⁺ in the perfusate (Romani et al., 1993).

Protein kinase C mediated Mg^{2+} uptake in cardiac myocytes

Since a natural cross-talk between cAMP and protein kinase C (pkC) activities exists in many cells (Nishizuka, 1984; Nishizuka, 1986), the observed Mg^{2+} accumulation in cardiac myocytes could be due to a decrease in adenylate cyclase activity, but also to the imbalance of cAMP/pkC equilibrium or/and the direct activation of pkC.

As we have reported elsewhere (Romani et al., 1992), a significant Mg^{2+} influx occurs across the sarcolemma of myocytes treated with agents which stimulate or mimic the action of pkC. Some of these results are reported in Figure 6A. This figure shows that phorbol 12,13-dibutyrate (PDBu), phorbol 12-myristate 13-acetate (PMA), 1-oleoyl 2-acetyl-*sn*-glycerol (OAG) and 1-stearoyl 2-arachidonoyl-*sn*-glycerol (SAG) induce a Mg^{2+} influx in cardiac ventricular myocytes which is qualitatively similar to that stimulated by carbachol. On the other hand myocytes incubated in the presence of the inhibitors of protein kinace C staurosporine or 1-(5-iso-quinoline-sulfonyl)-2-methylpiperazine dihydrochloride (H7) release amounts of cellular Mg^{2+} comparable to those extruded following the stimulation with NE, forskolin or permeant cAMP analogs (Fig. 6B).

These data, together with other published observations (Romani et al., 1992; and also Tab. 2), indicate an essential role of protein kinase C in inducing Mg^{2+} accumulation in cardiac cells. Also in the case of accumulation, the amount of Mg^{2+} transported across the sarcolemma is considerable, and corresponds to approx. 5–10% of the total cellular Mg^{2+} content. This observation and the absence of any significant change in cytosolic free Mg^{2+} (unpublished observations) imply that the Mg^{2+} transported across the plasma membrane is either rapidly redis-

Table 2. Magnesium transport in cardiac ventricular myocytes before and after down regulation of protein kinase C activity

	Incubation time (min)	
	0	5
After 2 h without PMA		
Control	42.96 ± 0.44	43.88 ± 0.56
10 µM norepinephrine	42.16 ± 0.68	52.92 ± 1.56
20 nM PMA	42.44 ± 0.64	35.24 ± 0.68
100 µM carbachol	42.04 ± 0.64	35.04 ± 0.44
After 2 h with PMA		
Control	43.56 ± 0.72	42.68 ± 1.04
10 µM norepinephrine	43.64 ± 0.20	52.20 ± 1.36
20 nM PMA	43.84 ± 0.60	42.92 ± 0.64
100 µM carbachol	42.52 ± 0.28	42.40 ± 0.12

Extracellular Mg^{2+} is expressed as nmol/million cells. Cells were incubated for 120 min in the absence or in the presence of 250 µM PMA. Data are mean ± S.E. of three different preparations.

A

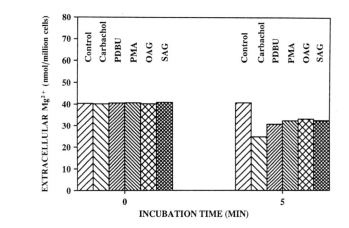

B

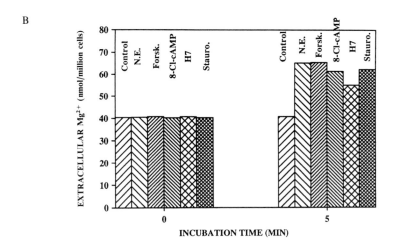

Figure 6. Mg^{2+} influx (Fig. 6A) and efflux (Fig. 6B) in cardiac ventricular myocytes. Cardiac ventricular myocytes were isolated according to the procedure of De Young et al., 1989. Cells were incubated in the *Mg^{2+}-free medium* reported in the legend to Figure 1, equilibrated with O$_2$:CO$_2$ (95:5 v/v) at 37°C. A final protein concentration of approx. 350 µg/ml was used. The stimulatory effect of 100 µM carbachol, 1 µM PDBU, 20 nM PMA, 20 µM OAG or 20 µM SAG on Mg^{2+} influx is reported in Figure 6A. The stimulatory effect of 10 µM nor-epinephrine, 50 µM forskolin, 100 µM 8-Cl-cAMP, 50 µM H7 or 50 nM staurosporine on Mg^{2+} efflux is reported in Figure 6B. At the times reported in the figures, aliquots of the incubation mixture were withdrawn in duplicate. After sedimentation of the cells in microfuge tubes, the Mg^{2+} content of the supernatants was measured by AAS. Data are means ± S.E. of four different preparations.

tributed within intracellular organelles or, alternatively, is in a form of complex with cytosolic components (protein or phosphonucleotide). Actually, it can be hypothesized that the activation of protein kinase C may result in the coordinated operation of two distinct transporters, located in the plasma membrane and in the membrane of intracellular organelles, respectively.

Intracellular mechanism(s) involved in Mg^{2+} extrusion

As previously described, the addition of NE or isoproterenol induces a time- and dose-dependent release of Mg^{2+} from cardiac cells through the activation of β-adrenoceptors and the increase in cellular cAMP. The efflux reaches a maximum within 5 min of stimulation, and then returns to basal levels. The magnitude of the efflux is not affected by changes in external Mg^{2+} concentration (Romani et al., 1993), and corresponds to approx. 5–10% of total cellular Mg^{2+} content.

The amount of Mg^{2+} mobilized by the addition of 1 μM NE in perfused rat hearts accounts for approx. 0.8–1 mM (Romani et al., 1993; Romani and Scarpa, 1990), a concentration which is equal or larger than the concentration of Mg^{2+} reported to be *free* in the cytosol of cardiac ventricular myocytes (see Tab. 1). This observation and the finding that no appreciable variations in cytosolic free Mg^{2+} follow the stimulation by adrenergic agonists (Romani et al., 1993; Murphy et al., 1991; Romani and Scarpa, 1990) suggest that the releasable Mg^{2+} is mobilized, at least partially, from an intracellular store.

The extensive investigation of this aspect in cardiac and liver cells (Romani et al., 1993; Romani et al., 1991) would indicate the mitochondria as the major pool from where Mg^{2+} is mobilized upon adrenergic stimulation (Romani et al., 1993; Romani et al., 1991).

We addressed this problem by permeabilizing cardiac ventricular myocytes with digitonin and incubating, thereafter, the cells in a *cytosol-like* medium (Romani et al., 1993). Under these experimental conditions, the control samples do not present variations in external Mg^{2+} level (Fig. 7) and the stimulation by NE, isoproterenol or forskolin was, as expected, ineffective in inducing Mg^{2+} efflux (data not shown). In contrast, as shown in Figure 7, the addition of nanomolar concentrations of cAMP, consistent with the level of second messenger reached under adrenergic stimulation (Morgan et al., 1983; Graham et al., 1987), elicits a time dependent Mg^{2+} release (approx. 10 nmol/10^6 cells over 6 min stimulation). Because of the permeabilization by digitonin, the sarcolemma is functionally abolished in terms of permeability, and several cytosolic components, including the adenylate cyclase, are lost from the cells. Hence, the Mg^{2+} extruded following the direct application of cAMP should be mobilized from an intracellular compartment(s). This result and the effectiveness of atractyloside and bongkrekate, two specific although structurally different inhibitors of mitochondrial adenine nucleotide translocase, to completely prevent

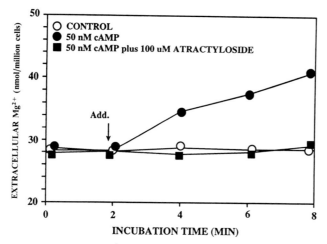

Figure 7. Mg^{2+} efflux from cardiac ventricular myocytes permeabilized by digitonin treatment and stimulated by 50 nM cAMP in the absence or in the presence of atractyloside. Isolated cardiac ventricular myocytes, prepared according to De Young et al., 1989, were permeabilized by digitonin treatment. Excess digitonin was removed by washing the cells three times with the following *cytosol-like medium*: KCl 100 mM, NaCl 3 mM, K-HEPES 10 mM, glucose 10 mM, KHCO$_3$ 10 mM, KH$_2$PO$_4$ 1.2 mM, MgCl$_2$ 0.8 mM, CaCl$_2$ 1 mM, EGTA 2 mM (pH 7.2), equilibrated with O$_2$:CO$_2$ (95:5 v/v), at 37°C. To perform Mg^{2+} determinations, the cells were incubated in the same medium, but in the adsence of added Mg^{2+}. ATP 1 mM, plus phosphocreatine 10 mM and 10 U/ml creatine phosphokinase as regenerating system, was added to the incubation system to prevent energy deficiency. Mg^{2+} content in the medium (contaminant plus carry-over) was measured by AAS and found to be approx. 2–10 µM. Where indicated, 50 nM cAMP was added to the system. At selected times, aliquots of the incubation mixture were withdrawn in duplicate. The cells were sedimented in microfuge tubes and the Mg^{2+} content of the supernatant was measured by AAS. Atractyloside (100 µM) was used as a selective inhibitor of mitochondrial adenine nucleotide translocase. Data are means ± S.E. of five different preparations.

cAMP or ADP induced Mg^{2+} efflux (Fig. 7 and also Romani et al., 1991) strongly argue for a mobilization of Mg^{2+} from the mitochondrial compartment.

Similar experiments were performed in isolated cardiac mitochondria. As Figure 8 shows, the addition of 50 nM cAMP (free acid form) to suspensions of isolated mitochondria induces a sizable Mg^{2+} release, having the same time dependence observed in permeabilized cells (see Figs 7 and 8 for a comparison). Also, in this experimental model the efflux could be induced by ADP and inhibited by the presence of atractyloside or bongkrekate in the incubation medium (unpublished observations).

Interestingly, only nanomolar concentrations of cAMP are required to induce Mg^{2+} efflux from permeabilized cells or isolated mitochondria, while ADP exerts the same effect at micromolar concentrations. In both cases, the amount of cAMP or ADP required to activate the efflux is consistent with the range of concentrations these agents attain in the

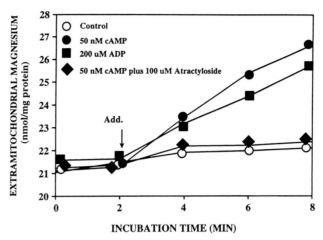

Figure 8. Mg^{2+} efflux from isolated cardiac mitochondria stimulated by 50 nM cAMP or 200 μM ADP, in the absence or in the presence of atractyloside. Isolated heart mitochondria were prepared according to Vinogradov et al., 1972. Their ADP/O$_2$ ratio was assessed polarographically (Estabrook, 1967). After isolation, mitochondria were resuspended at the final concentration of 40 mg protein/ml in the following medium: surcose 200 mM, KCl 10 mM, MgCl$_2$ 0.8 mM, TRIS 10 mM (pH 7.2) and kept in ice until used. To perform Mg^{2+} determinations, mitochondria were incubated, at the final concentration of 350 μg/ml, in the medium previously described, but in the absence of added Mg^{2+}, at room temperature. Mitochondria were stimulated by addition of 50 nM cAMP or 200 μM ADP. At selected time, aliquots of the incubation mixture were withdrawn in duplicate and the mitochondria were sedimented in microfuge tubes (10 000 × g × 2 min). Mg^{2+} content of the supernatants was measured by AAS. Atractyloside (100 μM) was used as a selective inhibitor of adenine nucleotide translocase. Data are means ± S.E. of five different preparations. In this as in the previous experiments, protein was measured according to the procedure of Bradford, 1976.

cytosol of cardiac myocytes (De Young et al,. 1989) under stimulatory conditions. Preliminary observations (Romani et al., 1991; Marfella et al., 1992) suggest that cAMP, via a direct binding, activates adenine nucleotide translocase and induces the extrusion of equivalent amounts of ATP and Mg^{2+} from the mitochondria.

Conclusions

The sequence of mechanisms involved in Mg^{2+} redistribution in cardiac cells, following an hormonal stimulation, can be summarized as follows:

a. The release of norepinephrine or epinephrine activates the β-adrenergic receptors of cardiac cells and, through the adenylate cyclase, enhances the cellular level of cAMP.
b. Cyclic AMP binds to mitochondrial adenine nucleotide translocase and induces the extrusion of Mg^{2+} and ATP from mitochondria.

c. Mg^{2+} accumulation, instead, appears to be mediated by the activation of muscarinic receptors, possibly through the activation of protein kinase C pathway and/or the down-regulation of the cAMP pathway.

Several aspects of the whole picture need to be further elucidated. In fact, it is uncertain if Mg^{2+} extrusion across the plasma membrane of cardiac cells depends on the increase in cellular cAMP level or, alternatively, on the increase in cytosolic ATP·Mg^{2+} content following their extrusion from mitochondria. In addition, no clear evidence has been provided about the nature of the extrusion mechanism located at the sarcolemma level, though the operation of an amiloride-sensitive Na$^+$/Mg^{2+} exchanger is inferred (Romani et al., 1993; Vormann and Gunther, 1987).

As for the influx pathway, the presence of physiological concentrations of external Na$^+$ and Ca^{2+} is required for a proper Mg^{2+} accumulation in cardiac myocytes (Romani et al., 1993). However, we lack detailed information of the nature of the inwardly directed transporter and the cellular compartment(s) involved in Mg^{2+} redistribution.

References

Blatter, L.A. and McGuigan, J.A.S. (1986) Free intracellular magnesium concentrations in ferret ventricular muscle measured with ion selective microelectrodes. *Quart. J. Exp. Physiol.* 71: 467–473.

Bode, D.C. and Brunton, L.L. (1989) Adrenergic, cholinergic, and other hormone receptors on cardiac myocytes. *In:* H.M. Piper and M.D. Isenberg (eds): *Isolated Adult Cardiomyocytes*, vol. I. CRC Press Inc., Boca Raton, Florida, pp. 163–201.

Bradford, M.M. (1976) A rapid and sensitive method for the quantitation of microgram quantities of protein utilizing the principle of protein-dye binding. *Biochemistry* 72: 248–254.

De Young, M.B., Giannattasio, B. and Scarpa, A. (1989) Isolation of calcium-tolerant atrial and ventricular myocytes from adult rat heart. *Meth. Enzym.* 173: 662–676.

Estabrook, R.W. (1967) Mitochondrial respiratory control and the polarographic measurements of ADP:O ratios. *Meth. Enzym.* 10: 41–47.

Fry, C.H. (1986) Measurement and control of intracellular magnesium ion concentration in guinea pig and ferret ventricular myocardium. *Magnesium* 5: 306–316.

Garfinkel, L., Altschuld, R.A. and Garfinkel, D. (1986) Magnesium in cardiac energy metabolism. *J. Mol. Cell. Cardiol.* 18: 1003–1013.

Graham, S.N., Herring, P.A. and Arinze, I.J. (1987) Age-associated alterations in hepatic β-adrenergic receptor/adenylate cyclase complex. *Am. J. Physiol.* 253: E277–E282.

Grubbs, R.D., Snavely, M.D., Hmiel, S.P. and Maguire, M.E. (1989) Magnesium transport in eukaryotic and prokaryotic cells using magneium-28 ion. *Meth. Enxym.* 173: 546–563.

Gupta, R.K., Gupta, P. and Moore, R.D. (1984) NMR studies of intracellular metal ions in intact cells and tissues. *Annu. Rev. Biophys. Bioenerg.* 13: 221–246.

Hess, P., Metzger, P. and Weingart, R. (1982) Free magnesium in sheep, ferret, and frog striated muscle at rest measured with ion selective microelectrodes. *J. Physiol.* 333: 173–188.

Jakobs, H.K., Aktories, K. and Schultz, G. (1979) GTP-dependent inhibitio of cardiac adenylate cyclase by muscarinic cholinergic agonists. *Naunyn-Schmiedeberg's Arch. Pharmacol.* 310: 113–119, 1979.

Marfella, C., Romani, A., Fatholahi, M. and Scarpa, A. (1992) cAMP induces Mg^{2+} efflux from rat liver and heart mitochondria by interacting with the adenine nucleotide translocase. *FASEB J.* 6: 2229.

Masuda, T., Dobson, G.P. and Veech, R.L. (1990) The Gibbs-Donnan near-equilibrium system of heart. *J. Biol. Chem.* 265: 20321–20334.

Morgan, N.G., Blackmore, P.F. and Exton, J.H. (1983) Age-related changes in the control of hepatic cycli AMP levels by $\alpha 1$- and $\beta 2$-adrenergic receptors in male rats. *J. Biol. Chem.* 258: 5103–5109.

Murphy, E., Freundenrich, C.C., Levy, L.A., London R.E. and Lieberman, M. (1989) Monitoring cytosolic free magnesium in cultured chicken heart cells by use of the fluorescent indicator furaptra. *Proc. Natl. Acad. Sci. USA* 86: 2981–2984.

Murphy, E., Freundenrich, G.G. and Lieberman, M. (1991) Cellular magnesium and Na/Mg exchange in heart cells. *Annu. Rev. Physiol.* 53: 273–287.

Nishizuka, Y. (1984) The role of protein kinase C in cell surface signal transduction and tumour promotion. *Nature* 308: 693–698.

Nishizuka, Y. (1986) Studies and perspectives of protein kinase C. *Science* 233: 305–312.

Page, E. and McCallister, L.P. (1973) A quantitative electron microscopic description of heart muscle cells: Application to normal, hypertrophied and thyroxin-stimulated hearts. *Am. J. Cardiol.* 31: 172–181.

Paradise, N.F., Beeler, G.W., Jr. and Visscher, M.B. (1978) Magnesium net fluxes and distribution in rabbit myocardium in irreversible contracture. *Am. J. Physiol.* 234: C115–C121.

Polimeni, P.I. and Page, E. (1973) Magnesium in heart muscle. *Circ. Res.* 4: 367–374.

Polimeni, P.I. and Page, E. (1974) Further observations on magnesium transport in rat ventricular muscle. *In:* N.S. Dhalle (ed.): *Recent Advances in Cardiac Cells and Metabolism.* University Park Press, Baltimore, pp. 217–232.

Romani, A., Dowell, E.A. and Scarpa, A. (1991) Cyclic AMP-induced Mg^{2+} release from rat liver hepatocytes, permeabilized heaptocytes, and isolated mitochondria. *J. Biol. Chem.* 266: 12376–12384.

Romani, A., Marfella, C. and Scarpa, A. (1993) Regulation of magnesium uptake and release in the heart and in isolated ventricular myocytes. *Circ. Res.* 72: 1139–1148.

Romani, A., Marfella, C. and Scarpa, A. (1992) Regulation of Mg^{2+} uptake in isolated rat myocytes and hepatocytes by protein kinase C. *FEBS Lett.* 296: 135–140.

Romani, A. and Scarpa, A. (1992) cAMP control of Mg^{2+} homeostasis in heart and liver cells. *Magnesium Res.* 5: 131–137.

Romani, A. and Scarpa, A. (1990) Hormonal control of Mg^{2+} transport in the heart. *Nature* 346: 841–844.

Romani, A., Secard, C., Fatholahi, M. and Scarpa, A. (1990) Magnesium transport across cardiac cells is regulated by the β-adrenergic receptors via cAMP. *FASEB J.* 4: 166.

Vinogradov, A., Scarpa, A. and Chance, B. (1972) Calcium and pyridine nucleotide in mitochondrial membranes. *Arch. Biochem. Biophys.* 152: 642–654.

Vormann, J. and Gunther, T. (1987) Amiloride-sensitive net Mg^{2+} efflux from isolated perfused rat hearts. *Magnesium* 6: 220–224.

Weiner, N. (1985) Drugs that inhibit adrenergic nerves and block adrenergic receptors. *In:* G.A. Goodman, S.L. Goodman, T.W. Rall and F. Murray (eds): *The Pharmacological Basis of Therapeutics.* Macmillan, New York, pp. 181–214.

Wu, S.T., Pieper, G.M., Salhany, J.M. and Eliot, R.S. (1981) Measurement of free Mg^{2+} in perfused and ischemic arrested heart muscle. A quantitative phosphorus 31 nuclear magnetic resonance and multiple equilibria analysis. *Biochemistry* 20: 7399–7403.

Biochemistry of Cell Membranes
ed. by S. Papa & J. M. Tager
© 1995 Birkhäuser Verlag Basel/Switzerland

Signal transduction by growth factor receptors

C. Battistini[1], S. Penco[1] and P.M. Comoglio[2]

[1]*Pharmacia — Research and Development, Nerviano (MI)*
[2]*Dept. of Biomedical Sciences, University of Torino, School of Medicine, Italy*

Introduction

Signal transduction involves an intracellular cascade of biochemical events that follow the interaction between extracellular growth factors and their membrane receptors, ending in the switch of nuclear mechanisms controlling the proper biological responses.

The cytoplasmic signal transduction pathway from the receptor to the nucleus includes a complex system of interactions, still largely unknown, between several proteins that are recruited to the inner surface of the plasma membrane and then activated. A greatly simplified scheme can be drawn for mitogenesis, where an activated receptor transmits a signal to a distinct set of membrane anchored proteins. Among these, the *Ras* protein, the enzymes phospholipase-Cγ and phosphatidyl-3-kinase appear to have, through a multistep pathway, a pivotal role in sending the signal to the nucleus.

Signal transduction is involved in the onset and progression of human cancer. There are several human tumors in which activated proto-oncogenes have been detected. In general, proto-oncogenes encode normal proteins involved in the cell signaling. Point mutations activate the oncogenes, giving rise to constitutively activated signaling proteins causing uncontrolled cell growth. In this perspective, a good knowledge of the signaling mechanism is essential for creating new therapeutic strategies for more efficient approaches in cancer treatment. An approach aimed at interfering with signal transduction can be advantageous when comparing with intervention at a receptor level by specific inhibition of extracellular ligand binding. Indeed, the failure to inhibit proliferation by means of a single receptor antagonist is due to the involvement of more than one membrane receptor in the mitogenic signal. Since different receptors can activate partially overlapping intracellular signal pathways, it is likely that targeting at critical steps of these internal pathways will result in a better chance of success for antitumoral therapy. Moreover, this approach is mandatory when the uncontrolled cellular growth is caused by constitutively activated intra-cellular signal transducers.

From growth factor receptors to the nucleus

Signaling from the activated receptor to the nucleus involves several proteins interacting along possibly redundant pathways, more or less interconnected, triggered by two initial events: (1) protein tyrosine-phosphorylation; (2) protein-protein recognition, by physical associa-tion between specific domains. Assuming *Ras* stimulation as a pivotal event in this intracellular signaling network, the downstream events (from *Ras* to the nucleus) can be distinguished from the upstream events (from the growth factor to Ras) and we can focus our attention

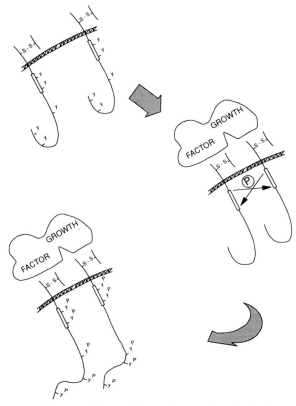

Figure 1. Activation of a growth factor by heterodimerization. In the absence of the ligand, receptors float as individual molecules in the plane of the plasma membrane. The ligand induces oligomerization, bringing adjacent molecules into close contact, and induces cross-phosphorylation (P) of tyrosine residues (Y). The phosphorylated residues of the receptor tail generate docking sites for intracellular signal transducers.

on the latter ones, which occur mainly at the interface between the plasma membrane and the periphery of the soluble cytoplasm.

The main events occurring between the growth factor receptor and activation of the "*Ras* system" are the tyrosine-phosphorylation of the receptors and the recognition of specific phospho-tyrosine residues by cytoplasmic transducers.

A growth factor receptor is a protein spanning the cell membrane, with a cytoplasmic portion including a tyrosine kinase domain and several tyrosine residues that can behave as auto-phosphorylation sub-strates. As first step, the growth factor activates its receptor by oligomerization (Schlessinger, 1993). The receptors are then brought close to each other, allowing "autophosphorylation" or, more likely, reciprocal phosphorylation of several tyrosine residues of the cytoplasmic portion of the adjacent molecule (Fig. 1). This seems to be a general mechanism to elicit an intracellular response by an extracellular ligand. Once the cytoplasmic region of the receptors has one or several sites phosphorylated, molecular recognition can occur between the phospho-rylated tyrosines and the appropriate domains of a spectrum of other proteins. These proteins are "cytoplasmic transducers"; which amplify the signal triggered by the ligand-bound receptor and transfer the message to the nucleus. There, transcription regulation influences the expression of specific genes and hence ultimately elicits biological re-sponses: cell proliferation in the case of growth factors. An incomplete list of the most studied cytoplasmic transducers is shown in Figure 2.

Several cytoplasmic effectors have a conserved domain named SH2 (Src Homology region 2). These short (approximately 100 amino acid long) domains are non-catalytic regions that are responsible for binding

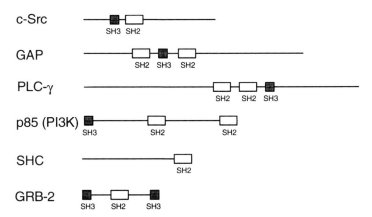

Figure 2. Schematic drawing of six cytoplasmic signal transducers endowed with SH2 (open boxes) and SH3 (filled boxes) domains. c-Src, GAP, and PLCγ are catalytic transducers. p85, SHC, and GRB-2, lacking enzymatic activity, behave as "adaptors", linking non SH2 catalytic molecule to receptors by a phosphorylation-dependent mechanism.

to phosphotyrosine containing aminoacid sequences. The SH2 domains are often found to be accompanied by another conserved domain named SH3 (Src Homology region 3). SH3 domains are also non catalytic, with affinity for protein stretches rich in proline. Both SH2 and SH3 domains are thought to have a pivotal role in the signal transduction for intramolecular recognition (Pawson and Gish, 1992).

The SH2 domains

The SH2-containing proteins can be subdivided into two groups. One includes proteins having an intrinsic enzymatic activity, hence with catalytic domains (kinase domains, phosphatase domains or others) besides the SH2. This group includes proteins like *Src*, *Abl*, *Syc*, PTP1C, PLCγ, GAP, *Vav*. The second group is comprised of SH2-containing proteins devoid of any known catalytic domain. The latter have the function of "adaptors", recruiting other cytoplasmic proteins endowed with catalytic properties to the receptor. For example, p85 provides a link between the activated (phosphorylated) receptor and the catalytic subunit of phosphatidylinositol-3′-kinase (P13K). Another relevant example of adaptor is *Grb2* (growth factor receptor binding protein 2) that links phosphorylated receptors to a protein thas has a guanine nucleotide exchange activity on *Ras*. Other examples of SH2-containing adaptors are *Shc*, *Crk*, and *Nck*; whose partners are largely unknown.

The SH2 domains are able to recognize the phosphorylated tyrosine residues of the receptors or of other downstream proteins. The SH2 structure is modular in nature, *i.e.* its carboxy- and amino-ends are very close in space, suggesting that the tertiary structure is remarkably independent of the rest of the protein. The general architecture of the domain and the phosphotyrosine recognition site have been clarified by crystal structure analysis of the Src SH2 domain, complexed with low or high affinity-bound phosphopeptides, and in free form (Waksman et al., 1992, 1993). The full picture emerged from studies of the crystal structure of *Lck* SH2, complexed with a high affinity phosphotyrosyl peptide (Eck et al., 1993), from nuclear magnetic resonance analysis of the free SH2 domains of the Abl tyrosine kinase (Overduin et al., 1992a,b) and from studies of the p85 subunit of P13K (Booker et al., 1992). The core element of the SH2 structure is an antiparallel β-sheet, sandwiched between two α-helices and a protruding small β-sheet (Fig. 3).

The multiple alignment of as many as 67 SH2 domains reveals highly conserved residues that are likely to participate in interactions with phosphotyrosine and in protein folding, and variable residues, some of which are responsible for a certain degree of specificity for the sequence

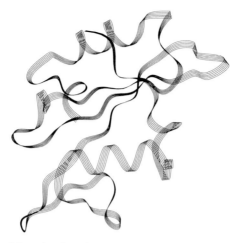

Figure 3. Schematic ribbon drawing of a Src homology 2 (SH2) protein domain. The core element of the SH2 structure is an antiparallel β-sheet, sandwiched between two α-helices and a protruding small β-sheet. The overall shape of the SH2 domain resembles a flattened hemisphere whose cross-sectional face provides the platform for binding tyrosine phosphory-lated peptides.

surrounding the phosphotyrosine of the bound protein (Russel et al., 1992). Two phosphorylated pentapeptides, Tyr(P)-Val-Pro-Met-Leu and Tyr(P)-Leu-Arg-Val-Ala, showing low affinity binding to Src SH2 domain, form complexes that crystallize: x-ray studies singled out the SH2 residues interacting with the ligand phosphotyrosine moiety (Waksman et al., 1992). The shape of the SH2 domain resembles a flattened hemisphere whose cross-sectional face provides the platform for peptide binding. Each pentapeptide makes contact with the SH2 domain by the side chains of three residues: the phosphotyrosine and the aminoacid in position +1 and +3 toward the carboxy-terminal end. The phosphotyrosine residue is firmly secured to the binding site by multiple interactions between its phenyl ring, its phosphate group and the side chains of aminoacids of the SH2 moiety, namely, Arg[155], Arg[175], Ser[177], Thr[179], and Lys[203] in the Src SH2. By contrast, the interactions at positions +1 and +3 of the phosphopeptide are quite loose and do not allow a clear definition of the basis for sequence-specific recognition, at least in the molecules studied. Similarly, in a crystalline complex formed by the Src SH2 domain and a high affinity 11-residue phosphopeptide derived from a sequence of the hamster middle T antigen, four central residues. Tyr(P), Glu(+1), Glu(+2), Ile(+3), make the primary strong interactions (Waksman et al., 1993). Comparable results have been obtained in crystallographic structure of the complex between the same peptide and the SH2 domain of the protein kinase *Lck*, a Src-like tyrosine kinase (Eck et al., 1993).

The SH3 domains

The other widely occurring protein-protein recognition domain in signal transduction is the Src-homology 3 (SH3) domain (Pawson et al., 1993). This structural motif is present in several SH2 containing signal transducers and in a number of cytoskeletal proteins. The SH3 domain contains approximately 60 amino acids and forms a compact structure made of five to eight antiparallel β-strands forming two orthogonal β-sheets. The crystal structure of the SH3 domain of the cytoskeletal protein *spectrin* has been determined (Musacchio et al., 1993); the solution structures of the SH3 domains of *Src*, of phospholipase C-γ and of p85 have been determined by multidimensional nuclear magnetic resonance (Yu et al., 1993; Kohda et al., 1993; Booker et al., 1993; Koyama et al., 1993). In all three cases the tertiary structure is very similar, despite a low homology at the level of primary structure. Indeed α-spectrin, c-Src and p85 SH3 domains have sequence identities as low as 18%. The adaptor p85 has SH3 domains of its own class, identified by the presence of a unique 15 amino acid insert with helix structure that may influence binding specificity.

Compared to the large amount of available information for SH2 domain functions, relatively little is known about SH3 domains. Two proteins (3BP1 and 3BP2) that bind specifically to the SH3 domain of the *Abl* kinase have been isolated (Ren et al., 1993). The binding sites of these molecules consist of a nine and, respectively, ten amino acid stretch rich in proline residue (e.g., APTMPPPLPP). A peptide composed of ten prolines binds poorly to the *Abl* SH3; thus, amino acids other than proline are critical for binding. Moreover, a peptide that binds strongly to *Abl* SH3, binds weakly to *Src* or *Grb2* SH3 domains and it is almost ineffective when tested with other SH3 moieties, demonstrating that there is specificity (Ren et al., 1993).

The ligand-binding sites of *Src* and p85 SH3 domains have been identified with the aid of appropriate proline-rich synthetic peptides and nuclear magnetic resonance. In the case of *Src* SH3, the perturbed residues define a slightly curved hydrophobic depression on the surface of the protein that is lined with side chains of aromatic amino acids (Yu et al., 1992). The ligand-binding site of p85 consists of a hydrophobic surface flanked by two charged loops (Booker et al., 1993).

The picture which emerges about the functions of SH3 domains appears to be a role in linking the signaling pathways of tyrosine kinases to guanine nucleotide binding proteins, such as Ras and small G-proteins (Kohda et al., 1993; Bar-Sagi et al., 1993). Moreover, the SH3 domains seem to be involved in the sub-cellular localization of signaling molecules, namely in the association with cytoskeletal proteins. Indeed, an isolated PLC-γ SH3 domain micro-injected into fibroblasts was localized to the actin cytoskeleton, while in a similar experiment the

SH3 domains of Grb2 direct its cellular localization to membrane ruffles (Bar-Sagi et al., 1993). Grb2 is an adaptor protein containing a central SH2 domain flanked by two SH3 domains at the amino- and the carboxy-terminal ends. One of the SH3 domains of Grb2 binds to *SoS*, the GDP-GTP exchanger that activates *Ras* (Lowenstein et al., 1992). In this context, it is of relevance that also *Ras* is preferentially localized in membrane ruffles. The similar cellular distribution of Grb2 and Ras is consistent with the idea that Grb2 plays a role in the control of Ras signaling.

The docking sites of growth factor receptors for SH2-transducers

SH2 containing intracellular signal transducers are known to recognize and bind the intracellular domain of a number of different growth factor receptors at specific tyrosine phosphorylated sites. The analysis of such interactions showed that receptors can bind, with reasonable affinity, a variety of SH2-signal transducers through a number of different specific docking sites.

The interaction between a tyrosine phosphorylated site of a protein and the SH2 domain of a cytoplasmic transducer can be investigated in several ways, but there are two principal methods. One is the mutation of the selected tyrosine to phenylalanine followed by the analysis of the residual binding with specific molecules, and of the loss of specific biological or biochemical responses. The second is based on the synthesis of short phosphopeptides with sequences derived from the receptor motif containing the putative tyrosine-phosphorylated docking site. These peptides are used to inhibit the binding *in vitro* between the selected cytoplasmic transducer and the phosphorylated receptor. For example the PDGF β-receptor contains two tyrosine phosphorylated sites that recognize selectively the SH2 domains of p85, namely Tyr^{740} and Tyr^{751}, one for the SH2 domain of *Gap*, Tyr^{771}, and two for the SH2 domains of phospholipase C-γ, Tyr^{1009} and Tyr^{1021} (Heldin, 1992; Kashishian et al., 1992). Other examples, listing the docking site of other receptors for the SH2-containing p85 adaptor, are shown in Table 1.

With a notable exception (see below), the tyrosine phosphorylated docking sites are rather selective for a given signal transducer. A synthetic phosphopeptide mimicking the docking site of the PDGF receptor for p85, namely containing Tyr^{740} and Tyr^{751}, inhibits the association between the receptor and p85, without affecting the binding to either *Gap* or to PLC-γ (Escobedo et al., 1991). Detailed studies on the binding between PDGF receptor, p85 and *Gap* have shown that specificity arises from the aminoacid sequence downstream from the receptor phosphotyrosine, as the upstream sequence is not necessary for recognition. Moreover, it has been ascertained that a pentapeptide is

Table 1. p85 Binding sites

Receptor	Sequence	Residue no.
Insulin-R	Y^{P}THMN	Y1334
Steel-R	Y^{P}MDMK	Y721
CSF-1-R	Y^{P}VEMR	Y723
PDGF-R-α	Y^{P}MDMS	Y731
PDGF-R-β	Y^{P}VPML	Y751
HGF-R	Y^{P}VHV	Y1349
	Y^{P}VNV	Y1356

sufficient to prevent the association between the receptor and the selected SH2-signal transducer (Fantl et al., 1992).

Some signal transducer molecules contain two SH2 domains. This is the case of p85 that has an SH2 proximal to the carboxy-terminus (SH2-C) and a second one proximal to the amino-terminus (SH2-N). Both of them take part in recognition of the PDGF receptor binding to either the phosphorylated Tyr^{740} or Tyr^{751}. These residues are embedded in a common aminoacid motif: Tyr(P)-Met/Val-X-Met (Kashishian et al., 1992). The p85 SH2-C domain binds with high affinity to the PDGF receptor site including Tyr^{740}; the SH2-N domain binds the site including Tyr^{751} site with a hundred-fold lower affinity (Klippel et al., 1992).

The HGF receptor "supersite"

As previously described, as a rule, different docking sites interact rather selectively with individual transducers. The notable exception is the docking site of the Hepatocyte Growth Factor (HGF) receptor. HGF, also known as "Scatter Factor," is a powerful extracellular signaling molecule known to induce a pleiotropic response in target cells. It is a powerful mitogen for epithelial and endothelial cells, and also promotes cell dissociation and extracellular matrix invasion ("scattering"). Moreover, during embryonal development HGF acts as a "morphogen," controlling the three-dimensional organization of some epithelial organs (for a review, see Comoglio, 1993). The high-affinity receptor for HGF is p190^{MET}, a membrane protein encoded by the *MET* proto-oncogene. The receptor is a heterodimer comprised of an extracellular 50 kDa subunit (α-chain) linked through a disulfide bond to a 145 kDa subunit (β-chain) spanning the membrane (Giordano et al., 1989). The intracellular part of the β-chain is comprised of a tyrosine kinase domain and several phosphorylated tyrosines.

As in the case of other tyrosine kinase receptors, stimulation by HGF causes the dimerization of the receptor and the intermolecular auto-

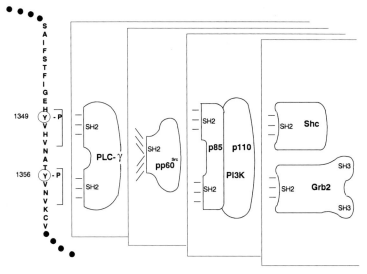

Figure 4. The HGF receptor "supersite". The multifunctional docking site is made of a tandemly arranged degenerate sequence YVH/NV (squared brackets). Phosphorylation of the site mediates intermediate to high-affinity interactions with multiple SH2-containing signal transducers, including PI 3-kinase, phospholipase-C-γ, pp60$^{c\text{-}Src}$, Shc and the GRB-2/SoS complex. The same bidentate motif is conserved in the evolutionarily related receptors encoded by the oncogenes *SEA* and *RON*, suggesting that in all members of the HGF-receptor family signal transduction is channeled through a multifunctional binding site.

phosphorylation (Fig. 1). Tyr^{1235} is a major site of p190^{MET} autophosphorylation both in vitro and in vivo. However, phosphorylation of this tyrosine residue, is apparently not involved in docking signal transducers: it rather enhances the kinase activity of p190^{MET}, hence playing a positive regulatory role (Longati et al., 1994). On the other hand, similarly to the EGF and insulin receptors, a negative regulation is provided through serine phosphorylation of the receptor justamembrane domain by protein kinase C (Gandino et al., 1994).

Like the receptors for PDGF, EGF, and other growth factors, in the ligand-bound state the HGF receptor is able to bind several cytoplasmic transducers. Indeed, the pleiotropic response induced by HGF results from activation of multiple signaling pathways. It has previously been shown that HGF stimulation enhances the activity of PI 3-kinase (Graziana et al., 1991; Ponzetto et al., 1993), of a *Ras* guanine nucleotide exchanger (Graziani et al., 1993), and of a tyrosine phosphatase (Villa Moruzzi et al., 1993). HGF stimulates PLCγ and MAP-kinase (an event associated with activation of the *Ras* pathway), and activates the *Src* kinase (Ponzetto et al., 1994).

Two phosphotyrosines located in the HGF receptor C-terminal tail within the sequence Y*^{1349}VHVNATY*^{1356}VNV, have been recently identified as intermediate-affinity sites for the p85 subunit of PI 3-kinase (Ponzetto et al., 1993). Surprisingly, this phosphotyrosine pair also mediates association of the receptor with PLCγ, *Src*, and the GRB-2/ SoS complex (Fig. 4; Ponzetto et al., 1994).

Comparison of the sequence YVH/NV with the optimal binding motifs listed by Songyang et al. (1993) indicates that the docking site of the HGF receptor is a sort of "supersite," a degenerated sequence potentially permissive for more than one SH2 domain. Using synthetic phosphopeptides and a BIAcore biosensor to measure intermolecular binding, we confirmed that the SH2 domains of p85, PLCγ, and *Src*, interact directly with either version of the "supersite" (YVHV or YVNV). GRB-2, which has a strong requirement for asparagine in the +2 position (Songyang et al., 1993), interacts specifically with the sequence YVNV. All bindings are characterized by fast association and dissociation rates. When the kinetic parameters measured for the HGF/ SF receptor sites were compared with those for the respective optimal sequences, all the SH2 domains studied showed similar affinity, with the exception of p85 which showed a K_d one order of magnitude higher than that for the optimal sequence YXXM (Ponzetto et al., 1993).

The kinetic binding parameters of the bidentate HGF receptor "supersite" are consistent with reversible association with a spectrum of rapidly exchanging signaling molecules. The close spacing of the two phosphotyrosines in the "bidentate" motif suggests that reiteration of the degenerate consensus may play a role in increasing binding affinities.

Site-directed mutagenesis of *TPR-MET*, the oncogenic counterpart of the receptor (Park et al., 1986), proved that phosphorylation of the bidentate "supersite" mediates biological function. Substitution of both sites with phenylalanine left the kinase activity unchanged, but completely abolished the transforming ability. Individual mutations inhibited transformation to a different extent. Substitution of Y^{1356}, which has the unique ability over Y^{1349} of binding GRB-2, drastically reduced the biological response. This result underscores the importance for neoplastic transformation by activated tyrosine kinases of the link with the Ras pathway provided by the GRB-2 adaptor (see also Pendergast et al., 1993). However, Y^{1356} alone is not sufficient to bring about the full transforming potential of the truncated HGF receptor protein. Mutation of Y^{1349} significantly reduced the number of foci with respect to wild type *TPR-MET*. This indicates that maximal signaling efficiency results from cooperation between the two tyrosines (Ponzetto et al., 1994).

The "supersite" motif Y-hydrophobic-X-hydrophobic-$(X)_3$-Y-hydrophobic-N-hydrophobic is conserved in the otherwise divergent C-terminal regions of the putative receptors *Sea* and *Ron*, two recently identified HGF receptor-related cDNAs (Huff et al., 1993; Ronsin et al., 1993, Gaudino et al., 1994). Synthetic phosphopeptides derived from the *Sea* and *Ron* multifunctional docking site binds to all the above listed SH2-containing transducers, showing that all members of the HGF receptor family use their own version of the "supersite" for signaling.

References

Albright, C.F., Giddings, B.W., Liu, J., Vito, M. and Weinberg, R.A. (1993) Characterization of a guanine nucleotide dissociation stimulator for a ras-related GTPase. *EMBO J.* 12: 339–347.

Bar-Sagi, D., Rotin, D., Batzer, A., Mandiyan, V. and Schlessinger, J. (1993) SH3 Domains direct cellular localization of signaling molecules. *Cell* 74: 83–91.

Booker, G.W., Breeze, A.L., Downing, A.K., Panayotou, G., Gout, I., Waterfield, M.D. and Campbell, I.D. (1992) Structure of an SH2 domain of the p85α subunit of phosphatidyl-inositol-3-OH kinase. *Nature* 358: 684–687.

Booker, G.W., Gout, I., Downing, A.K., Driscoll, P.C., Boyd, J., Waterfield, M.D. and Campbell, I.D. (1993) Solution structure and ligand-binding site of the SH3 domain of the p85α subunit of phosphatidylinositol 3-kinase. *Cell* 73: 813–822.

Comoglio, P.M. (1993) Structure, biosynthesis and biochemical properties of the HGF receptor in normal and malignant cells. *In:* I.D. Goldberg and E.M. Rosen (eds): *Hepatocyte Growth Factor-Scatter Factor (HGF-SF) and the c-Met receptor.* Birkhäuser Verlag Basel, Switzerland, pp. 131–165.

Domchek, S.M., Auger, K.R., Chatterjee, S., Burke, T.R., Jr. and Shoelson, S.E. (1992) Inhibition of SH2 domain/phospho-protein association by a nonhydrolyzable phospho-nopeptide. *Biochemistry* 31: 9865–9870.

Eck, M.J., Shoelson, S.E. and Harrison, S.C. (1993) Recognition of a high-affinity phospho-tyrosyl peptide by the Src homology-2 domain of p56[lck]. *Nature* 362: 87–91.

Egan, S.E., Giddings, B.W., Brooks, M.W., Buday, L., Sizeland, A.M. and Weinberg, R.A. (1993) Association of Sos Ras exchange protein with Grb2 is implicated in tyrosine kinase signal transduction and transformation. *Nature* 363: 45–51.

Escobedo, J.A., Kaplan, D.R., Kavanaugh, W.M., Turck, C.W. and Williams, L.T. (1991) A phosphatidylinositol-3 kinase binds to platelet-derived growth factor receptors through a specific receptor sequence containing phosphotyrosine. *Mol. Cell. Biol.* 11: 1125–1132.

Fantl, W.J., Escobedo, J.A., Martin, G.A., Turck, C.W., del Rosario, M., McCormick, F. and Williams, L.T. (1992) Distinct phosphotyrosines on a growth factor receptor bind to specific molecules that mediate different signaling pathways. *Cell* 69: 413–423.

Gandino, L., Longati, P., Medico, E., Prat, M. and Comoglio, P.M. (1994) Phosphorylation of Ser[985] negatively regulates the hepatocyte growth factor receptor kinase. *J. Biol. Chem.* 269: 1815–1820.

Gaudino, G., Follenzi, A., Naldini, L., Collesi, C., Santoro, M., Gallo, K.A., Godowski, P.J. and Comoglio, P.M. (1994) *RON* is a heterodimeric tyrosine kinase receptor activated by the HGF-homolog MSP. *EMBO J.* 13: 3524–3532.

Giordano, S., Ponzetto, C., Di Renzo, M.F., Cooper, C.S. and Comoglio, P.M. (1989) Tyrosine kinase receptor indistinguishable from the *c-Met* protein. *Nature* 339: 155–156.

Graziani, A., Gramaglia, D., Cantley, L.C. and Comoglio, P.M. (1991) The tyrosine phospho-rylated hepatocyte growth factor receptor associated with phosphatidyl inositol 3-kinase. *J. Biol. Chem.* 266: 22087–22090.

Graziani, A., Gramaglia, D., dalla Zonca, P. and Comoglio, P.M. (1993) Hepatocyte growth factor/scatter factor stimulates the Ras-guanine nucleotide exchanger. *J. Biol. Chem.* 268: 9165–9168.

Hara, M., Akasaka, K., Akinaga, S., Okabe, M., Nakano, H., Gomez, R., Wood, D., Uh, M. and Tamanoi, F. (1993) Identification of Ras farnesyltransferase inhibitors by microbial screening. *Proc. Natl. Acad. Sci. USA* 90: 2281–2285.

Heldin, C.-H. (1992) Structural and functional studies on platelet-derived growth factor. *EMBO J.* 11: 4251–4259.

Huff, J.L., Jelinek, M.A., Borgman, C.A., Lansing, T.J. and Parsons, J.T. (1993) The protooncogene *c-sea* encodes a transmembrane protein-tyrosine kinase related to the

Met/hepatocyte growth factor/scatter factor receptor. *Proc. Natl. Acad. Sci. USA* 90: 6140–6144.

James, G.L., Goldstein, J.L., Brown, M.S., Rawson, T.E., Somers, T.C., McDowell, R.S., Crowley, C.W., Lucas, B.K., Levinson, A.D. and Marsters, J.C., Jr. (1993) Benzodiazepine peptidomimetics: potent inhibitors of Ras farnesylation in animal cells. *Science* 260: 1937–1942.

Kashishian, A., Kazlauskas, A. and Cooper, J.A. (1992) Phosphorylation sites in the PDGF receptor with different specificities for binding GAP and PI3 kinase in vivo. *EMBO J.* 11: 1373–1382.

Kavanaugh, W.M., Klippel, A., Escobedo, J.A. and Williams, L.T. (1992) Modification of the 85-kilodalton subunit of phosphatidylinositol-3 kinase in platelet-derived growth factor-stimulated cells. *Mol. Cell. Biol.* 12: 3415–3424.

Klippel, A., Escobedo, J.A., Fantl, W.J. and Williams, L.T. (1992) The C-terminal SH2 domain of p85 accounts for the high affinity and specificity of the binding of phosphatidylinositol 3-kinase to phosphorylated platelet-derived growth factor β receptor. *Mol. Cell. Biol.* 12: 1451–1459.

Kohda, D., Hatanaka, H., Odaka, M., Mandiyan, V., Ullrich, A., Schlessinger, J. and Inagaki, F. (1993) Solution structure of the SH3 domain of phospholipase C-γ. *Cell* 72: 953–960.

Kohl, N.E., Mosser, S.D., deSolms, S.J., Giuliani, E.A., Pompliano, D.L., Graham, S.L., Smith, R.L., Scolnick, E.M., Oliff, A. and Gibbs, J.B. (1993) Selective inhibition of ras-dependent transformation by a farnesyltransferase inhibitor. *Science* 260: 1934–1937.

Koyama, S., Yu, H., Dalgarno, D.C., Shin, T.B., Zydowsky, L.D. and Schreiber, S.L. (1993) Structure of the PI3K SH3 domain and analysis of the SH3 family. *Cell* 72: 945–952.

Li, B.-Q., Kaplan, D., Kung, H. and Kamata, T. (1992) Nerve growth factor stimulation of the Ras-guanine nucleotide exchange factor and GAP activities. *Science* 256: 1456–1459.

Longati, P., Bardelli, A., Ponzetto, C., Naldini, L. and Comoglio, P.M. (1994) Tyrosines[1234–1235] are critical for activation of the tyrosine kinase encoded by the *MET* protooncogene (HGF receptor). *Oncogene* 9: 49–57.

Marshall, C.J. (1993) Protein prenylation: a mediator of protein-protein interactions. *Science* 259: 1865–1866.

Marshall, M.S. (1993) The effector interactions of p21[ras]. *Trends Biochem. Sci.* 18: 250–254.

Martegani, E., Vanoni, M., Zippel, R., Coccetti, P., Brambilla, R., Ferrari, C., Sturani, E. and Alberghina, L. (1992) Cloning by functional complementation of a mouse cDNA encoding a homologue of CDC25, a *Saccharomyces cerevisiae* RAS activator. *EMBO J.* 11: 2151–2157.

McNamara, D.J., Dobrusin, E.M., Zhu, G., Decker, S.J., Saltiel, A.R. (1993) Inhibition of binding of phospholipase Cγ1 SH2 domains to phosphorylated epidermal growth factor receptor by phosphorylated peptides. *Int. J. Pept. Protein Res.* 42: 240–248.

Montminy, M. (1993) Trying on a new pair of SH2s. *Science* 261: 1694–1695.

Musacchio, A., Noble, M., Pauptit, R., Wierenga, R. and Saraste, M. (1992) Crystal structure of a Src-homology 3(SH3) domain. *Nature* 359: 851–855.

Olivier, J.P., Raabe, T., Henkemeyer, M., Dickson, B., Mbamalu, G., Margolis, B., Schlessinger, J., Hafen, E. and Pawson, T. (1993) A drosophila SH2-SH3 adaptor protein implicated in coupling the sevenless tyrosine kinase to an activator of Ras guanine nucleotide exchange, Sos. *Cell* 73: 179–191.

Overduin, M., Mayer, B., Rios, C.B., Baltimore, D. and Cowburn, D. (1992a) Secondary structure of Src homology 2 domain of c-Abl by heteronuclear NMR spectroscopy in solution. *Proc. Natl. Acad. Sci. USA* 89: 11673–11677.

Overduin, M., Rios, C.B., Mayer, B.J., Baltimore, D. and Cowburn, D. (1992b) Three-dimensional solution structure of the src homology 2 domain of c-abl. *Cell* 70: 697–704.

Panayotou, G., Bax, B., Gout, I., Federwisch, M., Wroblowski, B., Dhand, R., Fry, M.J., Blundell, T.L., Wollmer, A. and Waterfield, M.D. (1992) Interaction of the p85 subunit of PI 3-kinase and its N-terminal SH2 domain with a PDGF receptor phosphorylation site: structural features and analysis of conformational changes. *EMBO J.* 11: 4261–4272.

Park, M., Dean, M., Cooper, C.S., Schmidt, M., O'Brien, S.J., Blair, D.G. and Vande Woude, G.F. (1986) Mechanism of *MET* oncogene activation. *Cell* 45: 895–904.

Pawson, T. and Gish, G.D. (1992) SH2 and SH3 domains, from structure to function. *Cell* 71: 359–362.

Pelicci, G., Lanfrancone, L., Grignani, F., McGlade, J., Cavallo, F., Forni, G., Nicoletti, I., Grignani, F., Pawson, T. and Pelicci, P.G. (1992) A novel transforming protein (SHC) with an SH2 domain is implicated in mitogenic signal transduction. *Cell* 70: 93–104.

Pendergast, A.M., Quilliam, L.A., Cripe, L.D., Bassing, C.H., Dai, Z., Li, N., Batzer, A., Rabun, K.M., Der, C., Schlessinger, J. and Gishizky, M.L. (1993) BCR-ABL-induced oncogenesis is mediated by direct interaction with the SH2 domain of the GRB-2 adaptor protein. *Cell* 75: 175–185.

Philips, M.R., Pillinger, M.H., Staud, R., Volker, C., Rosenfeld, M.G., Weissmann, G., Stock, J.B. (1993) Carboxyl methylation of Ras-related proteins during signal transduction in neutrophils. *Science* 259: 977–980.

Piccione, E., Case, R.D., Domchek, S.M., Hu, P., Chaudhuri, M., Backer, J.M., Schlessinger, J. and Shoelson, S.E. (1993) Phosphatidylinositol 3-kinase p85 SH2 Domain Specificity defined by direct phosphopeptide/SH2 domain binding. *Biochemistry* 32: 3197–3202.

Polakis, P. and McCormick, F. (1993) Structural requirements for the interaction of p21ras with GAP, exchange factors, and its biological effector target. *J. Biol. Chem.* 268: 9157–9160.

Ponzetto, C., Bardelli, A., Maina, F., Longati, P., Panayotou, G., Dhand, R., Waterfield, M.D. and Comoglio, P.M. (1993) A novel recognition motif for phosphatidyl-inositol 3-kinase binding mediates its association with the hepatocyte growth factor/scatter factor receptor. *Mol. Cell. Biol.* 13: 4600–4608.

Ponzetto, C., Bardelli, A., Zhen, Z., Maina, F., dalla Zonca, P., Giordano, S., Graziani, A., Panayotou, G. and Comoglio, P.M. (1994). A multifunctional docking site mediates signalling and transformation by the hepatocyte growth factor/scatter factor (HGF/SF) receptor family. *Cell* 77: 261–271.

Ren, R., Mayer, B.J., Cicchetti, P. and Baltimore, D. (1993) Identification of a ten-amino acid proline-rich SH3 binding site. *Science* 259: 1157–1161.

Ronsin, C., Muscatelli, F., Mattei, M.G. and Breathnach, R. (1993) A novel putative receptor protein tyrosine kinase of the *Met* family. *Oncogene* 8: 1195–1202.

Rozakis-Adcock, M., McGlade, J., Mbamalu, G., Pelicci, G., Daly, R., Li, W., Batzer, A., Thomas, S., Brugge, J., Pelicci, P.G., Schlessinger, J. and Pawson, T. (1992) Association of the Shc and Grb2/Sem5 SH2-containing proteins is implicated in activation of the Ras pathway by tyrosine kinases. *Nature* 360: 689–692.

Russell, R.B., Breed, J. and Barton, G.J. (1992) Conservation analysis and structure prediction of the SH2 family of phosphotyrosine binding domains. *FEBS Lett.* 304: 15–20.

Satoh, T., Nakafuku, M. and Kaziro, Y. (1992) Function of Ras as a molecular switch in signal transduction. *J. Biol. Chem* 267: 24149–24152.

Schlessinger, J. (1993) How receptor tyrosine kinase activate Ras. *Trends Biochem. Sci.* 18: 273–275.

Shoelson, S.E., Sivaraja, M., Williams, K.P., Hu, P., Schlessinger, J. and Weiss, M.A. (1993) Specific phospho-peptide binding regulates a conformational change in the PI 3-kinase SH2 domain associated with enzyme activation. *EMBO J.* 12: 795–802.

Simon, M.A., Dodson, G.S. and Rubin, G.M. (1993) An SH3-SH2-SH3 protein is required for p21Ras1 activation and binds to sevenless and Sos proteins in vitro. *Cell* 73: 169–173.

Skolnik, E.Y., Lee, C.-H., Batzer, A., Vicentini, L.M., Zhou, M., Daly, R., Myers, M.J., Jr., Backer, J.M., Ullrich, A., White, M.F. and Schlessinger, J. (1993a) The SH2/SH3 domain-containing protein GRB2 interacts with tyrosine-phosphorylated IRS1 and Shc: implications for insulin control of ras signalling. *EMBO J.* 12: 1929–1936.

Songyang, Z., Shoelson, S.E., Chaudhuri, M., Gish, G., Pawson, T., Haser, W.G., King, F., Roberts, T., Ratnofsky, S., Lechleider, R.J., Neel, B.G., Birge, R.B., Fajardo, J.E., Chou, M.M., Hanafusa, H., Schaffhausen, B. and Cantley, L.C. (1993) SH2 Domains recognize specific phosphopeptide sequences. *Cell* 72: 767–778.

Travis, J. (1993) Novel anticancer agents move closer to reality. *Science* 260: 1877–1878.

Valius, M. and Kazlauskas, A. (1993) Phospholipase C-γ1 and phosphatidylinositol 3-kinase are the downstream mediators of the PDGF receptor's mitogenic signal. *Cell* 73: 321–334.

Vojtek, A.B., Hollenberg, S.M. and Cooper, J.A. (1993) Mammalian Ras interacts directly with the serine/threonine kinase raf. *Cell* 74: 205–214.

Villa Moruzzi, E., Lapi, S., Prat, M., Gaudino, G. and Comoglio, P.M. (1993) A protein tyrosine phosphatase activity associated with the hepatocyte growth factor/scatter factor receptor. *J. Biol. Chem.* 268: 18176–18180.

Waksman, G., Kominos, D., Robertson, S.C., Pant, N., Baltimore, D., Birge, R.B., Cowburn, D., Hanafusa, H., Mayer, B.J., Overduin, M., Resh, M.D., Rios, C.B., Silverman, L. and Kuriyan, J. (1992) Crystal structure of the phosphotyrosine recognition domain SH2 of v-*src* complexed with tyrosine-phosphorylated peptides. *Nature* 358: 646–653.

Waksman, G., Shoelson, S.E., Pant, N., Cowburn, D. and Kuriyan, J. (1993) Binding of a high affinity phospho-tyrosyl peptide to the Src SH2 domain:crystal structures of the complexed and peptide-free forms. *Cell* 72: 779–790.

Yu, H., Rosen, M.K., Shin, T.B., Seidel-Dugan, C., Brugge, J.S., Schreiber, S.L. (1992) Solution structure of the SH3 domain of Src and identification of its ligand-binding site. *Science* 258: 1665–1668.

Zhang, K., Papageorge, A.G. and Lowy, D.R. (1992) Mechanistic aspects of signaling through Ras in NIH 3T3 Cells. *Science* 257: 671–674.

Biochemistry of Cell Membranes
ed. by S. Papa & J. M. Tager
© 1995 Birkhäuser Verlag Basel/Switzerland

Cell surface associated mucins: Structure and effects on cell adhesion

J. Hilkens, J. Wesseling, H.L. Vos, S.L. Litvinov, M. Boer
S. van der Valk, J. Calafat, C. Patriarca, E. van de Wiel-van Kemenade[1]
and C. Figdor[1]

Divisions of Tumor Biology and [1]Immunology, The Netherlands Cancer Institute, Plesmanlaan 121, NL-1066 CX Amsterdam, The Netherlands

Summary. Cell surface associated mucins are a class of transmembrane glycoproteins that protrude high above the cell surface and can mask other cell surface molecules, thereby interfering with the function of the latter molecules. In particular, adhesion molecules are affected, which has important consequences for cellular processes that are dependent on adhesion such as cytolysis by immune cells and probably also invasion and metastasis. Episialin is such a cell-associated mucin which is highly overexpressed on certain carcinomas and it can serve as a prototypic cell associated mucin. Here, we discuss its structure, its effect on cellular adhesion *in vitro* and *in vivo*, and the consequences for LAK cell and CTL mediated cytolysis. In addition, we discuss the biological impact of other related molecules.

Introduction

The mucins can be divided into the classic mucins that are secreted by specialized cells and form the mucus layer in, for example, the gastro-intestinal tract, and the cell surface associated mucins. In contrast to the former mucins, cell surface associated mucins are anchored in the membrane, and are produced by a variety of epithelial cell types. The common characteristic of both types of molecules is that a large part of the protein backbone consists of an array of threonine, serine and proline rich tandem repeats which are heavily O-glycosylated and con-stitute the actual mucin domain. Episialin (also designated EMA, PEM, CA 15-3 antigen, etc.), encoded by the MUC1 gene, is one of the best characterized cell surface associated mucins and can serve as the proto-type for this class of molecules. Here, we will extensively review episialin with special emphasis on its structure and the consequences of this structure for cell-cell and cell-matrix adhesion and cellular processes that involve cellular adhesion. We will also briefly discuss other cell surface associated mucins in this respect.

Discussion

The structure of episialin

Episialin is a heavily glycosylated molecule with an apparent molecular mass of over 400 kDa. It is a transmembrane molecule with a relatively large extracellular domain and a cytoplasmic domain of 69 amino acids (Ligtenberg et al., 1990; Wreschner et al., 1990; Gendler et al., 1990; Lan et al., 1990). The extracellular domain consists mainly of a region of nearly identical repeats of 20 amino acids. These repeats together with adjacent degenerated repeats contain many potential attachment sites for O-linked glycans and constitute the mucin-like domain which comprises more than half of the polypeptide backbone. The number of repeats and thus the size of the molecule is highly variable among different individuals as a result of genetic polymorphism. The number of repeats in all tested allelic forms was estimated to vary between 30 and 90.

The prolines and other helix breaking amino acids present in the mucin-like domain provide the molecule with an extended structure with many β-turns (Jentoft, 1990). The extended structure of the mucin-like domain is very rigid as a result of the numerous O-linked glycans attached to it. As a result, episialin protrudes high above the cell surface. We have estimated that the size of the molecule varies between 200 and 500 nm depending on the allele (Hilkens et al., 1992). Electron microscopy of episialin purified from ZR-75-1 breast cancer cells indeed shows a long thread-like structure that has the size predicted for episialin from this cell line. We observed that episialin also has this structure when present at the plasma membrane of these cells (Calafat et al., submitted).

Biosynthesis of episialin

The molecule is synthesized as one large precursor polypeptide with a few N-linked glycans attached to it (Fig. 1). This precursor is immediately cleaved while still in the endoplasmic reticulum (Hilkens et al., 1988). The cleavage separates the major part of episialin, the mucin-like domain, from the remainder of the molecule, which contains the cytoplasmic and transmembrane domains and approximately 65 amino acids of the extracellular domain. The two domains form a stable non-covalently bound complex, and thus the mucin domain remains associated with the membrane (Ligtenberg et al., 1992a). During maturation the molecule acquires the numerous O-linked glycans which are still incompletely sialylated upon reaching the cell surface. We have found that cell surface associated episialin is constitutively internalized

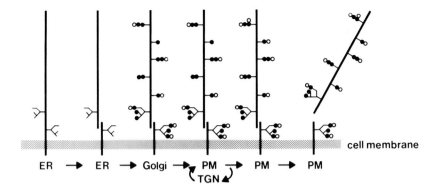

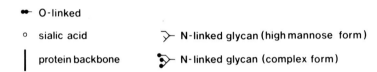

Figure 1. Schematic representation of the processing of episialin. The molecule is synthesized and cleaved in the endoplasmic reticulum (ER). Subsequently, it is translocated to the Golgi where the O-linked glycans are added and then transported to the plasma membrane (PM). The molecule completely matures by recycling to the trans-Golgi-network (TGN). During this process sialic acid residues are added to the glycans. Finally, the molecule is released from its cytoplasmic anchor and shed from the cell surface.

and recycled. During recycling the molecule matures completely by further addition of sialic acid residues, most likely in the trans-Golgi network (Litvinov and Hilkens, 1993). Constitutive recycling may be needed to maintain the fully sialylated state, which may keep the mucin-like domain in its most extended shape.

Several hours after synthesis, episialin is shed ($t_{\frac{1}{2}}$ is approximately 16 h) from the cell surface and can be detected in the medium of cultured cells (Litvinov et al., submitted). Whether shedding is due to dissociation of the complex, which would release the largest subunit into the extracellular environment, or whether an additional proteolytic cleavage is involved, remains to be established.

Localization of episialin in normal tissues

Episialin is mainly localized at the apical side of glandular epithelial cells. Immunohistochemical studies using various monoclonal antibodies (mAbs) revealed that episialin is also present at the luminal surfaces of certain cell layers lining other body cavities such as the mesothelium and in trophoblasts. Certain hematopoietic cells also show episialin expression, most notably a low proportion of the plasma cells.

Expression level of episialin is increased in cancer cells relative to normal epithelium

It has been reported that the expression of episialin is increased in tumor relative to normal epithelial cells as assessed by mAbs (Kufe et al., 1984; Croghan et al., 1983). However, due to the enzymatic detection routinely used in immunohistology and the differences in cellular localization (generally restricted to the apical surface in normal cells and present across the entire cell surface in cancer cells) it is difficult to quantitatively evaluate the immunohistochemical staining. Moreover, since the binding of most mAbs against episialin is affected by glycosylation, and episialin in normal and tumor tissues is differentially glycosylated, immunohistochemical staining does not necessarily represent the true expression level. Therefore, we have quantified the apparent overexpression of episialin in carcinomas at the mRNA level by *in situ* hybridization using an antisense RNA probe corresponding to a nonrepetitive region of the episialin cDNA. Episialin mRNA expression in almost all breast carcinomas was found to be more than ten times higher than in adjacent normal glandular epithelium.

We have shown that at least three different epitopes of episialin are absent or only present at a very low level on normal colonic epithelium and tubular adenomas. However, in 61% of the adenomas with severe dysplasia, and in almost all carcinomas of the colon a strong immunostaining was found (Zotter et al., 1987). The level of episialin mRNA in normal as well as in colon carcinomas was too low to be detected by *in situ* hybridization. Therefore, we cannot distinguish between an increased expression level of episialin and unmasking of the epitopes as a result of altered glycosylation in the carcinomas.

Episialin is also present in a number of non-epithelial malignancies such as 30% of sarcomas and brain tumors, and 20% of lymphomas, whereas none of the tissues of origin of these malignancies showed any episialin expression by immunohistochemical means (Zotter et al., 1988).

Episialin is a serum marker for breast cancer monitoring

As was discussed above, the mucin domain of episialin has a half-life at the surface of cells *in vitro* of approximately 16 h, and subsequently it is shed into the medium. The mucin domain of the molecule is very resistant to proteolytic degradation and can remain in the medium for several days. Shed episialin can also be detected in serum of healthy individuals. In the serum of breast cancer patients the level of episialin can be strongly elevated and there is a clear correlation with the stage of the disease. More importantly, episialin levels also correlate with the course of the disease, and can be used for early detection of recurrent breast cancer. Several commercial tests have been developed using mAbs in sandwich assays. The CA 15-3 assay using our mAb 115D8 is currently the most widely used serum assay for breast cancer (Hilkens, 1992).

Episialin expression and cellular adhesion

Overexpression of this rod-like molecule protruding high above the plasma membrane, as we observed in for instance breast carcinoma cells, may have a strong effect on cellular interactions. To investigate the effect of overexpression of episialin, we have transfected episialin cDNA under the control of the CMV promoter into various cell lines. We have compared the aggregation capacity of the transfectants, generated from SV40 transformed mammary epithelial cells (HBL-100) and melanoma cells (A375), with that of the episialin negative controls. The transfectants overexpressing episialin showed a strongly reduced aggregation capacity (Ligtenberg et al., 1992b). Episialin-negative revertant cells, bulk selected with the cell sorter to avoid as much as possible clonal outgrowth of cell types that had accidentally lost their adhesion molecules, regained their normal aggregation properties. Mouse L cells transfected with both E-cadherin and episialin cDNA, showed that episialin could overrule the intercellular adhesion mediated by E-cadherin. The presence of episialin on only one of a pair of aggregating cells is already sufficient to interfere with this process and results in sorting out of episialin positive and negative cells.

A proportion of the cells from several clones transfected with episialin cDNA were growing in suspension. A detailed investigation using adhesion assays demonstrated a reduced adherence of the transfectants to various extracellular matrix components (Wesseling et al., 1995). The anti-adhesion effect could be reversed by mAb induced activation of the integrins, indicating that there is a balance between adhesion mediated by integrins and anti-adhesion mediated by episialin. This balance can be tilted either way depending on the amount of active adhesion receptors and of episialin. MAbs directed against the mucin domains of episialin induced clustering of the episialin molecules at one side of the cell (capping) as a result of the presence of multiple identical epitopes in the molecule. Capping of episialin caused restoration of the normal adhesion properties of the cells. After capping, the adhesion molecules are evidently again available to their ligands, indicating that the adhesion molecules are not modified, and thus shielding of the integrins is the most likely mechanism of the anti-adhesion effect of episialin (Wesseling et al., 1995).

The anti-adhesion properties of episialin are probably the result of the unique extended and rigid structure of the molecule. As we have discussed above, episialin is towering 200–500 nm above the plasma membrane, whereas most proteins at the cell surface remain within the glycocalyx which is approximately 10 nm thick, Therefore, we postulate that episialin prevents cellular adhesion by masking adhesion molecules, and that a high density of episialin, as we observed on the surface of many carcinoma cells, can severely disturb the interaction of cell surface proteins with macromolecules on adjacent cell membranes. In addition, the abundance of the negatively charged sialic acid residues contributes to some extent to the anti-adhesion properties of episialin (Ligtenberg et al., 1992b). Thus the anti-adhesion effect could, in part, be caused by charge repulsion. Alternatively, the sialic acids could merely contribute to the rigidity of episialin. Anti-adhesion effects, induced by charge repulsion, have been postulated previously. For instance, the polysialylated embryonic form of N-CAM decreases adhesion of certain malignant cells and of neural cells during embryogenesis by charge repulsion and widens the intercellular spaces (Rutishauser et al., 1988; Yang et al., 1992).

The adhesion studies described above were carried out with tissue culture cell lines that do not grow in a polarized fashion. On normal differentiated glandular epithelial cells, episialin is only expressed at the apical side. Therefore, episialin can normally not interfere with adhesion molecules, which are only present at the basolateral side of the cell. However, at the apical side the function of episialin may be to prevent non-specific protein-protein interactions between glycoproteins on opposite membranes which could cause inadvertent adhesion of opposite apical membranes and disturb glandular function. In this way, episialin might facilitate the formation and maintenance of the lumen present in glandular structures and other tissue structures such as the high endothelial venules.

Episialin affects cellular adhesion in tumors in vivo

The episialin levels found (by FACS analysis) in transfectants that showed reduced adherence or aggregation were similar to those in human carcinoma cell lines. In many poorly differentiated adenocarcinomas the cellular polarization is lost. In these tumors the molecule is also found at those parts of the plasma membrane that are facing the stroma and adjacent cells, where it can interfere with adhesion. As a model to study the anti-adhesion effect of episialin *in vivo*, we subcutaneously injected episialin transfected melanoma cells into nude mice. Electron microscopy of the tumors that developed revealed strongly reduced cell-cell contacts in those regions of the plasma membrane where the molecule is abundantly present. In the tumor of an episialin-negative revertant we could not observe any loss of cell-cell contacts (Calafat et al., submitted). This implies that the presence of episialin destabilizes cell-cell interactions also *in vivo* and may be an important cofactor in metastasis. Moreover, episialin affects cellular adhesion to extracellular matrix components, which is known to be another important event during metastasis. We conclude that the balance between various adhesion processes can be disturbed when cellular polarization is lost and episialin is (over)expressed at the entire cell surface, and thus this mucin-like glycoprotein may enhance invasiveness and increase the efficiency of the metastatic process.

Episialin expression at the cell-stroma boundaries in breast carcinomas induced cleft formation

As discussed above, EM studies of melanoma xenografts overexpressing episialin showed that episialin prevents normal cell-cell contacts. Moreover, we have previously shown by electron microscopy on frozen sections of breast carcinomas that at those regions of the plasma membrane where the molecule is abundantly present, the adjacent membranes do not contact each other (Hilkens et al., 1984). We have tested whether similar spaces could also be visualized in human breast carcinomas by immunohistology on routinely fixed section. For this purpose, we investigated a series of formalin fixed paraffin embedded breast carcinomas by immunohistology. Several different staining patterns were observed, and usually more than one pattern was present in a single tumor. The first pattern is characterized by a polarized luminal staining of duct-like structures in well differentiated tumors. The second pattern shows a non-polarized membrane staining across the entire cell surface. The third pattern is characterized by a strong cytoplasmic staining. The fourth pattern is characterized by membrane staining mainly confined to cell-stroma boundaries of tumor cell clusters (Fig. 2). The latter type is most interesting because integrins are reported to be present at cell-stroma boundaries, yet no cell-stroma contact could be observed in those regions where episialin was expressed (Koukoulis et al., 1991; Pignatelli et al., 1992) and clefts were present at high frequency between the tumor nests and the stroma, suggesting that episialin actually prevents cell-stroma interactions. Thus, the presence of episialin can also destabilize cellular adhesion of human breast carcinomas, suggesting that upregulation of episialin may be important in invasion and metastasis.

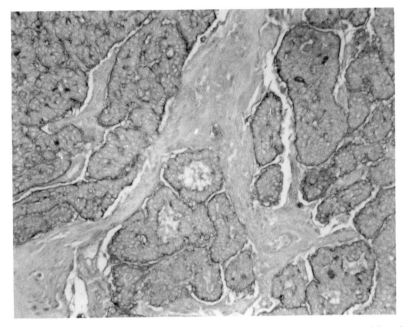

Figure 2. Cleft formation at tumor stroma boundaries. The figure shows a ductal invasive breast carcinoma immunostained with mAb 214D4 directed against episialin in an indirect immunoperoxidase test using peroxidase conjugated goat-anti-mouse and DAB as substrate. Reduced tumor-stroma contacts were present at those tumor margins where episialin is present (magnification 125 ×).

Episialin expression on target cells affects the susceptibility to LAK cell and allogenically stimulated T cell mediated cytolysis

Episialin prevents cell-cell adhesion also if only one of two aggregating cells expresses the mucin. Therefore, episialin is likely to prevent conjugate formation between cytotoxic effector cells and episialin positive target cells. Indeed, episialin transfected A375 melanoma cells showed a strongly diminished capacity to form conjugates with IL-2 activated large lymphocytes (LAK cells) and with allogeneic T lymphocytes, stimulated with A375 cells (CTL) relative to episialin negative revertants.

A375 cells can be efficiently lysed by LAK cells. We have measured the lysis of episialin transfected A375 melanoma cells and episialin-negative revertants by the LAK cells and CTL with time, in a ^{51}Cr release assay (Van de Wiel-van Kemenade et al., 1993). The kinetics of lysis of episialin-negative cells by the LAK cells was comparable to that of K562 cells, the standard target of LAK cells, whereas lysis of the episialin-positive A375 transfectants was more slow.

From these results, we can conclude that episialin transfected cells are partially protected against lysis *in vitro*, and may therefore escape immune destruction *in vivo*. The less efficient killing of episialin express-ing cells might be crucial to the survival of metastasizing cells. The effect of episialin on immune reactions proves that episialin expression has important implications for the cell that go beyond just the inhibition of intercellular adhesion.

Anti-adhesive effects of other mucin-like proteins

Anti-adhesive properties similar to those of episialin/MUC1 have also been observed for other mucin-like molecules, showing that there is a more general characteristic of these molecules. The best characterized example is the CD43 molecule, also known as leukosialin or sialophorin. The extracellular domain of this protein is almost entirely composed of a mucin-like domain consisting of semi-repetitive se-quences, which are heavily glycosylated. CD43/leukosialin has a length of about 45 nm (Cyster et al., 1991) which is much smaller than the observed length of episialin/MUC1, but is still larger than any adhesion molecule. The anti-adhesion property of leukosilian has been suggested by an experiment in which HeLa cells were transfected with leukosialin cDNA. The transfectants show a decreased adherence to T lymphocytes (Ardman et al., 1992). This is exactly the phenotype we have observed for episialin/MUC1 transfectants. Moreover, addition of mAbs directed against leukosialin on lymphocytes induces aggregation of these cells (Cyster and Williams, 1992). This effect has initially been interpreted by postulating that leukosialin is a receptor and binding of these mAbs mimics ligand binding causing activation of adhesion molecules by downstream signaling. Recently, it has been suggested that these mAbs induce cell-cell aggregation by crosslinking the cells when both legs of the antibody bind to leukosialin on different cells (Cyster and Williams, 1992). However, we have added mAbs to leukosialin expressing cells and found that these mAbs can induce patching of the antigen, suggest-ing that patching of leukosialin uncovers the adhesion molecules analogous to what we have shown for episialin.

Leukosialin is present on hematopoietic cells such as T cells and granulocytes, and is shed from these cells during activation *in vitro*, at the same time that integrins are upregulated (Rieu et al., 1992; Bažil and Strominger, 1993; Nathan et al., 1993). This suggests that leukosialin prevents interactions mediated by integrins or other adhe-sion molecules, when the cells are in the resting state. This view is reinforced by the observation that T-cell lines not expressing leukosialin spontaneously aggregate (Manjunath et al., 1993). In addition, it has been shown that leukosialin accumulates in the cleavage furrow during cell division (Yonemura et al., 1993). This might be interpreted as another anti-adhesion function in view of the data discussed above, because presence of leukosialin would prevent adhesive interactions in the cleavage furrow, thus ensuring the complete separation of the cells following division.

Another mucin-like molecule, CD24, has been implicated in preventing adhesion. Interestingly, CD34 is also conversely regulated with respect to integrins on endothelial cells (Delia et al., 1993); CD34 expression is increased when integrin expression is decreased. Recently, it has been shown that certain glycoforms of CD34 on endothelial cells also present ligands to selectins indicating that this molecule can also be involved in adhesion (Baumhueter et al., 1993).

Effects of other mucin-like molecules on tumor cells

Two additional membrane-associated mucin-like molecules, epiglycanin and ascites sialoglycoprotein-1 (ASGP-1), have been shown to affect cellular interactions of rodent carcinoma cells. Epiglycanin is produced by TA3 Ha mouse mammary carcinoma cells as one of the major components of the cellular surface. The epiglycanin and ASGP-1 genes have as yet not been cloned. Epiglycanin and episialin have many similarities to each other; both are highly glycosylated, and have long rod-like structures. TA3 Ha cells overexpressing epiglycanin grow readily in suspension, again suggesting an anti-adhesive effect. The addition of mAbs directed against epiglycanin caused capping of the mucin and adherence of the cells (H. Kemperman and E. Roos, personal communication) similar to our observations with episialin. Moreover, epiglycanin has been implicated in masking of the histocompatibility antigens, thereby rendering the cells allotransplantable (Codington et al., 1973, 1978; Miller et al., 1982). Fung and Longenecker (1991) observed immunosuppressive properties of this protein. The molecule has not been detected in any adult mouse tissue, but may be present in fetal mouse tissue (Beppu et al., 1987), suggesting a function during embryogenesis. Since epiglycanin is only found in a mouse mammary tumor it may be involved in a certain stage of mammary gland development.

ASGP-1 shares many properties with episialin with respect to biosynthesis and processing (Carraway et al., 1992). It is made from a single precursor, which is separated in a mucin-like domain (ASGP-1) and a cell-anchoring domain (ASGP-2) by proteolytic cleavage. It is abundantly present on the surface of the rat mammary carcinoma cell line 13762. Resistance of these cells to lysis by spleen lymphocytes and NK cells was shown to correlate with the expression level of this sialomucin (Sherblom and Moody, 1986; Bharathan et al., 1990; Moriarty et al., 1990). Moreover, studies on different metastatic clones of this rat mammary tumor cell line have established a positive correlation between their metastatic potential and the expression level of ASGP-1

(Steck and Nicolson, 1983). Since all cell-associated mucins can be covered by complex O-linked glycans including sialylated and sulfated Lewis-x structure, depending on the tissue of origin, they may in certain cell types serve as receptors for selectins, and thus these mucins may serve as adhesion molecules, too.

Conclusion

All properties of mucin-like domains are based on the extreme length and relative rigidity of these domains. Thus in normal cells, mucins may facilitate duct formation or maintenance of ductal structures, by preventing protein-protein interactions at opposite apical membranes. In carcinoma cells, the same anti-adhesive effects may be involved in dissemination of the carcinoma cells from their surroundings, thus enhancing their ability to metastasize, and the same anti-adhesion effect will also increase their chance to survive immune destruction. The anti-adhesive function is only effective when the mucin-like protein is overexpressed on the entire cell surface, so that most adhesion molecules are masked. It should be emphasized that overexpressing a mucin-like molecule is by no means the only way to obtain a less adherent cell. Lowering the effective amount of the adhesion molecules E-cadherin or integrins at the cell surface clearly has the same effect (Behrens et al., 1989). However, the mucin-like molecules nonspecifically affect all adhesion molecules, and therefore selection of cells that have upregulated a cell-associated mucin is likely to occur more frequently than that of cells that have downregulated a series of adhesion molecules to increase motility and thus the invasive potential.

Acknowledgements
This work was supported by the Dutch Cancer Society project number NKI 91-16 and NKI 93-523. Dr. C. Patriarca was on leave from the Department of Pathology, Ospedale S. Paolo, University of Milan, Italy.

References

Ardman, B., Sikorski, M.A. and Staunton, D.E. (1992) CD43 interferes with T-lymphocyte adhesion. *Proc. Natl. Acad. Sci. USA* 89: 5001–5005.
Baumhueter, S., Singer, M.S., Henzel, W., Hemmerich, S., Renz, M., Rosen, S.D. and Lasky, L.A. (1993) Binding of L-selectin to the vascular sialomucin CD34. *Science* 262: 436–438.
Bažil, V. and Strominger, J.L. (1993) CD43, the major sialoglycoprotein of human leukocytes, is proteolytically cleaved from the surface of stimulated lymphocytes and granulocytes. *Proc. Natl. Acad. Sci. USA* 90: 3792–3796.
Behrens, J., Mareel, M.M., VanRoy, F.M. and Birchmeier, W.N. (1988) Dissecting tumor cell invasion: Epithelial acquire invasive properties after the loss of uvomorulin-mediated cell-cell adhesion. *J. Cell Biol.* 108: 2435–2447.
Beppu, M., Codington, J.F., Lasky, R.D. and Jeanloz, R.W. (1987) Epiglycanin-immunoreactive glycoproteins in mouse fetal tissues and fetal cells in culture. *J. Natl. Cancer Inst.* 78: 1169–1175.

Bharathan, S., Moriarty, J., Moody, C.E. and Sherblom, A.P. (1990) Effect of tunicamycin on sialomucin and natural killer susceptibility of rat mammary tumor ascites cells. *Cancer Res.* 50: 5250–5256.

Calafat, J., Patriarca, C., Jansen, H., Peterse, H. and Hilkens, J. Structure of episialin and its effect on cell-cell and cell-matrix contacts *in vivo* assessed by electron microscopy and immunohistochemistry; *submitted.*

Carraway, K.L., Fregien, N., Carraway III, K.L. and Carothers-Carraway, C.A. (1992) Tumor sialomucin complexes as tumor antigens and modulators of cellular interactions and proliferation. *J. Cell Science* 103: 299–307.

Codington, J.F., Sanford, B.H. and Jeanioz, R.W. (1973) Cell-surface glycoproteins of two sublines of the TA3 tumor. *J. Natl. Cancer Inst.* 51: 585–591.

Codington, J.F., Klein, G., Cooper, A.G., Lee, N., Brown, M.C. and Jeanloz, R.W. (1978) Further studies on the relationship between large glycoprotein molecules and allotransplantability in the TA3 tumor of the mouse: studies on segregating TA3-Ha hybrids. *J. Natl. Cancer Inst.* 60: 811–818.

Croghan, G., Papsidero, L., Valenzuela, L., Nemoto, T., Penetrante, R. and Chu, T. (1983) Tissue distribution of an epithelial and tumor-associated antigen recognized by monoclonal antibody F36-22. *Cancer Res.* 43: 4980.

Cyster, J.G. and Williams, A.F. (1992) The importance of cross-linking in the homotypic aggregation of lymphocytes induced by anti-leukosialin (CD43) antibodies. *Eur. J. Immunol.* 22: 2565–2572.

Cyster, J.G., Shotton, D.M. and Williams, A.F. (1991) The dimensions of the T lymphocyte glycoprotein leukosialin and identification of linear protein epitopes that can be modified by glycosylation. *EMBO J.* 10: 893–902.

Delia, D., Lampugnani, M.G., Resnati, M., Dejana, E., Aiello, A., Fontanella, E., Soligo, D., Pierotti, M.A. and Greaves, M.F. (1993) CD34 expression is regulated reciprocally with adhesion molecules in vascular endothelial cells *in vitro. Blood* 81: 1001–1008.

Fung, P.Y.S. and Longenecker, B.M. (1991) Specific immunosuppressive activity of epiglycanin, a mucin-like glycoprotein secreted by a murine mammary adenocarcinoma (TA3-HA). *Cancer Res.* 51: 1170–1176.

Gendler, S.J., Lancaster, C.A., Taylor-Papadimitriou, J., Duhig, T., Peat, N., Burchell, J., Pumberton, L., Lalani, E.-N. and Wilson, D. (1990) Molecular cloning of human tumor-associated polymorphic epithelial mucin. *J. Biol. Chem.* 265: 15286–15293.

Hilkens, J., Ligtenberg, M.J.L., Vos, H.L. and Litvinov, S.V. (1992) The structure of cell-associated mucin-like molecules and their adhesion modulating property. *Trends Biochem. Sci.* 17: 359–363.

Hilkens, J. and Buys, F. (1988) Biosynthesis of MAM-6, an epithelial sialomucin; evidence for the involvement of a rare proteolytic cleavage step in the endoplasmic reticulum. *J. Biol. Chem.* 263: 4215–4222.

Hilkens, J., Buijs, F., Hilgers, J., Hageman, Ph., Calafat, J., Sonnenberg, A. and van der Valk, M. (1984) Monoclonal antibodies against human milk fat globule membranes detecting differentiation antigens of the mammary gland and its tumors. *Int. J. Cancer* 34: 197–206.

Hilkens, J., Buys, F. and Ligtenberg, M. (1989) Complexity of MAM-6, an epithelial sialomucin, associated with carcinomas. *Cancer. Res.* 49: 786–793.

Hilkens, J. (1992) CA 15-3 assay for detection of episialin; a serum marker for breast cancer. *In:* S. Sell (ed.): *Serological Cancer Markers.* Humana Press, Totowa, N.J., pp. 261–280.

Jentoft, N. (1990) Why are proteins O-glycosylated? *Trends in Biochem. Sci.* 15: 291–294.

Koukoulid, H.K., Virtanen, I., Korhonen, M., Laitnen, L., Quaranta, V. and Gould, V.E. (1991) Immunohistochemical localization of integrins in the normal, hyperplastic and neoplastic breast. *Am. J. Pathol.* 139: 787–799.

Kufe, D.W., Inghirami, G., Abe, M., Hayes, F., Justi, W.H. and Schlom, J. (1984) Differential reactivity of a novel monoclonal antibody (DF3) with human versus benign breast tumors. *Hybridoma* 3: 223.

Lan, M.S., Batra, S.K., Qi, W.-I., Metzger, R.S. and Hollingsworth, M.A. (1990) Cloning and sequencing of a human pancratic tumor mucin cDNA. *J. Biol. Chem.* 265: 15294–15299.

Ligtenberg, M.J.L., Vos, H.L., Gennissen, A.M.C. and Hilkens, J. (1990) Episialin, a carcinoma-associated mucin, is generated by a polymorphic gene encoding splice variants with alternative amino termini. *J. Biol. Chem.* 265: 5573–5578.

Ligtenberg, M.J.L., Kruijshaar, L., Bujis, F., van Meijer, M., Litvinov, S. and Hilkens, J. (1992a) Cell-associated episialin is a complex containing two proteins derived from a common precursor. *J. Biol. Chem.* 267: 6171–6177.

Ligtenberg, M.J.L., Buijs, F., Vos, H.L. and Hilkens, J. (1992b) Suppression of cellular aggregation by high levels of episialin. *Cancer Res.* 52: 2318–2324.

Linsley, P.S., Kallestad, J.C. and Horn, D. (1988) Biosynthesis of high molecular weight breast carcinoma associated mucin glycoproteins. *J. Biol. Chem.* 263: 8390–8397.

Litvinov, S.V. and Hilkens, J. (1993) The epithelial sialomucin, episialin, is sialylated during recycling. *J. Biol. Chem.* 268: 21364–21371.

Litvinov, S., Hageman, Ph., Buijs, F. and Hilkens, J. Episialin is released after complete maturation of its C-terminal domain; cell surface associated and released episialin differ in glycosylation; *submitted.*

Manjunath, N., Johnson, R.S., Staunton, D.E., Pasqualini, R. and Ardman, B. (1993) Targeted disruption of CD43 gene enhances T lymphocyte adhesion. *J. Immunol.* 151: 1528–1534.

Millar, S.C., Codington, J.F. and Klein, G. (1982) Further studies on the relationship between allotransplantability and the presence of cell surface glycoprotein epiglycanin in the TAS-MM mouse mammary carcinoma ascites cell. *J. Natl. Cancer Inst.* 68: 981–988.

Moriarty, J., Skelly, C.M., Bharathan, S., Moody, C.E. and Sherblom, A.P. (1990) Sialo-mucin and lytic susceptibility of rat mammary tumor ascites cells. *Cancer Res.* 50: 6800–6805.

Nathan, C., Xie, Q.-W., Halbwachs-Mecarelli, L. and Jin, W.W. (1993) Albumin inhibits neutrophil spreading and hydrogen peroxidase release by blocking the shedding of CD43 (sialophorin, leukosialin). *J. Cell Biol.* 122: 243–256.

Pignatelli, M., Cardillo, M.R., Hanby, A. and Stamp, G.W.H. (1992) Integrins and their accessory adhesion molecules in mammary carcinomas: Loss of polarization in poorly differentiated tumors. *Hum. Pathol.* 23: 1159–1166.

Rieu, P., Porteu, F., Bessou, G., Lesavre, P. and Halbwachs-Mecarelli, L. (1992) Human neutrophils release their major membrane sialoprotein, leukosialin (CD32), during cell activation. *Eur. J. Immunol.* 22: 3021–3026.

Rutishauser, U., Acheson, A., Hall, A.K., Mann, D.M. and Sunshine, J. (1988) The neural cell adhesion molecule (NCAM) as a regulator of cell-cell interactions. *Science* 240: 53–57.

Sherblom, A.P. and Moody, C.D. (1986) Cell surface sialomucin and resistance of natural cell-mediated cytotoxicity of rat mammary tumor ascites cell. *Cancer Res.* 46: 4543–4548.

Steck, P.A. and Nicholson, G.L. (1983) Cell surface glycoproteins of 13762NF mammary adenocarcinoma clones of differing metastatic potentials. *Exp. Cell Res.* 147: 255–267.

Van de Wiel-van Kemenade, E., Ligtenberg, M.J.L., de Boer, A.J., Buijs, F., Vos, H.L., Melief, C.J.M., Hilkens, J. and Figdor, C.G. (1993) Episialin (MUC1) inhibits cytotoxic lymphocyte-target cell interaction. *J. Immunol.* 151: 767–776.

Wesseling, J., van der Valk, S., Vos, H.L. and Hilkens, J. (1995) Episialin (MUC1) over-expression inhibits integrin mediated cell adhesion to extracellular matrix components. *J. Cell Biol.* 129; *in press.*

Wreschner, D.H., Hareuveni, M., Tsarfaty, I., Smorodinsky, N., Horev, J., Zaretsky, J., Kotkes, P., Weiss, M., Lathe, R., Dion, A. and Keydar, I. (1990) Human epithelial tumor antigen cDNA sequences: Differential splicing may generate multiple protein forms. *Eur. J. Biochem.* 189: 463–473, 1990.

Yang, P., Yin, X. and Rutishauser, U. (1992) Intercellular space is affected by the polysialic acid content of NCAM. *J. Cell Biol.* 116: 1487–1496.

Yonemura, S., Nagafuchi, A., Sato, N. and Tsukita, S. (1993) Concentration of an integral membrane protein, CD43 (leukosialin, sialophorin), in the cleavage furrow through the interaction of its cytoplasmic domain with actin-based cytoskeletons. *J. Cell Biol.* 120: 437–449.

Zotter, S., Lossnitser, A., Hageman, P.C., Delemarre, J.F.M., Hilkens, J. and Hilgers, J. (1987) Immunohistochemical localization of the epithelial marker MAM-6 in invasive malignancies and highly dysplastic adenomas of the large intesting. *Lab. Invest.* 57: 193–199.

Biochemistry of Cell Membranes
ed. by S. Papa & J. M. Tager
© 1995 Birkhäuser Verlag Basel/Switzerland

Role of β1 integrin cytoplasmic domain in signaling and cell adhesion

P. Defilippi, F. Balzac, F. Retta, C. Bozzo, A. Melchiorri[2], M. Geuna[1], L. Silengo and G. Tarone

Department of Genetic, Biology and Medical Chemistry; [1]Department of Biomedical Sciences and Oncology, University of Torino; [2]Istituto Nazionale per la Ricerca sul Cancro, Genova, Italy

Summary. We have investigated the intracellular signaling events generated during the interaction of integrins with extracellular matrix proteins in human endothelial cells (HEC). Within 30 s from adhesion to fibronectin, HEC showed increased tyrosine phosphorylation of a group of proteins with molecular mass of 100–130 and 70 kDa. Tyrosine phosphorylation of these proteins was triggered by several extracellular matrix ligands, including collagens type I and IV, laminin and vitronectin and could be mimicked by antibodies to $\alpha3\beta1$, $\alpha5\beta1$, and $\alpha6\beta1$ integrin complexes. Among the tyrosine phosphoproteins regulated by integrins, we identified the focal adhesion tyrosine kinase p125FAK. Inhibition of tyrosine phosphorylation with genistein during adhesion of HEC to fibronectin severly prevented the organization of focal adhesions and actin stress fibers, but did not inhibit adhesion *per se*. The role of β1 integrins in stimulating intracellular tyrosine phosphorylation was further investigated by analyzing a naturally occurring variant of β1 integrin (β1B) with a distinct cytoplasmic domain (Altruda et al., 1990). β1B, expressed in Chinese hamster ovary (CHO) cells, formed heterodimers with the α3 and α5 subunits and bind fibronectin in a RGD-dependent manner (Balzac et al., 1993). The $\alpha5\beta$1B complex, however, did not stimulate tyrosine phosphorylation of the 100–130 kDa proteins and did not localize at focal adhesions. Moreover, cells expressing β1B showed reduced migration in a Boyden chamber assay and poor spreading on fibronectin and laminin. This was not a generalized adhesive defect since cell spreading on vitronectin, a β3 integrin-dependent adhesion, was unchanged. These data show that β1 integrin cytoplasmic domain is critical in the stimulation of intracellular protein tyrosine phosphorylation; this signalling, moreover, is important in the organization of actin stress fibers and in the control of cell spreading and migration on matrix proteins.

Introduction

The interaction of cells with the extracellular matrix regulates several important cellular properties including proliferation and differentiation. Anchorage to a growth substratum is a prerequisite for fibroblast and epithelial cell growth in vitro. Cell adhesion in particular is required for the G1-S phase transition during the cell cycle (Guadagno and Assoian, 1991), but not for progression in G1 or in S phases. Cell differentiation is also regulated by matrix proteins. In keratinocytes adhesion to fibronectin inhibits differentiation (Adams and Watt, 1989) while in muscle cells fibronectin and its receptor play a positive role in differentiation (Menko and Boettiger, 1987). It is thus likely that cell-matrix interaction generates intracellular signaling that affects cellular response to different stimuli. The receptors involved in cell-matrix adhesion belong to the integrin family of adhesion molecules (for a review see Hynes, 1992). Integrins are heterodimers of α and β subunits connecting the extracellular environment to the contractile cytoskeleton inside the cell. For a long time it has been thought that the interactions mediated by integrins were purely mechanical. It is now emerging that integrins transduce signals from the outside of the cell to the cytoplasm. Evidence for such a property was first provided by showing that adhesion of cells to matrix proteins or to integrin antibodies modifies gene expression (Eierman et al., 1989; Werb et al., 1989). The integrin-mediated signaling pathways so far identified include the regulation of Na^+/H^+ antiporter (Schwartz et al., 1991), Ca^{2+} influx (Pardi et al., 1989; Miyauchi et al., 1991; Schwartz, 1993), stimulation of inositol lipid synthesis (McNamee et al., 1993) and protein tyrosine phosphorylation of a 130 kDa protein (Kornberg et al., 1991; Guan et al., 1991). This tyrosine phosphorylated protein was identified as the cytosolic tyrosine kinase p125[FAK] (Focal Adhesion Kinase) which is specifically localized in the focal adhesion (Schaller et al., 1992; Kornberg et al., 1992; Guan and Shalloway, 1992; Lipfert et al., 1992; Burridge et al., 1992).

We have analyzed the mechanisms of tyrosine phosphorylation mediated by integrins in human endothelial cells in culture. Moreover, by analyzing a natural variant of $\beta 1$ integrin with a distinct sequence at the cytoplasmic domain ($\beta 1B$), we show that activation of tyrosine phosphorylation is strictly correlated with actin stress fibers organization, cell spreading, and migration.

Materials and Methods

Cells, constructs and transfections

Human endothelial cells (HEC) were prepared from umbilical vein and cultured as described previously (Defilippi et al., 1991). Stable transfectants of Chinese hamster ovary (CHO) cells expressing the human integrin $\beta 1A$ or $\beta 1B$ were obtained as described previously (Balzac et al., 1993). Neomycin resistant clones were selected in HAM F12 medium with 10% fetal calf serum and 800 µg/ml of G418 (Gibco).

Adhesion experiments and detection of phosphotyrosine-containing proteins by western blotting

Tissue culture plates were coated with 10 µg/ml purified matrix proteins by overnight incubation at 4°C and post-coated with bovine serum albumin (Sigma) for 1 h at 37°C. Poly-L-lysine (PL) (Sigma) was used at 100 µg/ml. For the adhesion to integrin monoclonal antibodies (MAb), tissue culture plates were first coated overnight at 4°C with 10 µg/ml goat anti-mouse IgG (Sigma), post-coated with bovine serum albumin for 1 h at 37°C, and incubated with 10 µg/ml of purified MAbs for 2 h at 37°C. The following antibodies were used: MAb GOH3 to $\alpha 6$; MAb Sam1 to $\alpha 5$; MAb F4 to $\alpha 3$; MAb TS2/16 and MAb LM534 to human $\beta 1$; MAb 7E2 to hamster $\beta 1$; MAb B212 to $\beta 3$.

Cells at confluence were detached by EDTA treatment (5 mM), resuspended in prewarmed DMEM medium and seeded on coated culture plates for the indicated times. To eliminate the contribution of protein synthesis and secretion, cells were pretreated 2 h with 20 µM cycloheximide (Sigma) to prevent protein synthesis, and plated in the presence of 20 µM cycloheximide and 1 µM monensin (Sigma) to prevent secretion (Defilippi et al., 1991). At the end of the incubation, the cells were washed twice with a stop solution (5 mM EDTA, 10 mM NaF, 10 mM $Na_4P_2O_7$, 0.4 mM Na_3VO_4 in PBS), and detergent extracted in lysis buffer (1% NP-40, 150 mM NaCl, 50 mM Tris-Cl pH8, 5 mM EDTA, 10 mM NaF, 10 mM $Na_4P_2O_7$, 0.4 mM Na_3VO_4, 10 µg/ml leupeptin, 4 µg/ml pepstatin and 0.1 trypsin inhibitory unit/ml aprotinin) (all from Sigma).

Samples containing equal amounts of proteins were separated on 6% polyacrylamide gel electrophoresis in presence of SDS (SDS-PAGE) in reducing conditions. Proteins were transferred to nitrocellulose using a semi-dry apparatus (Novablot, Pharmacia) according to the manufacturer's instructions. Phosphotyrosine-containing proteins were visualized with 1 µg/ml phosphotyrosine MAbs (PY20, ICN Flow or PT66, Sigma) followed by anti-mouse IgG-peroxidase and a chemiluminescent substrate (ECL Amersham, UK).

Identification of phosphorylated focal adhesion kinase p125^{FAK}

Cell extracts were immunoprecipitated with MAb 2A7 to p125^{FAK} (gift of Dr. T. Parson). The immunocomplexes were bound to Protein-A Sepharose beads; the bound material was eluted by boiling beads in 1% SDS and subjected to SDS-PAGE. After transfer to nitrocellulose, blots were incubated with MAb to phosphotyrosine as described above.

Fluorescence microscopy

For immunofluorescence microscopy, cells were plated on glass coverslips coated with 10 µg/ml FN overnight, fixed in 3% paraformaldehyde and permeabilized with 0.1% Triton X-100. After washing, cells were incubated with one of the following antibodies: MAb PY20 to anti phosphotyrosine (10 µg/ml), MAb 2A7 to p125^{FAK} (5 µg/ml), MAb 7F9 to vinculin (1:5 dilution of hybridoma culture medium), MAb 7E2 to hamster β1 (5 µg/ml). The cells were then washed and exposed to the appropriate rhodamine-conjugated secondary antibodies. To obtain double-labeling MAb LM534 to β1 integrin was biotinilated according to Bayer and Wilchek (1980) and detected with rhodamine-conjugated extravidin (Sigma); fluorescein-labeled PT66 MAb (Sigma) to phospho-tyrosine was used in combination. Actin fibers were visualized with fluorescein-labeled phalloidin (Sigma). The coverslips were mounted in PBS:glycerol, 1:1, and viewed on an Olympus BH2-RFCA fluorescence microscope. Micrographs were taken with Kodak 400 film.

Boyden chamber migration assay

Cell migration was evaluated in the Boyden chamber assay (Albini et al., 1987). CHO cells and transfected clones were plated (2 × 10^5 cells/ml) on the 8 µm pore size filter coated with gelatine (5 µg/ml, Sigma) in the upper compartment. The lower compartment was filled with either RPMI medium containing 0.1% BSA as control, or with chemo-attractants consisting of RPMI with 25 µg/ml fibronectin. Cells were allowed to migrate 6 h at 37°C in a humidified atmosphere containing 5% CO_2. Cells on the lower surface of the filter were fixed, stained, and counted.

Results and Discussion

Protein tyrosine phosphorylation in HEC adhering to fibronectin

HEC were allowed to adhere to tissue culture dishes coated with either fibronectin (FN) or polylysine (PL), as a non-specific substrate and the detergent extracts were separated by SDS-PAGE, transferred to nitro-cellulose and processed for detection of phosphotyrosine-containing proteins. After 30 min of adhesion on fibronectin a strong increase in tyrosine phosphorylation of proteins migrating with an apparent molecular mass of 100–130 kDa and of a 70 kDa protein was detected (Fig. 1, lane b). In addition to these molecules three groups of tyro-sine phosphoproteins with molecular mass of 180–200 kDa, 85 kDa, and 60 kDa were detected. Their phosphorylation, however, was not stimulated after adhesion to fibronectin. The level of tyrosine phos-phorylation of the 100–130 kDa and 70 kDa proteins was rapidly downregulated when cells were detached from the culture dishes by EDTA or trypsin treatment (not shown), indicating that adhesion

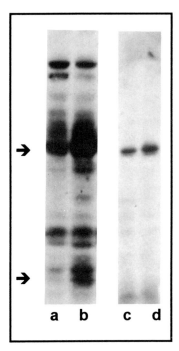

Figure 1. HEC adhesion to FN induces protein tyrosine phosphorylation. HEC were plated on PL (a, c) or FN (b, d) coated dishes for 30 min. Cell lysates were either directly run on SDS-PAGE (a, b) or immunoprecipitated with 2A7 MAb to p125^{FAK} and subsequently separated by electrophoresis (c, d). After western blotting phosphotyrosine proteins were visualized with MAb PY20, followed by peroxidase-labeled anti-mouse IgG and chemiluminescent substrate. Arrows indicate the 100–130 kDa and 70 kDa phosphoproteins.

to the growth substratum is required to maintain high level of phosphorylation.

We tested the possibility that the group of the 100–130 kDa proteins included the p125^{FAK}, a tyrosine kinase regulated during adhesion to fibronectin (Schaller et al., 1992; Guan and Shalloway, 1992; Kornberg et al., 1992). HEC plated on PL and on FN-coated dishes for 30 min were immunoprecipitated with the p125^{FAK} MAb 2A7. As shown in Figure 1 (lane c, d), MAb 2A7 immunoprecipitated a protein whose tyrosine phosphorylation was specifically stimulated in HEC adherent to FN showing that the p125^{FAK} in endothelial cells represents the major component of the 100–130 kDa phosphoproteins.

The time-course of protein tyrosine phosphorylation was determined by plating HEC on FN-coated dishes for different lengths of time. Increased phosphorylation of the 100–130 kDa proteins was detected 30 s after adhesion to FN. The phosphorylation increased very rapidly, reaching a maximum at 15 min, and persisted at high level for at least 3 h (Fig. 2). The phosphorylation of the 70 kDa protein increased

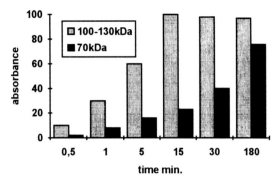

Figure 2. Time-course of protein tyrosine phosphorylation in HEC. HEC were centrifuged in cold medium on dishes coated with FN. Pre-warmed medium (37°C) was added and the cells were incubated at 37°C for the indicated times. Cell lysates were run on SDS-PAGE, transferred to nitrocellulose, and blotted with PY20 phosphotyrosine MAb. The absorbance of the bands corresponding to 100–130 kDa and 70 kDa phosphoproteins was determined by densitometric analysis. The absorbance values on the y axis are expressed in arbitrary units. Light gray bars: 100–130 kDa proteins; dark gray bars: 70 kDa protein.

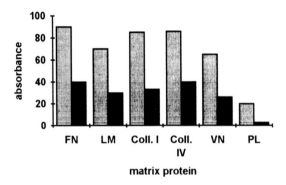

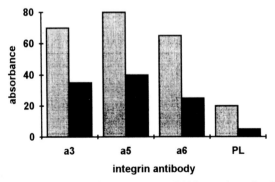

Figure 3. Tyrosine phosphorylation in response to ECM proteins, via different integrin complexes. HEC were plated for 30 min on dishes coated with PL, vitronectin, collagen I, collagen IV, laminin, fibronectin, MAb F4 to α3, MAb Sam1 to α5, MAb GOH3 to α6. Cell extracts were run on 8% SDS-PAGE, transferred to nitrocellulose and blotted with PY20

linearly within the first 3 h of adhesion: the rate of phosphorylation of the protein was slower compared to that observed with the 100–130 kDa molecules (Fig. 2).

Tyrosine phosphorylation of the 100–130 kDa and 70 kDa proteins is integrin-dependent

Plating of HEC on laminin, vitronectin, collagen I, and collagen IV induced tyrosine phosphorylation of the 100–130 kDa and of the 70 kDa proteins (Fig. 3) showing that different matrix proteins are able to elicit a similar tyrosine phosphorylation response. We next tested MAbs directed to α3, α5, and α6 integrin subunits that, in association with β1, form receptors for collagen, fibronectin, and laminin. Adhesion of HEC to each of the three MAbs adsorbed on culture dishes induced the tyrosine phosphorylation of the 100–130 and of the 70 kDa proteins (Fig. 3). Similar results were observed by plating cells on MAbs to integrin β3 that in association with αV forms the vitronectin receptor. These data show that the protein phosphorylation observed by plating cells on ECM proteins is mediated by the integrins expressed on HEC surface and demonstrate that integrin complexes consisting of different α and β subunits are able to elicit similar tyrosine phosphorylation signals.

Role of phosphotyrosine kinases in assembly of cytoskeletal and adhesion structures

Double fluorescence analysis of HEC plated for 3 h on FN-coated coverslips, showed complete co-localization of the tyrosine phosphorylated proteins with the β1 integrins in the focal adhesions (Fig. 4, panels a and b). Staining with MAb 2A7 showed that the p125^{FAK} kinase also co-distributes in these adhesive structures (not shown). To analyze the role of tyrosine phosphorylation in adhesion, endothelial cells were plated on FN-coated glass coverslips for 6 h in the presence of 100 μg/ml genistein, a known tyrosine kinase inhibitor (Akiyama et al., 1987). Under these conditions the phosphotyrosine level in cell extracts was inhibited by 50%, as determined by Western Blot analysis of HEC extracts. Treated cells adhere to the substratum, but did not spread

phosphotyrosine MAb. The absorbance of the bands corresponding to 100–130 kDa and 70 kDa phosphoproteins was determined by densitometric analysis. The absorbance values on the y axis are expressed in arbitrary units. Light gray bars: 100–130 kDa proteins; dark gray bars: 70 kDa protein.

completely; moreover, phosphotyrosine proteins, β1 integrin and vinculin did not organize in focal adhesions and remained concentrated in small dots at the cell periphery (Figure 4, panels c, d, c′). To test whether the inhibition of tyrosine kinases affected pre-organized adhesive structures genistein was also added at endothelial cells cultured on glass coverslips for 24 h. After 6 h treatment with 100 µg/ml genistein, the cells remained adherent but the phosphotyrosyl proteins were dispersed rather than localized to focal adhesion and the organization of preexistent actin stress fibers was disrupted (not shown). A weak reactivity for phosphotyrosyl proteins and β1 integrins was confined to a thin line at cell boundaries. These data suggest that tyrosine phosphorylation is a key event in the organization and maintenance of adhesion plaques structure.

Analysis of β1 variant to assess the cytoplasmic domain function in signaling and adhesion

We have previously identified a naturally occurring variant of β1 integrin (β1B) with a unique sequence in the cytoplasmic domain. In

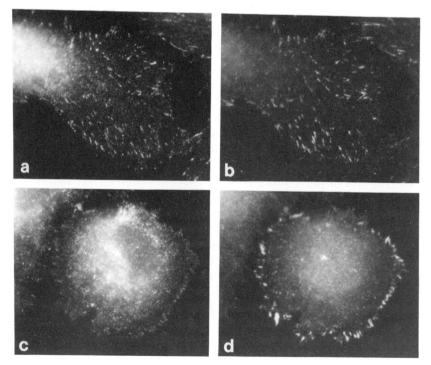

Figure 4(a–d)

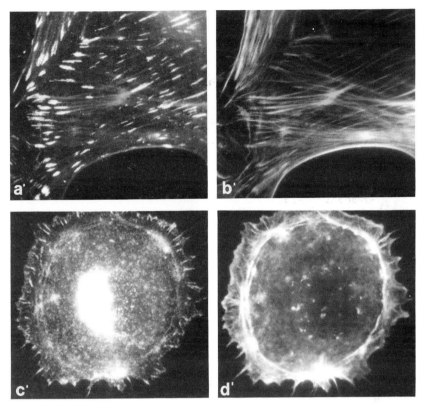

Figure 4. Inhibition of protein tyrosine kinases affects focal adhesion and cytoskeleton organization. To evaluate the role of tyrosine kinases during cell attachment, HEC were plated on FN-coated coverslips in the presence of the tyrosine kinase inhibitor genistein (100 µg/ml) for 6 h (c, d; c′, d′). Control untreated cells are shown in a, a′, b, and b′. Cells were double stained for: phosphotyrosine (a, c) and β1 (b, d), or vinculin (a′, c′) and F-actin (b′, d′).

this molecule a sequence of 11 amino acids replaces the last 22 COOH terminal amino acids characteristic of β1 (Altruda et al., 1990) (Fig. 5). β1B is preferentially expressed in keratinocytes and hepatocytes, while the conventional β1 (indicated as β1A) has a ubiquitous distribution (Balzac et al., 1993). We have taken advantage of this variant to investigate the role of the β1 integrin cytoplasmic domain in the activation of the tyrosine phosphorylation cascade. To do so, we expressed β1B in cells that do not express this variant by transfecting hamster CHO cells with human β1B cDNA under control of the early SV40 promoter. Transfectants with the human β1A isoform were also selected as a control (Balzac et al., 1993).

The transfected cells express the β1B at the cell surface, as determined by immunoprecipitation from cells labeled with lactoperoxidase cata-

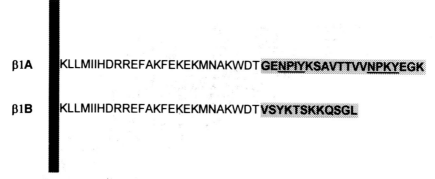

β1A   KLLMIIHDRREFAKFEKEKMNAKWDT GENPIYKSAVTTVVNPKYEGK

β1B   KLLMIIHDRREFAKFEKEKMNAKWDT VSYKTSKKQSGL

Figure 5. Amino acid sequence of β1A and β1B cytoplasmic domains. The vertical dark gray bar represents the membrane lipid bilayer. The sequences unique to the two molecules are shaded and in bold type (Altruda et al., 1990) The underlined residues in β1A are important for focal adhesion localization (Reszka et al., 1992).

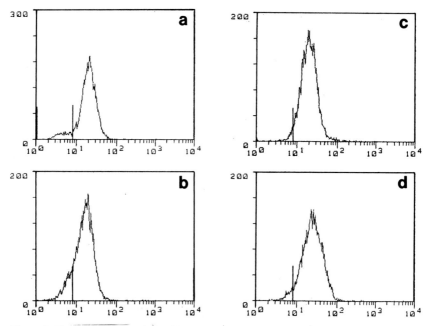

Figure 6. Flow cytometry analysis of human β1 integrin expression on transfected CHO cells. Cells of clones β1A22 (a), β1B-25 (b), β1B-18 (c), and β1B-50 (d) were detached with EDTA and incubated in suspension with the MAb LM534 specific for the transfected human β1, followed by FITC-labeled secondary antibody. Y axis: cell number; x axis: fluorescence intensity. The vertical bar represents the boundary between the negative values (on the left), and the positive values (on the right).

lyzed radio iodination and by fluorescence activated cell sorting (FACS) (Fig. 6). The transfected human and the endogenous hamster β1 integrins were identified with monoclonal antibodies that selectively react with the human or hamster protein. Three clones expressing different levels of β1B and one clone transfected with human β1A were selected. The level of expression of the transfected and endogenous β1 proteins at the cell surface was determined by measuring the radioactivity incorporated in the proteins after surface radioiodination and SDS-PAGE. The clones are indicated as β1B-18, β1B-25, β1B-50 and β1A-22; the numbers indicate the amount of β1B at the cells surface expressed as % of the total β1 (endogenous plus transfected).

β1B integrin does not trigger tyrosine phosphorylation of a 125 kDa protein

To test whether β1B was effective in activating the tyrosine phosphorylation pathway, cells were plated on plastic dishes coated with monoclonal antibodies specific for either the transfected human or the endogenous hamster β1. As shown in Figure 7, the transfected β1B did not significantly stimulate the tyrosine phosphorylation of the 125 kDa protein, while the endogenous β1A was active, showing that the signaling pathway was not altered in these cells. In clone β1A-22 both the endogenous and the transfected β1A integrins were equally able to stimulate tyrosine phosphorylation of a 125 kDa cytoplasmic protein, indicating that the human protein is active in the hamster cells and that the lack of function of β1B is related to the structural difference between the two integrin isoforms. Similar results were obtained with all three β1B transfected clones.

Expression of β1B results in reduced cell spreading

To test the adhesive properties of the β1B, transfected cells were placed on dishes coated with antibodies specific for the transfected human protein allowing adhesion via the variant integrin. As shown in Figure 8 (a, b), when the β1B isoform is involved in cell-substratum adhesion, cells attach but remain round and do not spread normally.

We then tested the adhesive properties of transfected CHO cells under conditions where both the endogenous and transfected β1 integrins mediate adhesion by plating cells on dishes coated with purified matrix proteins. Also in this case transfectants expressing β1B spread significantly less on fibronectin and on laminin substrata compared to clones expressing human β1A (Fig. 8). On the other hand, cell spreading on vitronectin, a β3 integrin-dependent adhesion, was comparable in β1A and β1B transfectants (not shown), indicating that expression of β1B

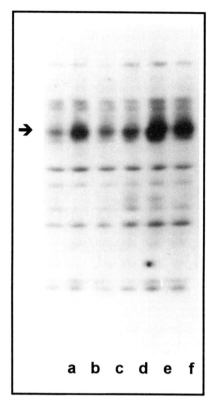

Figure 7. Tyrosine phosphorylation of p125 protein in β1A and β1B transfectants. β1A-22 and β1B-25 cells were plated on dishes coated with MAb 7E2 to the endogenous β1 (b, e) or with MAb TS2/16 to the transfected human β1 (c, f). Controls were obtained by plating cells on polylysine coated plates (a, d). After detergent extraction, equal amounts of proteins were separated by SDS-PAGE and transferred to nitrocellulose. Proteins containing phosphotyrosine were visualized by the monoclonal antibody PT66 and chemiluminescence reaction. The position of the 125kDa protein is shown by the arrow.

does not lead to a generalized spreading defect. The total number of cells adherent to dishes coated with fibronectin or laminin was not significantly different in clones β1B-18, β1B-25 and β1A-22 cells. Only clone β1B-50 showed reduced adhesion on both substrates, suggesting that reduction in adhesion occurs only when β1B expression exceeds a threshold level.

Expression of β1B results in reduced cell motility

Having established that β1B does not support cell spreading, we evaluated the motile properties of the transfected cells by measuring cell migration through a porous filter in a Boyden chamber. Soluble

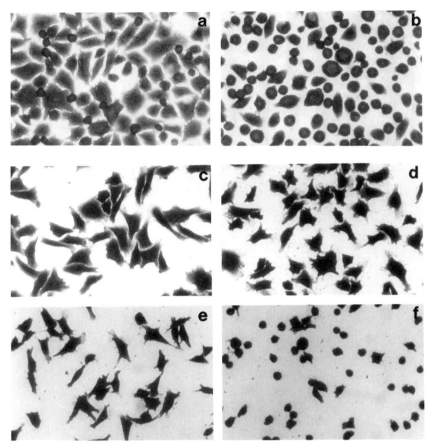

Figure 8. Adhesion and spreading of β1A and β1B transfected cells. β1A-22 (a) and β1B-50 (b) cells were plated on dishes coated with MAb TS2/16 to the transfected human β1 and allowed to adhere for 1 h at 37°C. β1A-22 (c, e) and β1B-50 (d, f) cells were plated on dishes coated with: 2 μg/ml of purified human plasma fibronectin (c, d); 10 μg/ml of purified mouse laminin (e, f) and allowed to adhere for 1 h at 37°C. Cells were photographed after fixation and staining.

fibronectin or medium conditioned by 3T3 cells was added in the bottom compartment as chemoattractant. As shown in Figure 9, β1B-25 and β1B-50 cells showed a significant reduction of migration compared to β1A-22 cells. Counting of the cells on the lower filter surface indicated that migration was inhibited by 60–75% depending on the clones.

β1B does not localize at focal adhesion in adherent cells

The subcellular localization of the β1B in transfected CHO cells, was analyzed by immunofluorescence. As shown in Figure 10, the β1B

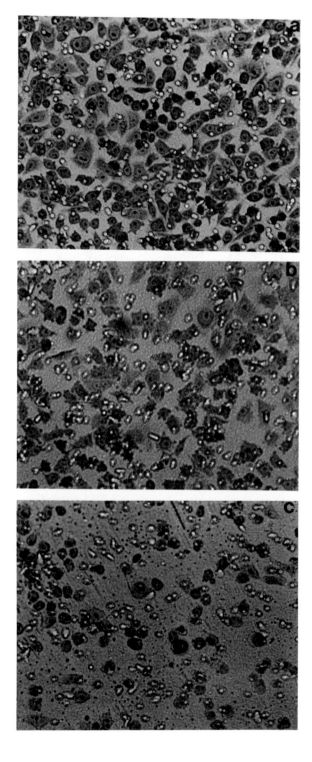

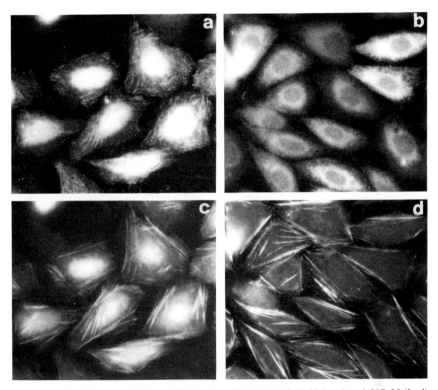

Figure 10. Localization of β1A and β1B in transfected cells. β1A-22 (a, c) and β1B-25 (b, d) cells, plated on fibronectin-coated glass coverslips, were fixed with formaldehyde and permeabilized with detergent. β1A and β1B were localized by staining with monoclonal antibody LM534 to human β1 followed by rhodamine labeled secondary antibody. Actin was visualized in the same cells by co-staining with fluorescein-labeled phalloidin.

molecule is uniformly distributed at the cell surface and does not concentrate at focal adhesions nor does it co-localize with stress fiber ends (Figure 10, panels b, d). Moreover, in β1B transfected cells, the endogenous hamster β1A localized to focal contacts (not shown), indicating that integrins in these cells retain the ability to cluster at cell substratum contacts. To verify whether human integrin molecules behave correctly in CHO cells, we then analyzed CHO cells transfected with human β1A. As shown in Figure 10, the human β1A co-localizes with actin stress fibers (panels a, c).

Figure 9. Migration of β1A and β1B transfected cells. β1A-22 (a), β1B-25 (b) and β1B-50 (c) cells were plated on the surface of a porous filter and allowed to migrate through the pores using fibronectin as chemoattractant. The pictures show the cells that had migrated through the pores after 6 h. The white round dots are the filter pores.

Conclusions

The data reported show that interaction of integrins with their extracellular matrix ligands triggers rapid increase of tyrosine phosphorylation of a group of proteins with an apparent molecular mass of 100–130 kDa and 70 kDa. The dominant component in the group of 100–130 kDa phosphoproteins was identified as the focal adhesion kinase p125^{FAK} (Kanner et al., 1990; Schaller et al., 1992). The 70 kDa protein may be related to paxillin, a protein localized in focal contacts, whose tyrosine phosphorylation during adhesion has been recently described (Burridge et al., 1992). Phosphorylation of these molecules persisted as long as the cells adhere to the substratum, but dropped to very low levels as soon as the cells are detached from the culture dish. This property may be related to the persistent occupancy of integrins by their ligands during adhesion.

Our data indicate that integrin-induced tyrosine phosphorylation of intracellular substrates is important in the organization of the actin skeleton and focal adhesions. Immunofluorescence experiments showed, in fact, that phosphotyrosyl proteins and p125^{FAK} are localized in adhesive structures, in correspondence to integrins and actin fiber ends. Moreover, inhibition of protein tyrosine phosphorylation by genistein, a tyrosine kinase inhibitor, leads to inhibition of the organization of integrins, vinculin, and actin stress fibers at the focal adhesion.

An interesting finding of our analysis is that several integrin heterodimers in the same cell can stimulate tyrosine phosphorylation of the 100–130 kDa and 70 kDa proteins during cell substrate adhesion. In fact, adhesion of cells to antibody-coated dishes showed that $\alpha 3/\beta 1$, $\alpha 5/\beta 1$, and $\alpha 6/\beta 1$ can promote tyrosine phosphorylation of the above components. This suggests that the $\beta 1$ subunit, sheared by all three integrin complexes, plays a major role in this signaling pathway. In addition, the $\alpha 3$, $\alpha 5$, and $\alpha 6$ subunits do not appear to affect the phosphorylation of a specific target molecule as far as the 100–130 kDa and the 70 kDa proteins are concerned.

The role of $\beta 1$ in the activation of tyrosine phosphorylation cascade was demonstrated by analyzing the natural variant $\beta 1B$ characterized by a unique sequence at the cytoplasmic domain. This variant did not trigger increased tyrosine phosphorylation of the cytoplasmic p125 kDa protein. The $\beta 1B$ variant was also ineffective in promoting cell spreading and migration and did not localize at focal adhesions.

All together, these data show that association with actin fibers at focal adhesion and signaling through tyrosine kinases are important for the adhesive function of $\beta 1$.

Acknowledgements
We are grateful to Dr. T. Parson (Univ. of Virginia, USA) for the gift of 2A7 Mab. We also thank Dr. R. Juliano (Univ. of North Carolina, USA) for the 7E2 Mab, Dr. L. Zardi (IST,

Genova, Italy) for the F4 Mab, and Dr. F. Sanchez-Madrid (Madrid, Spain), for the TS2/16 Mab. This work was supported by grants of the National Research Council "Progetto finalizzato ACRO 9202149PF39"; of the Italian Association for Cancer Research (AIRC) and of the European Economic Community, Biomed Project.

References

Adams, J.C. and Watt, F.M. (1989) Fibronectin inhibits the terminal differentiation of human keratinocytes. *Nature* 340: 307–309.

Albini, A., Allavena, G., Melchiorri, A., Giancotti, F., Richter, H., Comoglio, P.M., Parodi, S., Martin, G.R. and Tarone, G. (1987) Chemotaxis of 3T3 and SV3T3 cells to fibronectin is mediated through the cell attachement site in fibronectin and a fibronectin cell surface receptor. *J. Cell Biol.* 105: 1867–1872.

Altruda, F., Cervella, P., Tarone, G., Botta, C., Balzac, F., Stefanuto, G. and Silengo, L. (1990) A human integrin β1 with a unique cytoplasmic domain generated by alternative mRNA processing. *Gene* 95: 261–266.

Akiyama, T., Ishida, J., Nakagawa, S., Ogawara, H., Watanabe, S., Itoh, N., Shibuya, M. and Fukami, Y. (1987) Genistein, a specific inhibitor of tyrosine-specific protein kinases. *J. Biol. Chem.* 262: 5592–5595.

Balzac, F., Belkin, A., Koteliansky, V., Balabanow, Y., Altruda, F., Silengo, L. and Tarone, G. (1993) Expression and functional analysis of a cytoplasmic domain variant of the β1 integrin subunit. *J. Cell Biol.* 121: 171–178.

Bayer, E.A. and Wilchek, M. (1980) The use of avidin-biotin complex as a tool in molecular biology. *Methods Biochem. Anal.* 26: 1–45.

Burridge, C.A., Turner, C. and Romer, L. (1992) Tyrosine phosphorylation of paxillin and pp125FAK accompanies cell adhesion to extracellular matrix: a role in cytoskeletal assembly. *J. Cell Biol.* 119: 893–904.

Defilippi, P., Truffa, G., Stefanuto, G., Altruda, F., Silengo, L. and Tarone, G. (1991) Tumor necrosis factor α and interferon gamma modulate the expression of the vitronectin receptor (integrin β3) in human endothelial cells. *J. Biol. Chem.* 266: 7638–7645.

Eierman, D.F., Johnson, C.E. and Haskill, J.S. (1989) Human monocyte inflammatory mediator gene expression is selectively regulated by adherence substrates. *J. Immunol.* 142: 1970–1976.

Guadagno, T.M. and Assoian, R.K. (1991) G1/S control of anchorage-independent growth in the fibroblast cell cycle. *J. Cell. Biol.* 115: 1419–1425.

Guan, J.L. and Shalloway, D. (1992) Regulation of focal adhesion associated protein tyrosine kinase by both cellular adhesion and oncogenic transformation. *Nature*, 358: 690–692.

Guan, J.L., Trevethnick, J.E. and Hynes, R.O. (1991) Fibronectin/integrin interaction induces tyrosine phosphorylation of a 120 kD protein. *Cell Regul.* 2: 951–964.

Hynes, R.O. (1992) Integrins: versatility, modulation and signalling in cell adhesion. *Cell* 69: 11–25.

Kanner, S.B., Reynolds, A.B., Vines, R.R. and Parsons, J.T. (1990) Monoclonal antibodies to individual tyrosine kinase phosphorylated protein substrates of oncogene encoded tyrosine kinases. *Proc. Natl. Acad. Sci. USA* 87: 10222–10226.

Kornberg, L., Earp, H.S., Turner, C., Prokop, C. and Juliano, R.L. (1991) Signal transduction by integrins: increased protein phosphorylation caused by clustering of β1 integrins. *Proc. Natl. Acad. Sci. USA* 88: 8392–8396.

Kornberg, L., Earp, H.S., Parsons, J.T., Schaller, M. and Juliano, R.L. (1992) Cell adhesion or integrin clustering increases phosphorylation of a focal adhesion associated tyrosine kinase. *J. Biol. Chem.* 267: 23439–23442.

Lipfert, L., Haimovich, B., Schaller, M.D., Cobb, B.S., Parsons, J.T. and Brugge, J.S. (1992) Integrin-dependent phopshorylation and activation of the protein kinase pp125FAK in platelets. *J. Cell. Biol.* 119: 905–912.

McNamee, H.P., Ingberg, D.E. and Schwartz, M.A. (1993) Adhesion to fibronectin stimulates inositol lipid synthesis and enhances PDGF- induced inositol lipid breakdown. *J. Cell Biol.* 121: 673–678.

Menko, A.S. and Boettiger, D. (1987) Occupation of the extracellular matrix receptor, integrin, is a central point for myogenic differentiation. *Cell* 51: 51–57.

Miyauki, A., Alvarez, J., Greenfield, E., Teti, A., Grano, M., Colucci, S., Zambonin-Zallone, A., Ross, F.P., Teitelbaum, S.L., Cheresh, D. and Hruska, K.A. (1991) Recognition of osteopontin and related peptides by an $\alpha V/\beta 3$ integrin stimulate cell signals in osteoclasts. *J. Biol. Chem.* 266: 20396–20374.

Pardi, R., Bender, J.R., Dettori, C., Giannazza, E. and Engelman, E.G. (1989) Heterogeneous distribution and transmembrane signaling properties of lymphocyte function-associated antigen (LFA-1) in human lymphocyte subsets. *J. Immunol.* 143: 3157–3166.

Reszka, A.A., Hayashi, Y. and Horwitz, A.F. (1992) Identification of amino acid sequences in the integrin beta1 cytoplasmic domain implicated in cytoskeletal association. *J. Cell. Biol.* 117: 1321–1330.

Schaller, M.D., Borgman, C.A., Cobb, B.S., Vines, R.R., Reynolds, A.B. and Parsons, J.T. (1992) pp125FAK, a structurally unique protein tyrosine kinase associated with focal adhesion. *Proc. Natl. Acad. Sci. USA* 89: 5192–5196.

Schwartz, M.A. (1993) Spreading of human endothelial cells on fibronectin or vitronectin triggers elevation of intracellular free calcium. *J. Cell Biol.* 120: 1003–1010.

Schwartz, M.A., Lechene, C. and Ingber, D.E. (1991) Insoluble fibronectin activates the Na/H antiporter by clustering and immobilizing integrin $\alpha 5/\beta 1$, independent of cell shape. *Proc. Natl. Acad. Sci. USA* 88: 7849–7853.

Werb, Z., Tremble, P.M., Behrendtsen, O., Crowley, E. and Damsky, C.H. (1989) Signal transduction through the fibronectin receptor induces collagenase and stromelysin gene expression. *J. Cell Biol.* 109: 877–889.

Biochemistry of Cell Membranes
ed. by S. Papa & J. M. Tager
© 1995 Birkhäuser Verlag Basel/Switzerland

Structural analysis of membrane proteins

R. Jaenicke

*Department of Biophysics and Physical Biochemistry, University of Regensburg,
D-93040 Regensburg, Germany*

Overview

When Albert Szent-Györgyi, almost half a century ago, described the
cell as being more than a bag full of macromolecules, he referred to the
functional role of every piece of the cellular inventory and the implications of their organization as being divided into specialized compartments. The membrane "bag" Szent-Györgyi was alluding to has since
been shown to be one of the fundamental structural and functional
elements which not only surrounds the cell, but also divides it into a
whole variety of spatial substructures, such as the nucleus, mitochondria, chloroplasts, the endoplasmic reticulum and Golgi, lysosomes, etc.
Membranes close off specific regions, thus allowing the separation of
specialized tasks. The membranes are selective permeability barriers that
cause regulated potentials involved in energy conversion, substrate or
ion transport, and protein translocation. It is evident that the
metabolism requires the spatial separation of the synchronous energy
producing and consuming processes. In this context, membrane proteins
take care of the necessary specificity by gating or blocking fluxes of
charges or substrates (1). The degree of complexity becomes clear if we
consider, for example, the visual cycle. Figure 1 shows that in the
process of the light-induced interconversion of energy several membranes are involved in a single transduction process, and that cytoplasmic components play an essential role as mediators (2). As mentioned, the specificity depends on receptors, carriers, and enzymes. All
three are membrane proteins characterized by differences in their state
of association, as well as in their integration and orientation in the lipid
phase.

Considering the lipid part, specific lipid structures may function as
both membrane constituents and chemical signals. Regarding the lipid
membrane components, essential features such as variation in chain-
lengths, saturation, ionization, conjugation, association properties,
asymmetry, extremophilic anomalies, etc., have been studied in great

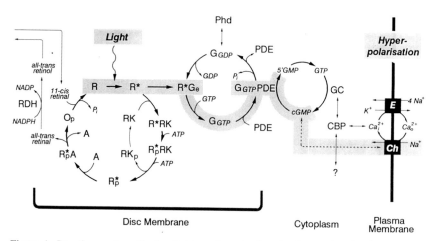

Figure 1. Reaction cycles of subunit interactions in the visual cascade, showing how the elementary processes are compartmentalized between disc membranes, cytosol, and plasma membrane. The four cycles are (from left to right) the rhodopsin cycle, the transducin cycle, with photoactivated rhodopsin R* as a catalyst of GDP/GTP exchange in the G protein, the guanine nucleotide cycles and the calcium cycle (cf. (2)).

detail (3–6). As far as chemical signals are concerned, specific lipids not only serve as tools of communication (e.g., as sex attractants or in connection with pollination), but also in the intricate network of the food-chain, where repellants for a specific guest may serve as attractants for a lethal parasite. In this context, the terpene family may serve as an example: more than 20 000 different derivatives have been discovered, ranging from simple lipid breakdown products to steroids (Fig. 2). They serve as ligands in the whole range of processes from chemotaxis of bacteria to the highly complex olfactory pathways observed in higher vertebrates. Minor changes in their configuration may have drastic effects on their biological efficiency, pointing to the high specificity at the receptor level, i.e., at the level of the three-dimensional structure of the corresponding membrane proteins.

The structural details of pheromone-receptor complexes remain to be elucidated. The same holds for the detailed topography of signal trans-duction cascades where increasing numbers of components are found to be involved in G-protein dependent elementary processes. G-proteins themselves show a wide variation of hetero-oligomeric quaternary struc-tures. For example, the β-adrenergic receptor system consists of at least 10 different polypeptide chains; again, membrane components and cytosolic factors cooperate in a complex way in the sensitization-seques-tration-desensitization cycle (8, 9).

Another example for a complex hierarchy of heterologous interac-tions, however in a totally different area, is the vesicular transport system: Starting from the coatomer family which combines with the

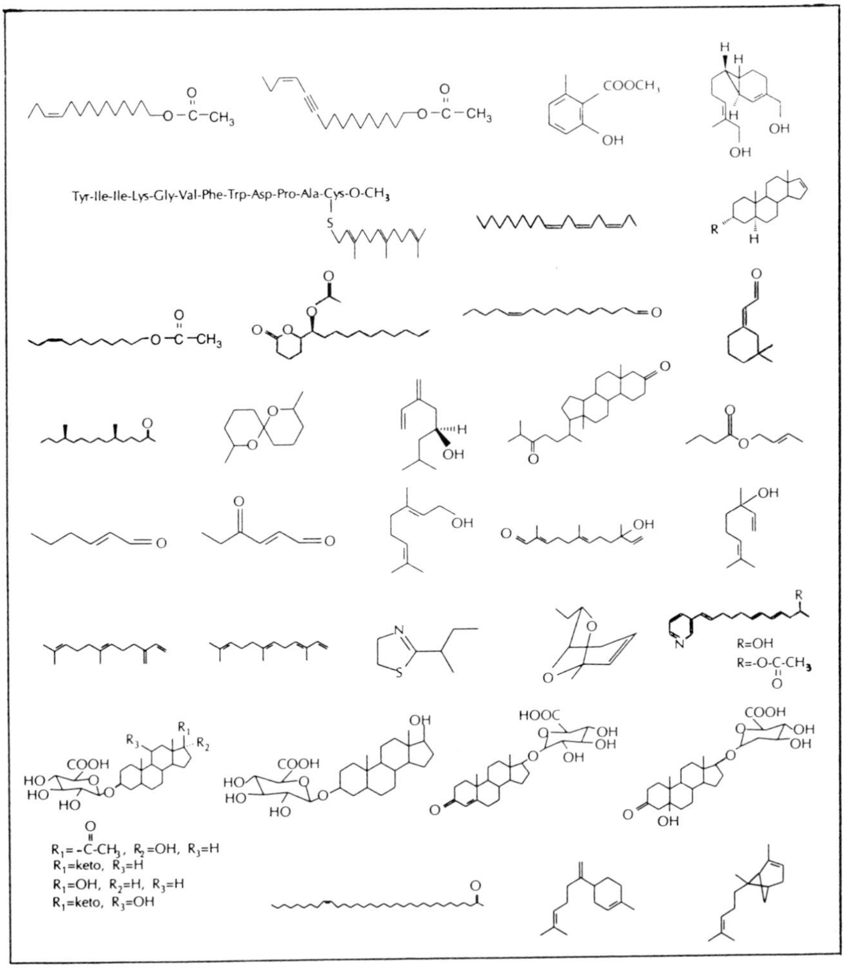

Figure 2. Lipidic pheromones. Modified from (11) where specific functions are discussed.

ADP-ribosylation factor Arf, the complex integrates into the Golgi membrane via the myristylic acid tail of Arf, so that altogether budding occurs, promoted only by the heterologous protein assembly of cytosolic Cops and Arf which induces the self-assembly of an NEM-sensitive fusion protein with SNAPs and SNARs. After targeting the complex to successive cisternae, ATP finally powers the fusion (10). Thus, from the point of view of cell biology, the overall mechanism seems resolved. For the structural chemist, a whole continent remains to be explored, because the central question regarding the mechanistic and topological details remains to be answered. The question refers to both the three-dimensional structure of the single components and their docking in forming the functional assembly. At both levels, the specific properties of the lipid phase, especially the asymmetry and the fluidity of the membrane are expected to be functionally important.

Reasons why membranes not only include specific kinds of lipids, but also topographical anisotropy have been related, first, to homoviscous adaptation to varying environmental conditions, second, to the communication across membrane barriers and, third, to specific interactions with proteins sharing the topographical orientation with the lipid framework. In this context, certain membrane-protein interactions may either be fostered or hindered by the presence of specific lipids (11).

As mentioned, the specific functions of membranes depend entirely on the protein moiety: neither energy conversion, nor substrate or ion transport could occur in the absence of specific light-harvesting complexes, ATPases, proton pumps or other transport systems. Compared to the vast number of high-resolution three-dimensional structures of soluble globular proteins, successful crystallization and structure determination of membrane proteins have been scarce (12–15). Most attempts to explore the structure-function relationship of membrane proteins is based on hydrophobicity considerations, making use of available experimental data which have shown that membrane-spanning elements of secondary structure may be α-helical or β-barrels. As taken

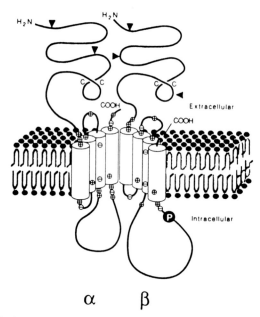

Figure 3. Hypothetical membrane topology of the GABA receptor from bovine cerebral cortex. Two copies of each of the subunits are assumed to be present in the α- and β-receptor complexes. For details, see (8).

from such "knowledge-based structure predictions", integral membrane proteins may transverse the bilayer membrane many times (Fig. 3) (16, 17). On the other hand, the fixation in the lipid phase may be restricted to a single fatty-acid chain or a phosphoinositol anchor. The type of interaction correlates with the functional requirements, e.g., in the case of photosynthetic reaction centers, bacteriorhodopsin, or the membrane form of the "variable surface glycoprotein (mfVSG) from Trypanosoma (18–20).

Among the three examples of successful structure determinations of membrane proteins reviewed in the following articles, annexin illustrates the dynamic potential of membrane proteins: as the result of crystallographic and patch-clamp data, it is shown that a peripheral membrane protein may serve as a regulated ion channel by means of an electroporation mechanism. Similarly, in the case of the bacterial porin, specific side-chain interactions of the integral membrane protein provide ion specificity and, at the same time, allow the protein to cope with membrane fluctuations, avoiding leakiness through tyrosine-membrane interactions. Finally, bacteriorhodopsin will serve as an example to illustrate how, in the case of a helical integral membrane protein, folding and association are determined and driven by side-to-side interactions of stable transmembrane helices.

Regarding the self-organization of membrane proteins, and their correct integration into the membrane, successful *in vitro* reconstitution has been accomplished for bacteriorhodospin, as well as the outer-membrane protein (OmpA, OmpH) from *Escherichia coli* (21–24). Obviously, no accessory proteins are required for the *in vitro* structure formation. Whether chaperones are essential *in vivo* is still unknown. On the other hand, targeting by signal sequences and fixation of receptors or antibodies in the membrane by hydrophobic anchor sequences is well-established. As indicated, the mode of insertion shows considerable diversity ranging from the integration of a single fatty acid (as in the case of Arf) or phosphoinositol anchors (as in the case of mfVSG) to multiple membrane-spanning α-helices or β-structures.

References

1. Petty, H.R. (1993) *Molecular Biology of Membranes: Structure and Function*. Plenum Press, New York, p. 404.
2. Hofmann, K.P. and Heck, M. Reaction cycles of the visual cascade. *In:* A.G. Lee (ed.): *Biomembranes*. JAI Press, Greenwich; *in press*.
3. Vance, D.E. and Vance, J.E. (1991) *Biochemistry of Lipids, Lipoproteins and Membranes. New Comprehensive Biochemistry*. Elsevier, Amsterdam, Vol. 20, p. 596; Watts, A. (1993) *Protein-Lipid Interactions*. Elsevier, Amsterdam, Vol. 25, p. 379.
4. Gurr, M.I. and Harwood, J.L. (1991) *Lipid Biochemistry: An Introduction*. 4th edition. Chapman & Hall, London, New York, p. 406.
5. Hadley, N.F. (1985) *The Adaptive Role of Lipids in Biological Systems*. Wiley, New York, p. 319.

 6. Langworthy, T.A. and Pond, J.J. (1986) Membranes and lipids of thermophiles. *In:* T.D. Brock (ed.): *Thermophiles: General, Molecular and Applied Microbiology.* Wiley, New York, pp. 107–135.
 7. Teal, P.E.A. and Tumlinson, J.H. (1992) Lipidic pheromones. *Curr. Opin. Struct. Biol.* 2: 475–481.
 8. Boege, F., Neumann, E. and Helmreich, E.J.M. (1991) Structural heterogeneity of membrane receptors and GTP-binding proteins and its functional consequences for signal transduction. *Eur. J. Biochem.* 199: 1–15.
 9. Lacal, J.C. and McCormick, J. (1993) *The Ras Superfamily of GTPases.* CRC Press, Boca Raton, p. 526.
10. Rothman, J.E. and Warren, G. (1994) Implications of the SNARE hypothesis for intracellular membrane topology and dynamics. *Curr. Biol.* 3: 220–223.
11. Marsh, D. (1992) Role of lipids in membrane structures. *Curr. Opin. Struct. Biol.* 2: 497–502.
12. Deisenhofer, J., Epp, O., Miki, K., Huber, R. and Michel, H. (1985) Structure of the protein subunits in the photosynthetic reaction centre of *Rhodopseudomonas viridis* at 3 Å resolution. *Nature* 318: 618–624.
13. Kühlbrandt, W. and Wang, D.N. (1991) Three-dimensional structure of plant light-harvesting complex determined by electron crystallography. *Nature* 350: 130–134; Kühlbrandt, W., Wang, D.N. and Fujiyoshi, Y. (1994) Atomic model of plant light-harvesting complex by electron crystallography. *Nature* 367: 614–621.
14. Weiss, M.S. and Schulz, G.E. (1992) Structure of porin refined at 1.8 Å resolution. *J. Mol. Biol.* 227: 493–509.
15. Michel, H. (1991) *Crystallization of Membrane Proteins.* CRC Press, Boca Raton, p. 224.
16. Fasman, G.D. and Gilbert, W.A. (1990) The prediction of transmembrane protein sequences and their conformation: An evaluation. *Trends Biochem. Sci.* 15: 89–92; Jähnig, F. (1990) Structure predictions of membrane proteins are not that bad. *Trends Biochem. Sci.* 15: 93–95.
17. von Heijne, G. and Manoil, C. (1990) Membrane proteins: From sequence to structure. *Prot. Engineering* 4: 109–112.
18. Huber, R. (1988) Beweglichkeit und Starrheit in Proteinen und Protein-Pigment-Komplexen. *Angew. Chem.* 27: 79–88.
19. Rehaber, P., Seckler, R. and Jaenicke, R. (1991) Intermolecular interactions involved in the association of the variable surface glycoprotein of *Trypanosoma brucei. Biol. Chem. Hoppe Seyler* 372: 593–598.
20. Lemmon, M.A. and Engelman, D.M. (1992) Helix-helix interactions inside lipid bilayers. *Curr. Opin. Struct. Biol.* 2: 511–518.
21. Popot, J.-L., Gerchman, S.-E. and Engelman, D.M. (1987) Refolding of bacteriorhodopsin in lipid bilayers: A thermodynamically controlled two-stage process. *J. Mol. Biol.* 198: 655–676.
22. Popot, J.-L. and Engelman, D. M. (1990) Membrane protein folding and oligomerization: The two stage model. *Biochemistry* 29: 4031–4037.
23. Dornmair, K., Kiefer, H. and Jähnig, F. (1990) Refolding of an integral membrane protein. OmpA of *Escherichia coli. J. Biol. Chem.* 265: 18907–18911.
24. Surrey, T. and Jähnig, F. (1992) Refolding and oriented insertion of a membrane protein into a lipid bilayer. *Proc. Natl. Acad. Sci. USA* 89: 7457–7461.

Biochemistry of Cell Membranes
ed. by S. Papa & J. M. Tager
© 1995 Birkhäuser Verlag Basel/Switzerland

Helix-helix interactions inside membranes

D.M. Engelman, B. Adair, A. Brunger, J. Hunt, T. Kahn, M. Lemmon,
K. MacKenzie and H. Treutlein

*Department of Molecular Biophysics and Biochemistry, Yale University, New Haven,
CT 06520-8114, USA*

Introduction

For several years, we have been studying the role of transmembrane
helices in the folding and oligomerization of membrane proteins. Our
work has included general conceptual notions, experimental studies of
reassembly and oligomerization, and, most recently, computational
protocols. In the following, we summarize the main points from our
work, and make reference in part to the work of others. Our main
purpose, however, is to present the past findings and current trends
toward the future that are embodied in our own efforts.

Transmembrane helices can be independently stable

Helical structure is known to be induced in polypeptides in nonaqueous
environments (1, 2), as is expected given the large free energy costs of
transferring an unsatisfied hydrogen bond donor or acceptor from an
aqueous to a nonpolar environment or of breaking such a bond in a
nonpolar environment. It is therefore expected that hydrogen bonds
must be satisfied in the transbilayer region. If a polypeptide has a
sufficiently hydrophobic sequence of amino acid sidechains, the hydrophobic effect will favor its partition into the nonaqueous region of a
lipid bilayer. Combining these concepts leads to the notion that nonpolar sequences will be stable as transbilayer structures in which backbone hydrogen bonds are systematically satisfied and the polar ends of
the helix are in more polar environments outside the bilayer (3–6).
Experimental observation of many integral membrane proteins shows
that, with the exception of some outer membrane proteins such as
bacterial outer membrane porins, transmembrane proteins can be understood as having bundles of hydrophobic α-helices (7–10). Some
studies have shown that subfragments of polytopic proteins retain

helical structure when they are separately reconstituted into lipid bilayers, as would be predicted if their separate helices were stable (11).

Helix-helix association in bilayers

If hydrogen bonding of the main chain and the hydrophobic effect are
accounted for in forming independent transmembrane helices, then
factors other than these must drive their association if they interact in
tertiary folding and oligomerization. These factors may include polar
interactions within the membrane, the constraints of interactions outside the bilayer (including helix-helix polypeptide links), and packing
effects. These factors have been contemplated in the proposal that
folding and oligomerization of membrane proteins involves two stages,
one in which hydrophobic α-helices are established across the lipid
bilayer and a second in which helices interact to form functional
transmembrane structures (6, 12). Work with bacteriorhodopsin (BR)
and with glycophorin A (GpA) constructs has supported the con-
tentions of this two-stage model (see below).

BR studies show that helix-helix interactions dominate in folding

In BR, the seven transmembrane helices of the structure interact very
closely and some interactions of helices are also involved in forming the
trimeric structure found in membranes (9). BR can be regenerated from
two chymotryptic fragments containing two and five helical transmembrane segments (11), and, more dramatically, from two separate helical
transmembrane peptide plus the five helix chymotryptic fragment (13).
Since the two separate helices were also studied as independent entities
and found to be transbilayer helices (14), the notions of the two-stage
concept are supported. Calorimetric studies (15) show that cleavage of
the B-C link or the A-B link, or removal of the retinal each act to
destabilize the structure somewhat, but do so in independent ways and
do not abolish the folding of the BR molecule. Since reformation of BR
occurs from fragments first reconstituted as separate transmembrane
helical structures, and then introduced into the same lipid bilayer,
viewing each fragment as a subunit suggests that helix-helix interactions
are a component of specific oligomerization in this system.

SDS gel assay of dimerization using a chimeric construct: GpA studies

Specific, stable side-by-side interactions between transmembrane α-
helices can occur in detergent environments that, to some extent, mimic

the properties of lipid bilayers. We have developed a system for the genetic and biphysical analysis of such interactions (20). The transmembrane domain of interest is fused to the C-terminus of staphylococcal nuclease (SN). The resulting chimera can be expressed at high levels in *E. coli* and is readily purified. Interactions can be studied by SDS PAGE and other methods. In our initial application, we have studied the single transmembrane α-helix of GpA, which mediates the SDS-stable dimerization of this protein. Vectors and methods are described in Lemmon et al. (20,21).

The chimeric protein runs as a dimer in SDS gels, as shown in Figure 1. Lane 1 in panel A shows the presence of substantial amounts of dimer in the chimeric construct in SDS. In lanes 2, 3, and 4, the pure transmembrane peptide is added and shown to compete for binding at the dimer interface, resulting in a large proportion of heterodimer between the chimeric construct and the transmembrane peptide as shown in lane 4 and panel B. In lanes 5–8, other transmembrane peptides are added, and are found not to disrupt the SN/GpA dimer. Thus, the dimerization arises from an interaction of the transmembrane part of the chimeric construct, and is sufficiently specific so that transmembrane domains from the *neu* oncogene, the EGFR, the HER-2 receptor, or helix A from BR do not compete. In panel B, a fluorescently labeled peptide was used in the competition experiment, and shown to migrate at a molecular weight just above the monomer molecular weight of the chimeric construct. This establishes that the band is in fact a peptide-chimera heterodimer.

Mutational analysis of GpA dimerization

The behavior of the chimeric protein on SDS gels has permitted a mutational analysis of requirements for dimerization. We explored C-terminal truncations and deletions of the extra-membranous region proximal to the fusion junction in order to define the part of the sequence responsible for the dimerization seen on SDS gels, and it is found that the dimerization potential depends on the transmembrane region from GpA and on the presence of a flexible linker with no particular sequence requirement.

Further experiments were conducted by saturation mutagenesis at each amino acid position in the transmembrane domain, using SDS gel migration as an assay (21). The effects of sequence alterations were assigned to four categories: no effect, a modest effect with significant dimer remaining, a stronger effect in which there remains relatively little dimer, or abolition of dimerization altogether resulting in the migration of all protein at the monomer position. Changes to charged amino acids produced strong effects, entirely abolishing all dimerization in almost all

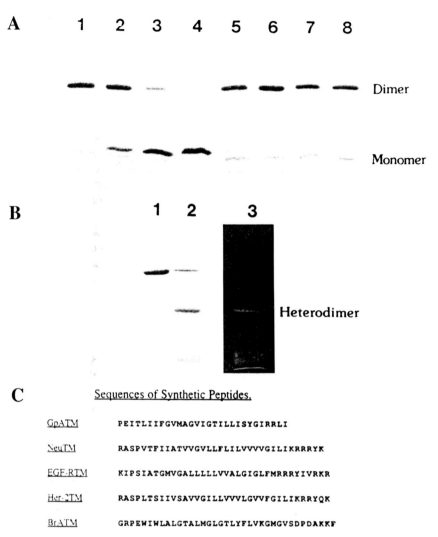

Figure 1. (A) 12.5% polyacrylamide SDS Phastgel. *Lane 1,* SN/GpA. *Lanes 2–4* include a 1-, 5-, and 10-fold molar excess of GpATM, respectively. *Lanes 5–8* contain a 10-fold molar excess of synthetic peptides of Neu, EGFR, HER-2, and BrA, respectively. Dimer and monomer positions are marked. (B) competition experiment performed with GpA transmembrane peptide labeled with Dansyl. *Lane 1* shows SN/GpA. *Lane 2* with addition of 10-fold molar excess of Dansylated TM peptide. *Lane 3* as *lane 2* but under UV illumination. The diffuse fluorescent band at the bottom of the gel corresponds to free Dansylated peptide.

cases. Such effects could result from influences on the detergent micelle arrangement, and are not easily interpreted. Terminations anywhere within the transmembrane region likewise abolish dimerization. The phenotypes of changes to moderately polar amino acids were more mixed. N, Q, P, and Y substitutions generally abolished dimerization,

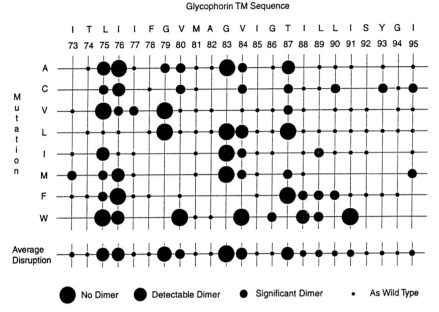

Figure 2. Effect of hydrophobic substitutions on dimerization of SN/GpA.

but G, S, and T were tolerated to some extent in some sequence locations.

More easily interpreted are the results of substitutions of less polar sidechains. Figure 2 shows the current state of these mutagenesis studies. As will be readily appreciated, this is a large body of work. Two striking messages emerge from the analysis. The first is that the dimeric interaction is extraordinarily specific, responding to modest changes in amino acid sidechain structure. The second is that this response varies along the sequence. Consideration, for example, of the three leucine residues in the transmembrane domain illustrates these points. At L75 all nonpolar substitutions disrupt the dimer. Many, including those to isoleucine or valine, have a large effect. However, the same substitutions at L89 and L90 are at most only slightly disruptive. Similarly, while all substitutions observed at G79 and G83 disrupt the dimer significantly, G86 and G94 will accommodate a variety of different residues with no significant effect. Thus, these results strongly suggest the intimate involvement of a subset of amino acid sidechains in the helix-helix interaction.

There is in general a periodicity in the ability of nonpolar sidechains to disrupt the dimer. The bottom of Figure 2 shows the average degree of disruption by nonpolar substitutions at each position in the transmembrane domain. This shows that there are four clear regions in which the disruptive potential of substitutions is strong compared with other locations, namely L75–76; G79–V80; G83–V84; and T87. A straightforward interpretation of this pattern would be that these residues comprise the interacting surfaces of α-helices in a supercoil. These putative sites of interaction are separated, on average, by almost four residues. This would be consistent with the dimer consisting of a pair of helices in a right-handed supercoil. Circular dichroism studies in lipid bilayers and in detergent environments show that both the isolated peptide and the transmembrane domain of the chimeric protein have strong helical character. Given the possibility that the interaction is between helices and that the interaction appears sensitive to small details in amino acid sidechain interactions, additional objectives arose. We wished to establish whether the interactions are of parallel helices, whether the interactions occur in lipid bilayers as well as in detergent environments, and whether theoretical descriptions can be created and tested against the experimental observations so that prediction of membrane protein folding and oligomerization can be facilitated. We have made progress in each of these directions.

The question of whether the dimers are parallel was addressed based on mutant complementation and on studies of the flexible link in the chimeric construct. When the link was shortened excessively, dimerization was abolished. Addition of alanines to replace the deleted residues progressively restored dimerization. This result can be explained if steric interference of the nuclease moiety in the chimera is acting to disrupt the dimer, which can only occur if it contains a parallel arrangement of helices. Further, we have recently found that dimerization can be restored in a mixture of L75V with I76A, in which dimerization was abolished for each substitution taken separately. The complementation suggests that Leu 75 and Ile 76 interact with one another, which is possible only in a parallel dimer (21).

GpA transmembrane peptides dimerize in lipid bilayers

To follow the behavior of peptides in a lipid bilayer environment, new methods needed to be developed. A great deal of effort was spent on exploring neutron-scattering methods. While encouraging data were obtained on BR in deuterated lipid vesicles, it became clear that this approach is impractical as a routine assay. Consequently, we turned to resonance energy transfer, and have now developed an assay based on mixing 2,6-Dansyl-labeled peptides with Dabsyl-labeled peptides in lipid environments. 2,6-Dansyl and Dabsyl are a good donor-acceptor pair for resonance energy transfer since they have a good spectral overlap and a calculated $R_0(2/3)$ of 34.9 Å. Dabsyl, however, does not fluoresce, so sensitized emission cannot be detected.

We have developed a protocol to measure the interaction of transmembrane peptides using this technique. Peptides to be labeled were prepared from chimeric protein as described in Lemmon et al. (20). Peptides were labeled in 50% $CHCl_3$ 50% MeOH, 1% triethyl amine with either 2-dimethylaminoaphthyl-6-sulfonyl chloride (2,6-Dansyl Cl) or 4-dimethylaminoazobenzene-4-sulfonyl chloride (Dabsyl Cl) for 24 h at 8 °C. Peptides were separated from unreacted dye on an LH-20 column in $CHCl_3$/MeOH and further separated from other peptides by reversed-phase HPLC (H_2O, acetonitrile, isopropanol). The peptides were then assayed for their ability to disrupt chimeric protein dimers on SDS gels and further assayed for self association on 20% gels. Analysis of previous labelings with 1,5-Dansyl indicated that the N-terminus and at least one of the C-terminal lysines were labeled. The peptides probably contain a number of different labeling isomers.

Peptides were reconstituted separately into DMPC vesicles by detergent dialysis. For the experiments, vesicles with labeled peptides were fused by another round of freeze-thaw-sonication and fluorescence measurements made on an SLM 8000C fluorometer. Measurements indicate that the peptides are associating in fluid DMPC membranes, as shown by a reduction in Dansyl fluorescence. Energy transfer is not diluted out with a 20-fold increase in DMPC to peptide ratio, indicating that transfer is due to actual peptide interactions, as opposed to transfer between near neighbors. Competition experiments have been done with unlabeled EGFR-TM, Neu-TM, and HER2-TM, none of which compete, and with unlabeled GpA-TM, which does compete (Fig. 3). By analyzing the variation of the observed quenching as a function of the mole ratio of donor-to-acceptor-labeled peptides, we have clearly established that the peptides associate in bilayers as dimers.

Simulated annealing approach

To study the interactions, we have recently turned to studies of the dimerization of glycophorin A using conformational searches by simulated annealing techniques with restraints (23). Simulated annealing was implemented through molecular dynamics simulations with geometric (24) and empirical energy functions (55). Our assumptions are that the transmembrane domains of glycophorin form dimers of α-helices with a parallel chain directionality and that the helices are in contact across the lipid bilayer. In the modeling, the helices were constrained with a half parabolic potential to prevent them from dissociating. Several initial conditions were chosen, including one in which the helices are straight and parallel. A search over a range of relative helix rotational angles resulted in the identification of two global energy minima (26). Modeling from initially parallel helices resulted in the surprising conclusion

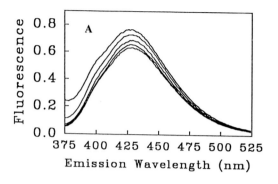

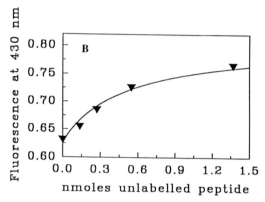

Figure 3. (A) Emission spectra of 0.11 nmoles 2,6-Dansyl- and 0.16 nmoles Dabsyl-GpA peptides in DMPC. Excitation wavelength at 325 nm. Top trace: peptides in separate vesicles; bottom trace: after vesicle fusion; intermediate traces: with increasing amounts of unlabeled peptide added, 0.14, 0.27, 0.55, 1.4 nmoles (lower trace to upper). (B) Data from panel A plotted as a function of unlabeled peptide added. The line is best fit to experimental data, assuming dimeric behavior.

that the best structure is a right-handed supercoil. To check our procedure, we did a similar calculation with GCN4, which is known to have
a left-handed supercoiled dimer, and found that the modeling leads to
the expected handedness (27, 28).

Figure 4 shows the result of a number of simulations using the
glycophorin transmembrane helix and the GCN4 helix dimers, revealing
a propensity for left-handed supercoiling (positive crossing angles) in
the GCN4 and right-handed supercoiling (negative crossing angles) in
the glycophorin cases. Figure 5 shows a comparison of our best theoretical models with the mutagenesis data described above and in (21). The
agreement for these right-handed supercoiled models is striking. We
were unable to generate a left-handed coiled-coil of GpA helices which
permitted good agreement with the mutagenesis data.

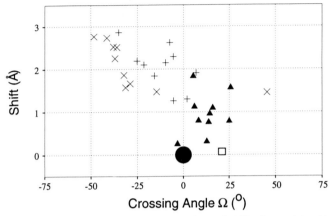

Figure 4. Scatterplot showing final crossing angles and shifts for models obtained starting from uncoiled dimers of ideal helices.

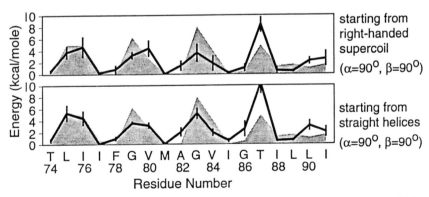

Figure 5. Comparison of average residue interaction energies with the mutational sensitivity profile of GpA.

Our findings support the hope of predicting folded transmembrane protein structure in the hydrophobic region of the lipid bilayer. It is conceivable that a more complete modeling of the bilayer environment and more accurate empirical energy functions could improve the modeling, and we intend to move in these directions. If a successful modeling of the basic transmembrane structure is accomplished, then the hope of modeling interactions with pharmacological agents is improved.

High-resolution NMR studies of transmembrane peptides

We have undertaken preliminary studies designed to establish whether atomic level structural information about detergent-solubilized peptide oligomers can be obtained using NMR. The detergent conditions chosen are those where the peptides (and the full-length chimeric proteins) have been shown to form sequence-specific oligomers under equilibrium conditions (SDS-PAGE).

Attempts to study synthetic peptides corresponding to GpA-TM in SDS micelles by standard homonuclear two-dimensional (2D) techniques met with considerable technical difficulties: not only does the large peptide/detergent micelle cause the peptide resonances to be rather broad, but the presence of residual proton signals from detergent (98% deuterated, 1–5% w/v) renders the aliphatic portion of the spectrum difficult to interpret. Due to the broad linewidths and poor dispersion of the amide protons (1.0 ppm), the "fingerprint" region of 2D COSY spectra are also problematic. We therefore generated ^{15}N-labeled peptides by proteolysis of the overexpressed SN/GpA fusion protein grown in minimal medium. 2D ^{15}N-^{1}H heteronuclear single quantum correlation experiments (HSQC) of a 2 mM peptide sample in 98% deuterated

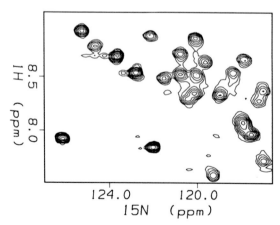

Figure 6. Portion of the amide region of an HSQC spectrum of a 2.0 mM sample of uniformly ^{15}N-labeled GpA-TM peptide in predeuterated SDS (3.5% w/v) and 90% H_2O/ 10% D_2O. The spectrum was acquired using a GE Omega-500 spectrometer and a minor modification of the pulse sequence of (30). 80 scans (1K data points) were collected at each of 128 t_1 increment for a ^{1}H (D1) spectral width of 6000 Hz; the spectral width in the ^{15}N (D2) dimension was 1333 Hz.

SDS (at 500 MHz) show that the ^{15}N chemical shift dispersion is sufficient to resolve amide protons from one another almost completely (Fig. 6). The nature of the heteronuclear experiments, which permits one to filter out signals from protons that are not coupled to heteronuclei, results in the elimination of the detergent resonances and demonstrates that assignment of the peptide resonances should be relatively straightforward using existing 2D and 3D heteronuclear techniques, and that a structure determination should be feasible.

The need to distinguish intra-monomer NOEs from inter-monomer NOEs is a particularly acute problem for symmetric parallel oligomers (as our results show the GpA peptides to be). Our approach to this problem is to use heteronuclear half-filter experiments to distinguish between protons that are on one peptide monomer or another. This will require the use of ^{13}C-labeled peptide, which will be produced biosynthetically.

Specific interactions between transmembrane α-helices

Oligomerization of transbilayer proteins *in vivo* is also thought to involve helix-helix interactions. The structures of the BR trimer (9) and the photosynthetic reaction center (10) show many helix-helix interactions at oligomeric interfaces. Another example is the interaction between the T-cell receptor α chain and CD3 δ (16, 17) in the assembly of the T-cell receptor complex. In this case, potentially charged groups appear to be important in the interaction; however, in other instances association of transmembrane domains without charged groups is also seen. One of these is the human erythrocyte sialoglycoprotein glycophorin A (GpA), which forms dimers stable under the conditions of SDS PAGE, and in which the interactions of its single transmembrane domain drive the association (18–20). The properties of the GpA transmembrane domain were studied by genetically fusing it to the carboxy terminus of staphylococcal nuclease. The resulting chimera forms a dimer in SDS, which is disrupted upon addition of a peptide corresponding to the transmembrane domain of GpA, but not upon addition of transmembrane domains from the EGF receptor, *neu* oncogene or BR. Deletion mutagenesis has been used to define the boundaries of the transmembrane domain responsible for dimerization, and site-specific mutagenesis has been used to explore the properties of the interactions in the dimer (20, 21). This work indicates that dimerization is driven largely by close association of the transmembrane portions of GpA in a parallel coiled-coil, and that the interactions are highly specific.

Similar interactions may by important in transmembrane signalling by growth factor receptors with a single transmembrane α-helix (12).

For a number of these, most notably the epidermal growth factor receptor (EGFR), specific interactions between transmembrane domains may provide part of the energy and specificity for dimerization. Indeed, a valine to glutamic acid mutation in the transmembrane domain of the *neu* oncogene product cauases this EGFR-like molecule to become constitutively active (22), and increases the proportion of it found as dimer (23).

In summary, the interactions of transmembrane helices can be highly specific and can have sufficient interaction energies in a lipid context to drive oligomerization.

Discussion

The idea of predicting the folded structure of a protein based on its amino acid sequence has been a theme in biochemistry for many years. While progress is being made, it remains one of the most challenging problems in modern biology, and, at the same time, one of the most important. If it were understood, the connection between the functional, 3D structure that is selected by evolutionary influences and the gene used to encode the information for that function would be placed on a chemically conceptualized basis.

It now appears that the folding problem may be solved in the case of membrane proteins that consist of many α-helices interacting with each other in a side-to-side fashion within a lipid bilayer environment. Studies of subfragments of bacteriorhodopsin strongly suggest that many of its helices form independently stable transbilayer structures as single helices or, perhaps, as hairpins. These interact spontaneously with each other to give the functional protein, even though some of the extra-membranous loops are not intact. Thus, the information for folding is largely contained in the interactions of previously formed helices. Studies of the dimerization of the transmembrane region of glycophorin have shown that, at least for such a simple case, current approaches to computational representation of chemistry lead to a relatively small number of possible structures, and only one of these agrees with the experimental evidence from mutagenesis. NMR studies are expected to yield detailed structural information. In future studies, it is hoped that a combination of continued computational developments, a broader base of cases studied by mutagenesis and physical biochemistry, and measurements of energies of association will lead to a refinement of the computational approaches so as to permit the determination of strongly preferred models for the chemical structures based solely on sequence information. Should this prove possible, it will constitute an important step toward understanding the chemistry of membrane protein folding.

References

1. Singer, S. (1962) Properties of proteins in non-aqueous solvents. *Adv. Prot. Chem.* 17: 1–68.
2. Singer, S. (1971) *In: Structure and Function of Biological Membranes.* Academic Press, New York, pp. 145–222.
3. Engelman, D.M. and Steitz, T.A. (1981) The spontaneous insertion of proteins into and across membranes: The helical hairpin hypothesis. *Cell* 23: 411–422.
4. Engelman, D.M., Steitz, T.A. and Goldman, A. (1986) Identifying non-polar transbilayer helices in amino acid sequences of membrane proteins. *Ann. Rev. Biophys. Chem.* 15: 321–353.
5. Popot, J.-L. and De Vitri, C. (1990) On the microassembly of integral membrane proteins. *Ann. Rev. Biophys. Biophys. Chem.* 19: 369–403.
6. Popot, J.L. and Engelman, D.M. (1990) Membrane protein folding and oligomerization: The two-stage model. *Biochemistry* 29: 4031–4037.
7. Henderson, R. and Unwin, P.N.T. (1975) Three-dimensional model of purple membrane obtained by electron microscopy. *Nature* 257: 28–32.
8. Henderson, R. (1977) The purple bacterium from Halobacterium halobium. *Ann. Rev. Biop. Bioeng.* 6: 87–109.
9. Henderson, R., Baldwin, R.J.M., Ceska, T.A., Zemlin, E., Beckman, E. and Downing, K.H. (1990) Model for the structure of bacteriorhodopsin based on high-resolution electron cryo-microscopy. *J. Mol. Biol.* 213: 899–929.
10. Deisenhofer, J. and Michel, H. (1991) High resolution structures of photosynthetic reaction centers. *Ann. Rev. Biophys. Biophys. Chem.* 20: 247–266.
11. Popot, J.L., Gerchman, S.-E. and Engelman, D.M. (1987) Refolding of bacteriorhodopsin in lipid bilayers. A thermodynamically controlled two-stage process. *J. Mol. Biol.* 198: 655–676.
12. Bormann, B.-J. and Engelman, D.M. (1991) Intramembrane helix-helix association in oligomerization and transmembrane signalling. *Ann. Rev. Biophys. Biophys. Chem.* 14: 223–242.
13. Kahn, T.W. and Engelman, D.M. (1992) Bacteriorhodopsin can be refolded from two independently stable transmembrane helices. *Biochemistry* 31: 6144–6151.
14. Hunt, J.F. (1991) Biophysical studies of the integral membrane protein folding pathway. *Biophys. J.* 59: 400A.
15. Kahn, T.W., Sturtevant, J.M. and Engelman, D. M. (1992) Thermodynamic measurements of the contributions of helix-connecting loops and of retinol to the stability of bacteriorhodopsin. *Biochemistry* 31: 8829–8839.
16. Manolios, N., Bonifacino, J.S. and Klausner, R.D. (1990) Transmembrane helical interactions and the assembly of the T-cell receptor complex. *Science* 249: 274–277.
17. Cosson, P., Lankfors, S.P., Bonifacino, J.S. and Klausner, R.D. (1991) Membrane protein association by potential intramembrane charge pairs. *Nature* 351: 414–416.
18. Furthmayr, H. and Marchesi, V.T. (1976) Subunit structure of human erythrocyte glycophorin A. *Biochemistry* 15: 1137–1144.
19. Bormann, B.-J., Knowles, W.J. and Marchesi, V.T. (1989) Synthetic peptides mimic the assembly of transmembrane glycoproteins. *J. Biol. Chem.* 264: 4033–4037.
20. Lemmon, M.A., Flanagan, J.M., Hunt, J.F., Adair B.D., Bormann, B.J., Dempsey, L.E. and Engelman, D.M. (1992) Glycophorin A dimerization is driven by specific interactions between transmembrane α-helices. *J. Biol. Chem.* 267: 7683–7689.
21. Lemmon, M.A., Flanagan, J.M., Treutlein, H.R., Zhang, I. and Engelman, D.M. (1992) Sequence specificity in the dimerization of transmembrane α-helices. *Biochemistry* 31: 12719–12725.
22. Bargmann, C.I., Hung, M.-C. and Weinberg, R.A. (1986) The *neu* oncogene encodes for an epidermal growth factor receptor-related protein. *Nature* 319: 226–230.
23. Weiner, D.B., Liu, J., Cohen, J.A., Williams, D.V. and Greene, M.L. (1989) A point mutation in the *neu* oncogene mimics ligand induction of receptor aggregation. *Nature* 319: 226–230.
24. Nilges, M. and Brunger, A. (1991) Successful prediction of the coiled-coil geometry of the GCN4 leucine zippen domain by simulated annealing. *Protein Engineering* 4: 649–659.
25. Brunger, A.T. (1991) Simulated annealing in crystallography. *Ann. Rev. Phys. Chem.* 42: 197–223.

26. Karplus, M. and Petsko, G.A. (1990) Molecular dynamics simulations in biology. *Nature* 347: 631–639.
27. Treutlein, H.R., Lemmon, M.A., Engelman, D.M. and Brunger, A.T. (1992) The glycophorin A transmembrane domain dimer: Sequence-specific propensity for a right-handed supercoil of helices. *Biochemistry* 31: 12726–12732.
28a. Jones, E.W. and Fink, G.R. (1982) *In*: J.N. Strathern and J.R. Brouch (eds): *The Molecular Biology of the Yeast Saccharomyces: Metabolism and Gene Expression.* Cold Spring Harbor Laboratory, Cold Spring Harbor, New York, pp. 181–299.
28b. Pathak, D. and Sigler, P.B. (1992) Updating structure-function relationships in the bZIP family of transcription factors. *Curr. Opin. Struct. Biol.* 2: 116–123.
29. O'Shea, E.K., Klemm, J.D., Kim, P.S. and Alber, T. (1991) X-ray structure of the GCN4 leucine zipper, a two-stranded, parallel coiled coil. *Science* 254: 539–544.
30. Tolman, J.R., Chung, J. and Prestegard, J.H. (1992) *J. Mag. Resonance* 98: 462–467.

Biochemistry of Cell Membranes
ed. by S. Papa & J. M. Tager
© 1995 Birkhäuser Verlag Basel/Switzerland

Annexin V: Structure-function analysis of a voltage-gated, calcium-selective ion channel

P. Demange[1], D. Voges, J. Benz, S. Liemann, P. Göttig, R. Berendes, A. Burger and R. Huber

Abt. Strukturforschung, Max-Planck-Institut für Biochemie, D-82152 Martinsried, Germany
[1]*Present address: Laboratoire de Pharmacologie et Toxicologie Fondamentales, 118 route de Narbonne, F-31062 Toulouse Cedex, France*

Introduction

The structural basis of ion channel function remained mostly a mystery for more than 140 years since the existence of water-filled pores through membranes was first postulated (Brücke, 1843). Despite intense efforts in the determination of the electrophysiological properties of ion channels (for a review see Hille, 1992), starting with the seminal work of Hodgkin and Huxley, the field still suffers from the lack of high-resolution structures of ion channel proteins. The first determination of the structure of a membrane protein, the photosynthetic reaction center (Deisenhofer et al., 1985), was a flash of hope for the crystallization of membrane-integrated ion channel proteins in order to understand their selectivity, conductance regulation, and voltage-gating. But still, no crystals of ion channel proteins diffracting x-rays to high resolution could be obtained. Therefore, channel model systems with known crystal structures, such as the porins (e.g., Weiss et al., 1991) or colicins (e.g., Parker et al., 1989), were valuable. Unfortunately, these proteins do not combine the main features of typical ion channels mentioned above. To compensate for this shortcoming extensive investigations by mutational analysis (see Catterall, 1988 or Miller, 1991 for reviews) and molecular modeling (see Guy and Seetharamulu, 1992; Durell and Guy, 1992) of ion channels and of pore-forming peptides (see Sansom, 1991 for a review) were carried out. In 1990, annexin V was structurally characterized as the first ion channel protein, thus paving the way, in combination with electrophysiological analysis of annexin V mutants, for an understanding of ion channel function in molecular terms.

The annexins are proteins that bind to negatively charged phospholipids in a calcium-dependent manner. They are found in a variety of cell types in higher and lower eukaryotes (Moss et al., 1991). The consensus sequence and the structure of their calcium-binding sites clearly distinguish them from the "EF-hand" family (Kretsinger, 1980). Annexins are amphipathic proteins distinct from soluble and integral membrane proteins, but sharing features of both; therefore they were termed "Janus-faced" proteins (Brisson et al., 1991). With the exception of the eight-repeat annexin VI, all members of the annexin family have primary sequences composed of four repeats. These repeats are homologous between each other as well as between all the annexins. In contrast, the N-termini of the annexins are variable in length and in sequence (Crompton et al., 1988; Barton et al., 1991).

Although the physiological role of the annexins remains to be determined, they are most likely involved in membrane fusion and exocytosis (Creutz, 1992). A variety of *in vitro* properties has so far been described for the annexins. Among them are the interaction with cytoskeletal proteins (Mangeat, 1988), anticoagulant activities (Römisch et al., 1990), and the inhibition of phospholipase A_2 to regulate inflammation (Davidson et al., 1990). It has also been shown that some members of the annexin family are expressed in a growth-dependent manner (Schlaepfer and Haigler, 1990; Fox et al., 1991) and are targets for cellular kinases *in vivo* (Moss et al., 1991). These results led to the proposal that annexins are involved in differentiation and mitogenesis. The annexins I, V, VI and VII form calcium selective ion channels in phospholipid bilayers (Pollard and Rojas, 1988; Rojas et al., 1990; Pollard et al., 1992; Karshikov et al., 1992; Berendes et al., 1993).

The determination of the three-dimensional structure of annexin V (Huber et al., 1990a, b) was the first step in understanding its ion channel activity. This review will focus on the structure-function analysis of the ion channel of human annexin V. We will briefly describe the structures of annexin V in crystalline and membrane-bound form and present our ideas about features of the annexin V-membrane complex. We will then give an account of the mutational analysis of human annexin V and finally outline the function of this ion channel protein.

The crystal structure of human annexin V

The annexin V molecule of 320 amino acids is almost completely α-helical (see Figs 1a and b). Viewed from the side (Fig. 1b), it has a slightly curved shape with the calcium binding sites located on the convex and the N-terminus on the opposite surface. The four repeats in the primary sequence form four compact domains each consisting of five α-helices wound into a right-handed super-helix. The four domains are arranged in a cyclic array. The domains I and IV and II and III, respectively, form tight modules with approximate twofold symmetry. In the center of the molecule lies a prominent hydrophilic pore, which we have suggested as the ion conduction pathway (Huber et al., 1992)

A

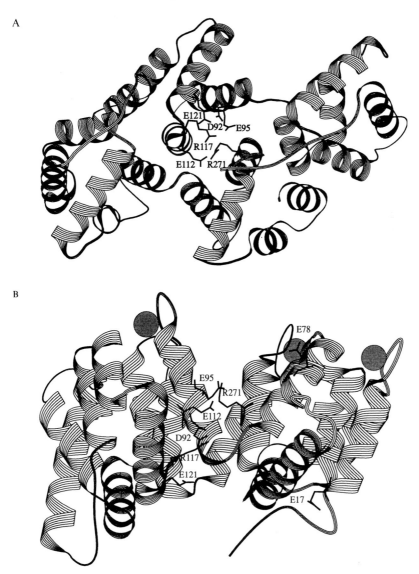

B

Figure 1. (a) Ribbon plot of the annexin V crystal structure as seen from the concave side. The membrane plane would lie behind the plane of this representation. Overlaid are important charged residues located within the central pore. (b) Ribbon plot as seen from the side of the molecule with the convex surface with the calcium binding sites at the top (calcium ions are depicted as grey circles) and the concave side with the N-terminus at the bottom. The mutated residues described in this review are overlaid on the plot.

for the following reasons: the invariant calcium sites 1 and 2, symmetrically arranged around the molecular axis, mediate an intimate contact with the membrane; almost all residues forming the central pore, which is highly hydrated, are conserved; half of all buried charged residues participate in formation of the pore; similarly, more than half of all buried waters are localized in the pore. We proposed therefore that the chain of waters through the central pore marks the ion conduction pathway through annexin V. Free diffusion through the pore is prevented by two salt-bridges across the pore. Ion permeation would therefore require some side chain rearrangements, leading to the suggestion that the salt-bridges are important constituents of the voltage-gates (Karshikov et al., 1992). These structural features are invariably found in different crystal forms of human annexin V (Lewit-Bentley et al., 1992), and in other annexins such as chicken annexin V (Bewley et al., 1993) and annexin I (Weng et al., 1993).

The main difference between the various crystal structures of wildtype annexin V determined so far (Huber et al., 1990a, b; Lewit-Bentley et al., 1992) is the relative orientation of the two modules consisting of domains I and IV and domains II and III, respectively. The two modules are covalently connected by the peptide linkers between domains I and II as well as III and IV, respectively. The rhombohedral forms of the native annexin V and the mutants have the same module-module angle which differs from hexagonal annexin V by about 3.5°.

An interesting new feature of the annexin V structure was discovered recently by analysis of yet other crystal forms (Concha et al., 1993; Berendes et al., 1993; Sopkova et al., in press; Burger et al., 1994), revealing a calcium-induced conformational change leading to the formation of a calcium-binding site in domain III absent in the other crystal forms studied earlier. In these new crystals the loop with Trp187 at its top is exposed and may penetrate the membrane surface up to the ester bond region, as supported by spectroscopic data (Meers, 1991; Meers and Mealy, 1993). In the previous crystal forms the Trp187 loop is completely buried within domain III.

The annexin V-membrane complex

Several water-soluble, channel-forming proteins are proposed to undergo profound structural changes when they integrate into the hydrophobic phase (e.g., Parker et al., 1989). In contrast, all experimental data available so far show that the annexins do not integrate into the phospholipid membrane but remain peripherally bound (Voges et al., 1994).

In previous crystallization studies of phospholipid-bound annexin V and electron microscopical analysis the orientation of the molecule relative to the membrane was unclear (Mosser et al., 1991; Brisson et al., 1991; Newman et al., 1991; Driessen et al., 1992). Therefore, we have analyzed negatively stained two-dimensional crystals bound to monolayers composed of acidic phospholipids by electron microscopy and image processing (see Figs 2a and b) and have determined the three-dimensional structures using reconstruction techniques (see Fig. 2d) (Voges et al., 1994).

These experiments show that the overall domain structure of annexin V remains unchanged after membrane binding compared to its structure in crystals (see Fig. 2). The molecule binds with its convex surface (containing the calcium-binding sites) to the membrane surface. Annexin V does not appreciably penetrate as its stain excluding height on the membrane equals its thickness. The shape of the membrane-bound molecule fits well to the high-resolution structure (Huber et al., 1990a, b) if the molecule composed of domains II and III is rotated by about 30° around an axis perpendicular to the central pore (see Fig. 2c). This "hinged domain motion," similar to the one proposed in the first paper on the annexin V structure (Huber et al., 1990), results in a nearly flat molecule bringing all four calcium-binding loops into one plane.

How can peripherally bound annexin V form highly conducting ion channels through membranes? Annexin V displays a strong electrostatic field on the membrane-binding surface (Karshikov et al., 1992). Since electric fields of similar magnitude can invoke formation of membrane pores (Neumann, 1988), it was proposed that annexin V disturbs adjacent membrane areas by a similar mechanism (electroporation) (Karshikov et al., 1992). We suggest that the synergistic action of the "electroporation phenomenon" (Karshikov et al., 1992), the membrane-penetrating Trp187, and the interactions between the calcium-binding loops and the phospholipid headgroups makes the membrane locally ion-permeable by generating either long-lived water-filled pores or transiently disordered less tightly packed areas. Our current picture of the annexin V-membrane complex is sketched in Figure 3, indicating disordered phospholipids.

In the following paragraph we describe experiments to demonstrate that the voltage-gating, the conductance regulation, and the ion selectivity filter of the protein membrane complex reside within the protein.

Mutational analysis of human annexin V

The crystal structure of human annexin V suggests a variety of interesting mutational sites. The most obvious target is the central pore of the molecule. We have substituted glutamic acid-95 for a serine residue in the center of the pore region (Berendes et al., 1993). The mutation resulted in a lower single-channel conductance for calcium and a strongly increased conductance for sodium and potassium (see Figs 4a and b). The mutant had also completely lost selectivity for Ca^{2+} versus Na^+ ions (see Fig. 4c), indicating that Glu95 is a crucial constituent of

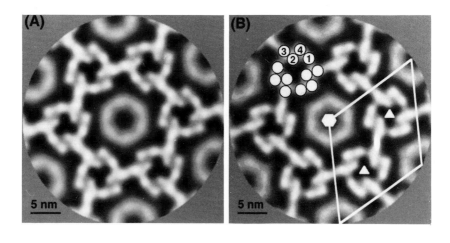

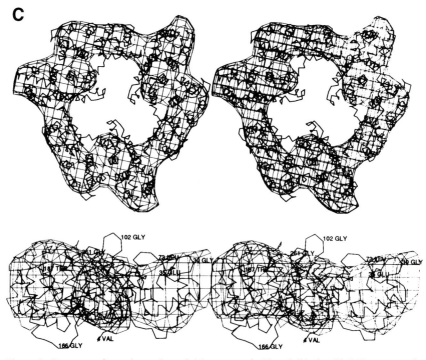

Figure 2. Structure of membrane bound (a) wt annexin V and (b) the Glu95Ser mutant in projection normal to the membrane as seen from the membrane side obtained by processing electron microscopic images of two-dimensional crystals (Berendes et al., 1993). All mutants form isomorphous crystals, with a p6 lattice and a periodicity of about 18 nm. The unit cell (outlined by the white diamond in (b)) contains two trimers of annexin V and a central ring located on the hexad. This ring was found to be a translationally and rotationally disordered annexin V trimer (Voges et al., 1994). A trimer from the high-resolution crystal structure of the R3 form can only be fitted into one trimer identified in the electron micrographs when seen from the side of annexin V containing the Ca²⁺ binding sites. (c) Top: Superposition of

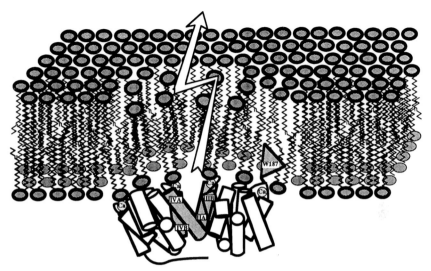

Figure 3. Current picture of the annexin V membrane complex. The membrane-bound annexin V molecule is shown in a cylinder representation, with the pore-forming four-helix bundle (helix assignments as in Huber et al., 1990a, b) in gray and the calcium ions as circles. The membrane-penetrating Trp187 (depicted as triangle), the electroporation phenomenon (outlined by a flash of lightning), and the interactions between the calcium binding loops and the phospholipid headgroups are proposed to lead to a local disorder of the phospholipids adjacent to the annexin V molecule.

the ion selectivity filter. There were only minor differences in the crystal structures of mutant and wild-type annexin V around the mutation site.

The change from an inward to an outward rectification after reversal of the Na-Ca gradient is predicted from the position of the mutated Glu95Ser residue in the center of the ion pathway through the annexin V pore (Fig. 5).

A high-affinity Ca^{2+} binding site within the pore is not seen in the crystal structure, but Glu95 might constitute a low-affinity Ca^{2+} binding site (Fig. 5). A variety of studies has shown that the selectivity for Ca^{2+} versus monovalent cations decreases with a decrease of negative charges within the putative pore-forming region (Heinemann et al., 1992; Kim et al., 1993; Ferrier-Montiel and Montal, 1993; Yang et al., 1993). Similarly, the Ca^{2+} permeability and selectivity of glutamate receptors is

the trimer found by x-ray structure analysis in the described orientation with a trimer cut-out of the EM-crystal structure of membrane bound annexin V perpendicular to the membrane. The superposition led to the assignment of the four domains, which are marked by white numbered circles in (b); bottom: The same two structures as seen parallel to the membrane. The dashed model created by rotating the module consisting of domains II and III fits better into the EM-contour than the totally unchanged high-resolution structure (solid lines) (Voges et al., 1994).

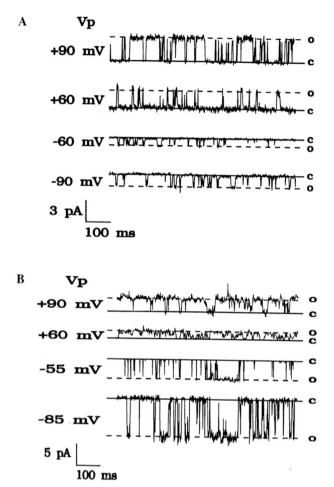

Figure 4. Single-channel current recordings of (a) wt annexin V and (b) the Glu95Ser mutant after incorporation into acidic phospholipid bilayers at different pipette potentials Vp (pipette solution: 50 mM Ca^{2+}; bath solution: 100 mM Na^+).

predominantly controlled by the residue occupying the so-called "Gln/ Arg site" (Burnashev et al., 1992; Dingledine et al., 1992; Egjeberg and Heinemann, 1993). Our electrophysiological analysis of the first struc- turally characterized ion channel allows one to conclusively identify amino acid Glu95 as an important constituent of the ion pathway and the selectivity filter of annexin V by virtue of its size, charge, and interactions with adjacent amino acid side chains (Berendes et al., 1993). In contrast, the substitution of glutamic acid-121 for an aspartate in the lower part of the central part of the channel pore (see Fig. 1b) did not change any of the ion channel characteristics of annexin V.

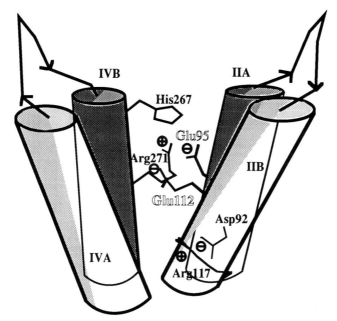

Figure 5. Cylinder representation of the central four-helix bundle forming the hydrophilic pore through annexin V. Some of the mutated residues (Glu95: selectivity filter; Glu112: voltage gate), the two salt-bridges and potential Ca^{2+}-ligating residues (ionizable groups are marked) are shown as derived from the high-resolution crystal structure.

Recent experiments with an annexin V mutant that had a glycine substituted for the salt-bridge forming residue glutamic acid-112 in the upper part of the central pore (see Figs 1b and 5) revealed a loss of the voltage-dependence of the channel gating (see Fig. 6a). These results indicate that the salt-bridge formed between Glu112 and Arg271 constitutes an important part of the voltage sensor of the ion channel annexin V (Göttig et al., unpublished data). Indeed, Karshikov et al. (1992) had already postulated that the salt-bridge Glu112/Arg271 is one of the two voltage-sensing gates within the central ion-conducting pore. This model, initially based on stereochemical and energetic considerations, is now strongly supported by the mutational analysis. The role of the second salt-bridge within the channel pore (formed by Arg117 and Asp92) remains to be investigated.

Hodgkin and Huxley developed in 1952 the concept that the electric field acts by moving charges that are part of the channel molecule, the so-called gating charges or voltage sensor (Hodgkin and Huxley, 1952). The models proposed for the known Na, K, and Ca channels suggest a highly charged helix called S4 as the voltage sensor (Stühmer et al., 1989; Papazian et al., 1991; Greeff, 1992). Similarly, for colicin the

A

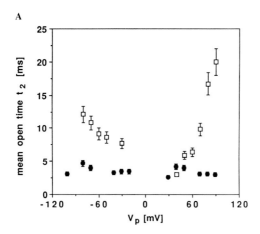

B

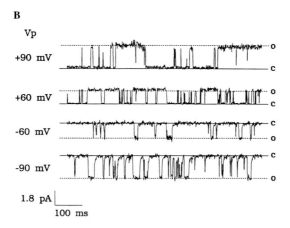

C

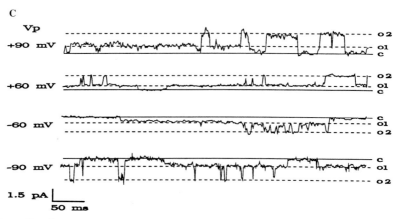

Figure 6(a–c)

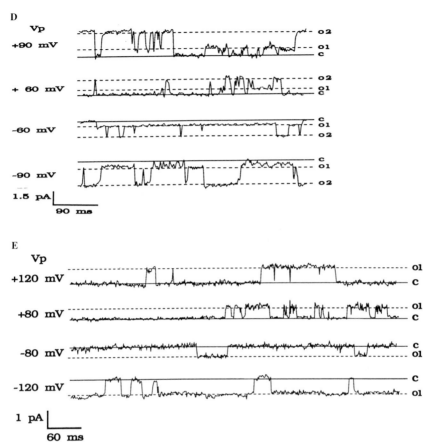

Figure 6. (a) Voltage-dependence of the large mean open time t_2 of wt annexin V (open squares) and the large conductance level of the Glu112Gly mutant (filled circles). The mutant has clearly lost the voltage-dependence of its channel gating (Göttig et al., unpublished data). Single-channel current recordings of (b) wt annexin V, (c) the Glu17Gly mutant, (d) the Glu78Gln mutant, and (e) the Glu17Gly/Glu78Gln mutant incorporated into acidic phospholipid bilayers at different pipette potentials Vp (recordings made using symmetrical pipette and bath solutions, both containing 50 mM Ca^{2+}) (Burger et al., 1994).

movement of a highly charged segment into the lipid bilayer under the influence of a transmembrane field is thought to be essential for the gating process (Merrill and Cramer, 1990). The studies presented here provide the first picture of a voltage sensor in an ion channel at atomic resolution.

We have also mutated charged amino acids located on the annexin V surface for glycine residues in order to examine effects of a change of the macro-dipole on the channel properties (Burger et al., 1994). The exchange of Glu17 located at the start of the first helix in domain I and far from the ion conduction pathway leads to the appearance of a

second conductance level of 9 pS in addition to the conductance level of
about 30 pS in the wildtype molecule (see Figs. 6b and c). This was also
the case for Glu78, which is part of a weak calcium-binding site (see
Fig. 6d). The removal of Glu17 and Glu78 produced a mutant retaining
only the smaller conductance level (see Fig. 6e).

Invariance of the three-dimensional structures in crystals suggests
that the changes in the single-channel characteristics of the Glu17Gly,
Glu78Gln and Glu17Gly/Glu78Gln mutant are probably due to
changes in dynamic rather than static properties. The mutations are far
removed from the central hydrophilic pore, which is the ion pathway.
They are located in different parts of the molecule, but their electro-
physical properties are very similar, suggesting a common mechanism
for their strong effects on channel conductance and kinetics. We suggest
the inter-module angle as the common denominator as it varies in the
different crystal structures of wt annexin V and its mutants (see Huber
et al., 1992; Lewit-Bentley et al., 1992) suggesting a preferred flexing
mode of the annexin molecule.

We propose that the inter-module angle, which strongly influences the
width of the central pore of annexin V, is the essential determinant of
molecular conformers with different channel conductances (Burger et
al., 1994). The equilibrium between these conformers is influenced by
the electrical charge constellation of the two modules. We suggest that
an alteration of the charge distribution of the module consisting of
domains I and IV between wt annexin V and the mutants Glu17Gly,
Glu78Gln and Glu17Gly/Glu78Gln may favor alternative conformers
with different intermodule angles. These conformers may have different
conductance levels in the membrane-bound form of the mutants. These
discrete conductance levels are likely to be associated with conformers
with defined inter-module angles not necessarily coincident with those
observed in the crystals, where crystal packing has a dominant influence.

The proposed gating model is in accordance with the absence of a
threshold voltage for ion-channel activity and the weak asymmetry
(rectification) of the voltage dependence of the channel openings in
annexin V.

Conclusions

Human annexin V is the first structurally and functionally characterized
ion channel protein. The combination of high-resolution crystal structures, electron microscopical images of membrane bound annexin V,
and single-channel analysis of the wildtype molecule and a variety of
mutants allows us to sketch a picture identifying the amino acid residues
performing the main tasks of an ion channel: Glu95 as an important

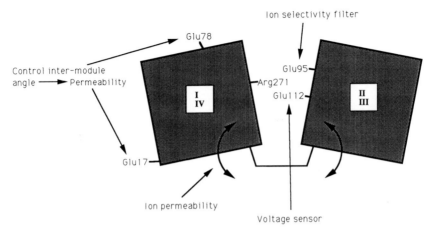

Figure 7. Scheme of the annexin V molecule depicting the amino acid residues performing the main tasks of an ion channel.

constituent of the ion pathway and the main selectivity filter, Glu112 as the main voltage-sensor controlling the gating of the channel, and Glu17 and Glu78 influencing the inter-module angle and thereby the ion permeability through the channel pore (see Fig. 7).

References

Barton, G.J., Newman, R.H., Freemont, P.S. and Crumpton, M.J. (1991) Amino acid sequence analysis of the annexin super-gene family of proteins. *Eur. J. Biochem.* 198: 749–760.

Berendes, R., Voges, D., Demange, P., Huber, R. and Burger, A. (1993) Structure-function analysis of the ion channel selectivity filter in human annexin V. *Science* 262: 427–430.

Bewley, M.C., Boustead, C.M., Walker, J.H., Waller, D.A. and Huber, R. (1993) Structure of chicken annexin V at 2.25 Å resolution. *Biochemistry* 32: 3923–3929.

Brisson, A., Mosser, G. and Huber, R. (1991) Structure of soluble and membrane-bound human annexin V. *J. Mol. Biol.* 220: 199–203.

Brücke, E. (1843) Beiträge zur Lehre von der Diffusion tropfbarflüssiger Körper durch poröse Scheidenwände. *Ann. Phys. Chem.* 58: 77–94.

Burger, A., Voges, D., Demange, P., Perez, C.R., Huber, R. and Berendes, R. (1994) Mutagenisis of human Annexin V, an *in vitro* voltage-gated calcium channel, provides information about the structural features of the ion pathway, the voltage sensor and the ion selectivity filter. *J. Mol. Biol.* 237: 479–499.

Burnashev, N., Monyer, H., Seeburg, P.H. and Sakmann, B. (1992) Divalent ion permeability of AMPA receptor channel is dominated by the edited form of a single subunit. *Neuron* 8: 189–198.

Catterall, W.A. (1988) Structure and function of voltage-sensitive ion channels. *Science* 242: 50–61.

Concha, N.O., Head, J.F., Kaetzel, M.A., Dedman, J.R. and Seaton, B.A. (1993) Rat annexin V crystal structure – Ca^{2+} induced conformational changes. *Science* 261: 1321–1324.

Creutz, C.E. (1992) The annexins and exocytosis. *Science* 258: 924–931.

Crompton, M.R., Moss, S.E. and Crumpton, M.J. (1988) Diversity in the lipocortin calpactin family. *Cell* 55: 1–3.

Davidson, F.F., Lister, M.D. and Dennis, E.A. (1990) Binding and inhibition studies on lipocortins using phosphatidylcholine vesicles and phospholipase-α2 from snake-venom, pancreas, and a macrophage-like cell-line. *J. Biol. Chem.* 265: 5602–5609.

Deisenhofer, J., Epp, O., Miki, K., Huber, R. and Michel, H. (1985) Structure of the protein subunits in the photosynthetic reaction centre of *Rhodopseudomonas viridis* at 3 Å resolution. *Nature* 318: 618–624.

Dingledine, R., Hume, R.I. and Heinemann, S.F. (1992) Structural determinants of barium permeation and rectification in non-NMDA glutamate receptor channels. *J. Neurosc.* 12: 4080–4087.

Driessen, H.P.C., Newman, R.H., Freemont, P.S. and Crumpton, M.J. (1992) A model of the structure of human annexin VI bound to lipid monolayers. *FEBS Lett.* 306: 75–79.

Durell, S.R. and Guy, H.R. (1992) Atomic scale structure and functional models of voltage-gated potassium channels. *Biophys. J.* 62: 238–250.

Egjeberg, J. and Heinemann, S.F. (1993) Ca^{2+} permeability of unedited and edited versions of the kainate selective glutamate receptor GluR6. *Proc. Natl. Acad. Sci. USA* 90: 755–759.

Ferrer-Montiel, A.V. and Montal, M. (1993) A negative charge in the M2 transmembrane segment of the neuronal a7 acetylcholine receptor increases permeability to divalent cations. *FEBS Lett.* 324: 185–190.

Fox, M.T., Prentice, D.A. and Hughes, J.P. (1991) Increases in p11 and annexin II proteins correlate with differentiation in the pc12 pheochromocytoma. *Biochem. Biophys. Res. Comm.* 177: 1188–1193.

Guy, H.R. and Seetharamulu, P. (1986) Molecular model of the action potential sodium channel. *Proc. Natl. Acad. Sci. USA* 83: 508–512.

Heinemann, S.H., Terlau, H., Stühmer, W., Imoto, K. and Numa, S. (1992) Calcium channel characteristics conferred on the sodium channel by single mutations. *Nature* 356: 441–443.

Hille, B. (1992) *Ionic channels of excitable membranes.* Sinauer Associates, Sunderland, MA, USA.

Huber, R., Römisch, J. and Pâques, E. (1990a) The crystal and molecular structure of human annexin V, an anticoagulant protein that binds to calcium and membranes. *EMBO J.* 9: 3867–3974.

Huber, R., Schneider, M., Mayr, I., Römisch, J. and Pâques, E. (1990b) The calcium binding sites in human annexin V by crystal structure analysis at 2.0 Å resolution. *FEBS Lett.* 275: 15–21.

Huber, R., Berendes, R., Burger, A., Schneider, M., Karshikov, A., Luecke, H., Römisch, J. and Pâques, E. (1992) Crystal and molecular structure of human annexin V after refinement. *J. Mol. Biol.* 223: 683–704.

Karshikov, A., Berendes, R., Burger, A., Cavalié, A., Lux, H.-D. and Huber, R. (1992) Annexin V membrane interaction: an electrostatic potential study. *Eur. Biophys. J.* 20: 337–344.

Kim, M.-S., Morii, T., Sun, L.-X., Imoto, K. and Mori, Y. (1993) Structural determinants of ion selectivity in brain calcium channel. *FEBS Lett.* 318: 145–148.

Kretsinger, R.H. (1980) Structure and evolution of calcium-modulated proteins. *CRC Crit. Rev. Biochem.* 8: 119–174.

Lewit-Bentley, A., Morera, S., Huber, R. and Bodo, G. (1992) The effect of metal binding on the structure of annexin V and implications for membrane binding. *Eur. J. Biochem.* 210: 73–77.

Mangeat, P.-H. (1988) Interaction of biological membranes with the cytoskeletal framework of living cells. *Biol. Cell* 64: 261–281.

Meers, P. (1990) Location of tryptophans in membrane-bound annexins. *Biochemistry* 29: 3325–3330.

Meers, P. and Mealy, T. (1993) Relationships between annexin V tryptophan exposure and calcium and phospholipid binding. *Biochemistry* 32: 5411–5418.

Miller, C. (1991) 1990: Annus mirabilis of potassium channels. *Science* 252: 1092–1096.

Moss, S.E., Edwards, H.C. and Crumpton, M.J. (1991) Diversity in the annexin family. *In:* C.W. Heizmann (ed.): *Novel calcium-binding proteins.* Springer-Verlag, Berlin, pp. 535–566.

Mosser, G., Ravanat, C., Freyssinet, J.-M. and Brisson, A. (1991) Subdomain structure of lipid-bound annexin-V resolved by electron image analysis. *J. Mol. Biol.* 217: 241–245.

Neumann, E. (1988) The electroporation hysteresis. *Ferroelectrics* 86: 325–333.

Newman, R.D., Leonard, K. and Crumpton, M.J. (1991) 2D crystal forms of annexin IV on lipid monolayers. *FEBS Lett.* 279: 21–24.

Parker, M.W., Pattus, F., Tucker, A.D. and Tsernoglou, D. (1989) Structure of the membrane-pore-forming fragment of colicin A. *Nature* 337: 93–96.

Pollard, H.B. and Rojas, E. (1988) Ca^{2+}-activated synexin forms highly selective, voltage-gated Ca^{2+}-channels in phosphatidylserine bilayer membranes. *Proc. Natl. Acad. Sci. USA* 85: 2974–2978.

Pollard, H.B., Guy, H.R., Arispe, N., de la Fuente, M., Lee, G., Rojas, E.M., Pollarrd, J.R., Srivastava, M., Zhang-Keck, Z.-Y., Merezhinskaya, N., Caohuy, H., Burns, L.A. and Rojas, E. (1992) Calcium channel and membrane fusion activity of synexin and other members of the Annexin gene family. *Biophys. J.* 62: 19–22.

Römisch, J., Schorlemmer, U., Fickenscher, K., Paques, E.P. and Heimburger, N. (1990) Anticoagulant properties of placenta protein-4 (annexin-v). *Thromb. Res.* 60: 355–366.

Rojas, E., Pollard, H.B., Haigler, H.T., Parra, C. and Burns, A.L. (1990) Calcium-activated endonexin II forms calcium channels across acidic phospholipid membranes. *J. Biol. Chem.* 265: 21207–21215.

Sansom, M.S.P. (1991) The biophysics of peptide models of ion channels. *Prog. Biophys. Molec. Biol.* 55: 139–235.

Schlaepfer, D.D. and Haigler, H.T. (1990) Expression of annexins as a function of cellular growth-state. *J. Cell. Biol.* 111: 229–238.

Sopkova, J., Renouard, M. and Lewit-Bentley, A. (1993) The crystal-structure of a new high-calcium form of annexin-V. *J. Mol. Biol.* 234: 816–825.

Voges, D., Berendes, R., Burger, A., Demange, P., Baumeister, W. and Huber, R. (1994) Three-dimensional structure of membrane-bound annexin V – a correlative electron microscopy-x-ray crystallography study. *J. Mol. Biol.* 238: 199–213.

Weng, X., Luecke, H., Song, I.S., Kang, D.S., Kim, S.-H. and Huber, R. (1993) Crystal structure of human annexin I at 2.5 Å resolution. *Prot. Sci.* 2: 448–458.

Weiss, M.S., Abele, U., Weckesser, J., Welte, W., Schiltz, E. and Schulz, G.E. (1991) Molecular architecture and electrostatic properties of a bacterial porin. *Science* 254: 1627–1630.

Yang, J., Ellinor, P.T., Sather, W.A., Zhang, J.-F. and Tsien, R.W. (1993) Molecular determinants of Ca^{2+} selectivity and ion permeation in L-type Ca^{2+} channels. *Nature* 366: 158–161.

Biochemistry of Cell Membranes
ed. by S. Papa & J. M. Tager

Structure and function of the membrane channel porin

G.E. Schulz

Institut für Organische Chemie und Biochemie der Albert-Ludwigs-Universität, Albertstr. 21, D-79104 Freiburg im Breisgau, Germany

Summary. The structure of a porin from the outer membrane of the phototrophic bacterium *Rhodobacter capsulatus* is described in detail. Three more porins have been structurally characterized showing, in general, the same features. Accordingly, the essential points of this report, i.e., the oligomeric structure, the secondary structure, the immersion in the membrane, the electrostatic separation effect, the chain-folding process, the voltage dependence and the observed specificity of some porins, apply for all the structurally known and probably also for a much larger group of porins.

Introduction

Prokaryotes have to protect themselves against adverse environments. For this purpose, Gram-negative bacteria have an outer membrane. This membrane contains channels that are permeable for nutrients. The channels consist of porins which are usually homotrimeric proteins with subunit sizes ranging from 30 to 50 kDa and solute exclusion limits around 600 Da (Benz and Bauer, 1988; Nikaido, 1992). The channels are well permeable for polar solutes, but exclude nonpolar molecules of comparable sizes. Porins have been subdivided into specific and non-specific ones. For a particular solute, specific porins show a comparatively large diffusion rate at low, and saturation effects at high concentrations. In contrast, nonspecific porins act like inert holes, their diffusion rates are always proportional to the solute concentration.

A rather general description of the porin architecture was derived from electron microscopy studies (Engel et al., 1985; Jap et al., 1991). The first porin crystals suitable for an x-ray analysis were obtained from *Escherichia coli* (Garavito and Rosenbusch, 1980). The first atomic structure was elucidated for the major porin of *Rhodobacter capsulatus* (Kreusch et al., 1991; Schiltz et al., 1991; Weiss et al., 1991a,b; Weiss and Schulz, 1992). An analysis of other crystals using molecular replacement with the *R. capsulatus* porin structure showed that the 16-stranded β-barrel fold of this porin is also present in quite a number of other porins (Pauptit et al., 1991). Now, three more porin structures are known in atomic detail (Cowan et al., 1992; Kreusch et al., 1994); they confirm the general picture derived from the first porin structure.

Results and discussions

Structure

After careful protein preparation the major porin of *R. capsulatus* yielded particularly well-ordered crystals (Kreusch et al., 1991). As shown by amino acid sequence analysis (Schiltz et al., 1991), one subunit consists of 301 amino acids. The crystal structure has been solved at 1.8 Å resolution and refined to a crystallographic R-factor of 18.6% using a 97% complete data set (Weiss and Schulz, 1992). The model of one subunit contains all polypeptide atoms, 274 water molecules, three calcium ions, three detergent molecules (octyltetra-oxyethylene, C_8E_4), and one bound ligand. Since this ligand could not be identified, it was modeled as C_8E_4.

Each porin subunit consists of a very large 16-stranded antiparallel β-barrel and three short α-helices. All β-strands are connected to their nearest neighbors. The loops at the bottom end of the barrel are short and this end is smooth. In contrast, the top end of the barrel is rough, containing much longer loops. The longest loop has 44 residues and runs into the interior of the barrel, where it constricts the pore to a small eyelet (Weiss and Schulz, 1992).

The chain fold of trimeric porin (Fig. 1) shows that the β-barrel height in the trimer center is lower than the barrel height facing the membrane. This center forms a three-pronged star composed of all three subunits resembling hub and spokes of a wheel. The mobilities of the atoms in this central star are the lowest of the whole molecule as

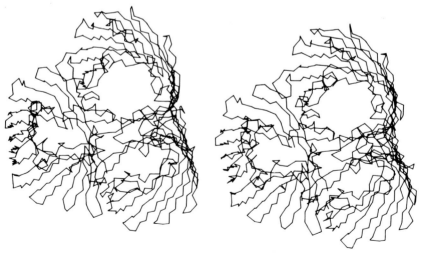

Figure 1. Stereo view of the C_α-backbone chain fold of the trimeric major porin from *Rhodobacter capsulatus* taken from Weiss et al. (1991). The view is from the external medium.

determined by the crystallographic temperature factors (Weiss and Schulz, 1992). The interior of this star is nonpolar; 18 phenylalanines (six from each subunit) interdigitate tightly, while the surface is polar. The star contains all six chain termini paired in salt bridges. Trimeric porin can thus be described as a rigid central core constructed like a water-soluble protein, that is surrounded by three β-barrel walls fencing off the membrane.

Orientation in the membrane

A cut through the center of a pore shows the shape of the channel as illustrated in Figure 2. The aggregation to trimers connects the rear β-barrel walls at the height levels indicated by dashed lines. The central part is low, giving rise to a common channel formed by all three subunits in the upper half. The three pore eyelets limiting the diffusion are located between the barrel wall *1* close to the molecular threefold-axis and the 44-residue loop *3* inside the barrel.

The longitudinal position of porin in the membrane is clearly defined by the nonpolar ring with a height of 24 Å that surrounds the trimer

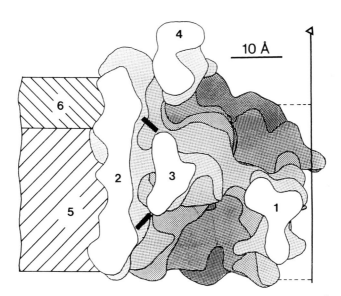

Figure 2. Shape of a *Rb. capsulatus* porin subunit represented by a sliced 6 Å resolution density map calculated from the high resolution model (Weiss et al., 1991b). The density level is at 1σ; the distance from the viewer is indicated by shading. The sectional areas **1, 2, 3, 4, 5** and **6** belong to: the β-barrel wall at the trimer interface, the β-barrel wall facing the membrane, the pore-size-defining 44-residue loop inside the β-barrel, a smallish domain facing the external medium, the nonpolar moiety of the membrane, and the LPS core, respectively. As indicated by bars, there is no free space between sectional areas **2** and **3**.

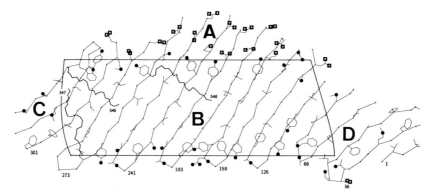

Figure 3. Projection of the outer surface and of the β-barrel onto a cylinder as taken from Weiss and Schulz (1992). All loops at the periplasmic side are indicated. Surface areas **A, B, C** and **D** contain the ionogenic side chains connecting to the LPS core, the nonpolar side chains facing the nonpolar interior of the membrane, the left-hand and the right-hand part of the interface, respectively. Ionogenic atoms are emphasized by squares and polar atoms by circles. The aromatic girdles are obvious.

and fits the nonpolar moiety of the membrane. The rough upper end of the β-barrel (Fig. 1) contains the larger loops with numerous charged side chains, whereas the smooth lower barrel end has only small loops with few polar residues. As shown by binding studies with antibodies and phages, the large polar loops face the external medium (Tommassen, 1988). Accordingly, the external medium is at the top of Figures 1 through 5.

This orientation agrees well with other features of the outer surface of the trimer (Fig. 3). While the nonpolar surface moiety faces the nonpolar interior of the lipid bilayer, the upper polar part with its numerous ionogenic side chains is most likely connected to the lipopolysaccharides (LPS) forming the external layer of the bacterial outer membrane. Presumably, the numerous Asp and Glu side chains are glued by divalent cations (like calcium and magnesium) to the carboxylate groups of the LPS cores. This network integrates porin efficiently in the tough bacterial protection layer of crosslinked LPS molecules, avoiding a fragile protein-membrane interface.

Furthermore, Figure 3 shows two girdles of aromatic residues along the upper and lower border lines between polar and nonpolar residues. The upper girdle contains mostly tyrosines pointing with their hydroxyl groups to the upper polar moiety of the membrane, while the lower girdle has phenylalanines pointing to the nonpolar membrane interior and also tyrosines pointing to the polar part of the periplasmic membrane layer. These patterns are obviously significant. They were also observed in three recently elucidated porin structures (Cowan et al., 1992; Kreusch et al., 1994).

Also, there are four type-II' β-turns at the lower smooth end of the barrel. These four turns point with their peptide amides toward the membrane where they can form hydrogen bonds to the phosphodiesters of the periplasmic layer of the bacterial outer membrane.

Packing in the crystal

The exceptionally well-ordered crystals of the *R. capsulatus* porin (Weiss and Schulz, 1992) have the packing scheme shown in Figure 4. It resembles the crystal packing of the photoreaction center (Deisenhofer and Michel, 1989) as the polypeptides form only polar contacts. Apart from the subunit interface, there exists merely one contact type, i.e., between head and tail, giving rise to an arrangement of the trimers similar to a cubic closest packing. This contact contains five hydrogen bonds and one salt bridge and is obviously strong. The closest lateral distance between trimers is 15 Å, and it is between nonpolar surfaces. Any conceivable contact through detergent molecules should therefore be very weak and should provide only a minor contribution to the crystal packing energy.

The molecular packing in the crystal corresponds to a stack of lipid bilayers containing porins, suggesting that such bilayers can also be formed in the crystal. This hypothesis could be tested by measuring the electric conductivity of the crystal along and perpendicular to the threefold rotation axis. Such measurements can also be done after changing the detergent in the crystal (possibly to lipids) or after crystallizing with other detergents, which would yield information about the detergent(lipid)-protein contact.

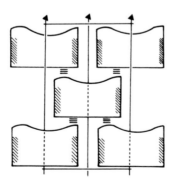

Figure 4. Crystal packing arrangement of trimeric porin molecules, the molecular threefold axes ($\triangle$) are crystallographic. The space group R3 requires only one contact type to form a three-dimensional network. This contact ($\equiv$) is head to tail and polar. There is no lateral contact between the nonpolar surfaces (cross-hatched), which are presumably covered by detergent.

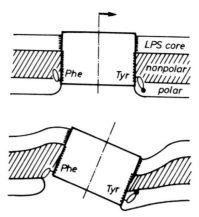

Figure 5. Sketch describing the suggested shielding role of phenylalanines and tyrosines during relative movements of protein and membrane.

The aromatic girdles

Since the Tyr(Trp)-Phe-pattern (Fig. 3) has been observed in all the four known porin structures (Weiss and Schulz, 1992; Cowan et al., 1992; Kreusch et al., 1994), it most likely reflects a function. I suggest that these bulky aromatic side chains prevent conformational damage of the protein on mechanical movements in the membrane, as indicated in Figure 5. Any transversal wave in the membrane or any knocking at a porin trimer immersed in the membrane causes rocking movements that would expose nonpolar protein surface to a polar membrane layer and polar protein surface to the nonpolar membrane interior. Both contact types give rise to a large surface tension that is likely to scramble the polypeptide conformation, which to a large part forms a single layer β-sheet where the residues can turn around easily. Since the aromatic side chains rotate around their C_α–C_β-bonds much faster than the trimer can rock, they should be able to shield the respective surfaces against the wrong counterpart (Fig. 5).

Folding pathway

Considering the chain fold of the homotrimer with the central rigid star, the prongs of which are connected by the three β-barrel walls, a folding pathway can be suggested: In a first step, the three-pronged star folds like a water soluble protein. Since this star contains all chain termini, the remaining chain parts form three large loops of about 200 amino acid residues each suspended between the three prongs. On contact with the nonpolar membrane interior, the three 200-residue loops arrange themselves to the actually observed most simple β-barrel topology with all strands antiparallel and connected to their next neighbors. This folding process is straight-forward as it requires no chain crossing-over or wrapping-around. On folding, the nonpolar side chains of the β-barrel residues orient themselves to the outside and face the nonpolar moiety of the membrane. Subsequently, the large 44-residue loop is inserted into the barrel. This loop supports the barrel at its center against the membrane pressure, and it defines the pore size.

Permeation properties

The pore eyelet
The eyelet of the channel is lined by ionogenic groups that segregate into positively and negatively charged rims (Fig. 6). The positive rim is closer to the trimer center and consists of half a dozen Arg, Lys, and His side chains. The negative rim is further out toward the circumference of the trimer, it contains about a dozen Asp and Glu residues mostly located in the 44-residue loop that is inserted into the β-barrel. These abundant negative charges are partially compensated by two bound Ca^{2+} ions which appreciably tighten the rim structure. As a consequence, the removal of these Ca^{2+} should change permeabilities. Actually, it had been observed that the analyzed porin permits the

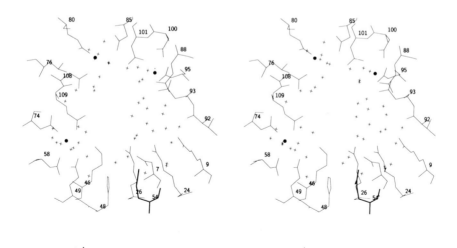

Figure 6. Stereo view of a pore eyelet as taken from Weiss and Schulz (1992). The pore eyelet is lined by negatively charged side chains at its upper rim and positively charged ones at the lower rim. There exists a strong transversal electric field between these charges as indicated by the rigidity of the neighboring arginines at the lower rim (see text). Water molecules (+) and bound Ca^{2+} (●) are given. The molecular threefold axis is indicated (▲).

diffusion of ATP only after the bacterium has been treated with EDTA (Carmeli and Lifshitz, 1989).

An electric separator

The juxtaposed positive and negative rims cause a transversal electric field across the eyelet. The field strength is best estimated from two arginines participating in the positive rim. These arginines are, at van der Waals distance, in good electron density, demonstrating that they are ridigly positioned in spite of their repelling positive charges (the crystals are at pH 7.2, the pK of an isolated Arg is 12.4). Such an arrangement is only possible if the charges are fixed by a strong electric field.

An X-ray structure analysis at 1.8 Å resolution allows for the assignment of reasonably well fixed water molecules (Weiss and Schulz, 1992). The eyelet contains quite a number of them (Fig. 6). They orient their dipoles in the electric field (Fig. 7). In the center of the eyelet there are mobile water molecules confined to a rather small cross-section of about 4 Å by 5 Å. They form a hole through which a molecule with the cross-section of an alkyl chain should be able to permeate.

The diffusion of a small polar solute is illustrated in Figure 8A. The solute is oriented by the transversal electric field formed by the charged rims of the eyelet and will remain oriented over the whole diffusion distance. The solute is oriented in the pore eyelet like a substrate on an enzyme. This reduces the entropy of the diffusion process in the same way as it is reduced for a chemical reaction by binding to an enzyme surface. Without tumbling, the activation energy barrier for permeation

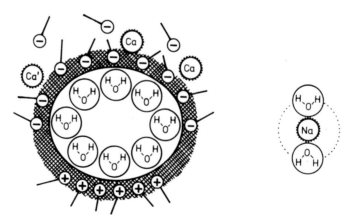

Figure 7. Sketch of the pore eyelet indicating a ring of bound water molecules that are oriented in the transversal electric field. The rim of negatively charged side chains is strengthened by two strongly (Ca) and one weakly (Ca′) bound Ca^{2+}. The depth of the pore eyelet is about 6 Å. For a hydrated ion, always half the hydration shell has an energetically unfavorable orientation in the electric field.

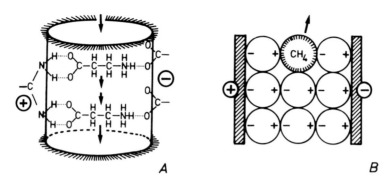

Figure 8. Diffusion through the pore eyelet. (A) Sketch of a permeating zwitterion β-alanine which remains oriented in the field. Its permeation is facilitated by entropy reduction (B) Sketch of the pore eyelet as a capacitor explaining the high activation energy barrier for the permeation of nonpolar solutes.

is lowered, giving rise to an appreciable acceleration of the diffusion of polar solutes.

Although the eyelet cross-section that is free of fixed water molecules is spacious enough for the penetration of alkyl chains, these still cannot permeate because they are blocked by the strong transversal electric field. Field and eyelet behave like a charged capacitor that stores an electric energy proportional to the high dielectric constant of 80 of oriented water molecules (Fig. 8B) multiplied by the volume. Any incoming molecule with a lower dielectric constant (the respective value for an alkyl chain is about 2), i.e., with a smaller dipole than water, causes a decrease of the stored capacitor energy and therefore generates a force expelling the intruder. Consequently, the transversal field acts as an electric separator, facilitating the permeation of polar solutes while blocking off nonpolar ones.

A special situation seems to arise for ions. When diffusing through the pore, half their hydration shell enters in an energetically unfavorable orientation (Fig. 7). In the $R.$ $capsulatus$ porin, there exists a possible separate pathway for positive ions at one corner of the eyelet. This pathway is outside the electric field and completely lined by carbonyl and carboxyl oxygens. This may explain the observed though low cation selectively of this porin. Since this channel is blocked by a weakly bound Ca^{2+} ion, its importance can be tested by measuring cation currents as a function of Ca^{2+} concentration.

Voltage dependence
There are observations indicating that some porins incorporated in black lipid films reduce their electric conductivity on application of a voltage of around 100 mV across this film (Jap and Walian, 1990), i.e., there occurs voltage closure. Such a behavior can be understood when

considering the shape of the pore (Fig. 2). The membrane channel has a rather large cross-section over its whole length except for the short distance of about 6 Å at the diffusion-defining eyelet at the center, where the cross section is small. This implies that any voltage across the membrane (say 100 mV) lies essentially across the 6 Å length of the eyelet (certainly more than 90 mV) forming a strong longitudinal electric field. Since the energy gained by moving a single charge (e.g., the protonated ε-amino group of a lysine) over 100 mV is 10 kJ/mol and thus equivalent to the energy of a hydrogen bond, the strong longitudinal electric field along the eyelet can tear off any charged group that can be moved over the 6 Å length of the eyelet (which is possible for lysine). The applied voltage thus disrupts the eyelet structure and diminishes the permeability. In conclusion, the pores are vulnerable at such high voltages. This does not exclude, however, that some porins may possess lids that move in a longitudinal electric field onto the pore, thus closing it. Such porins would have a defined and reproducible voltage dependence.

Porin specificity

Porins have been subdivided into two classes, specific and nonspecific ones (Nikaido, 1992). The structurally known *R. capsulatus* porin has been previously classified as nonspecific (Flammann and Weckesser, 1980). In the crystal structure, however, this porin is ligated by a small molecule (Fig. 9). Moreover, efficient binding of tetrapyrrols to this

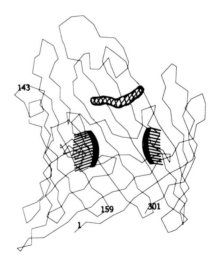

Figure 9. Sketch of the C$_\alpha$ backbone of one subunit with the pore eyelet and the observed solute binding site within the pore close to the eyelet. Note that the solute binds at the external side of the eyelet.

porin has been reported (Bollivar and Bauer, 1992). In the crystal, the bound ligand cannot be identified from the density. The binding site is formed between the 44-residue loop inserted into the barrel (where it forms the eyelet) and a smallish domain protruding to the external medium. Facing the external medium, this binding site can pick up solutes at very low concentrations from the environment. After being bound close to the eyelet (Fig. 9), the solute may subsequently dissociate and diffuse through the pore at a considerable rate. Binding and permeation should follow Michaelis-Menten kinetics if the bound solute hinders the permeation of another molecule of its kinds. As a consequence there should be saturation effects as observed within the class of specific porins. Following our observation of a ligand binding site, the R. capsulatus porin previously assigned to the class of nonspecific porins is actually both, specific and nonspecific, depending on the solute. Generalizing this observation, it is suggested that many more porins belong to both classes, but that the specific solutes have been detected only in very few cases.

Acknowledgments
I thank A. Kreusch, E. Schiltz, J. Weckesser, M.S. Weiss, and W. Welte for their essential contributions to the structure analysis of the R. capsulatus porin.

References

Benz, R and Bauer, K. (1988) Permeation of hydrophilic molecules through the outer membrane of gram-negative bacteria. *Eur. J. Biochem.* 176: 1–19.

Bollivar, D.W. and Bauer, C.E. (1992) Association of tetrapyrrole intermediates in the bacteriochlorophyll *a* biosynthetic pathway with the major outer-membrane porin protein of *Rhodobacter capsulatus*. *Biochem. J.* 282: 471–476.

Carmeli, C. and Lifshitz, Y. (1989) Nucleotide transport in *Rhodobacter capsulatus*. *J. Bacteriol.* 171: 6251–6525.

Cowan, S.W., Schirmer, T., Rummel, G., Steiert, M., Ghosh, R., Pauptit, R.A., Jansonius, J.N. and Rosenbusch, J.P. (1992) Crystal structures explain functional properties of two *E. coli* porins. *Nature* 358: 727–733.

Deisenhofer, J. and Michel, H. (1989) The photosynthetic reaction center from the purple bacterium *Rhodopseudomonas viridis*. *Science* 245: 1463–1473.

Engel, A., Massalski, A., Schindler, H., Dorset, D.L. and Rosenbusch, J.P. (1985) Porin channel triplets merge into single outlets in *Escherichia coli* outer membranes. *Nature* 317: 643–645.

Flammann, H.T. and Weckesser, J. (1984) Porin isolates from the cell envelope of *Rhodopseudomonas capsulate*. *J. Bacteriol.* 159: 410–412.

Garavito, R.M. and Rosenbusch, J.P. (1980) Three-dimensional crystals of an integral membrane protein: an initial x-ray analysis. *J. Cell. Biol.* 86: 327–329.

Jap, B.K. and Walian, P.J. (1990) Biophysics of the structure and function of porins. *Quart. Rev. Biophys.* 23: 367–403.

Jap, B.K., Walian, P.J. and Gehring, K. (1991) Structural architecture of an outer membrane channel as determined by electron crystallography. *Nature* 350: 167–170.

Kreusch, A., Neubüser, A., Weckesser, J. and Schulz, G.E. (1994) The structure of the membrane channel porin from *Rhodopseudomonas blastica* at 2.0 Å resolution. *Protein Sci.* 3: 58–63.

Kreusch, A., Weiss, M.S., Welte, W., Weckesser, J. and Schulz, G.E. (1991) Crystals of an integral membrane protein diffracting bo 1.8 Å resolution. *J. Mol. Biol.* 217: 9–10.

Nikaido, H. (1992) Porins and specific channels of bacterial outer membrane. *Mol. Microbiol.* 6: 435–442.

Pauptit, R.A., Schirmer, T., Jansonius, J.N., Rosenbusch, J.P., Parker, M.W., Tucker, A.C., Tsernogiou, D., Weiss, M.S. and Schulz, G.E. (1991) A common channel-forming motif in evolutionarily distant porins. *J. Struct. Biol.* 107: 136–145.

Schiltz, E., Kreusch, A., Nestel, U. and Schulz, G.E. (1991) Primary structure of porin from *Rhodobacter capsulatus*. *Eur. J. Biochem.* 199: 587–594.

Tommassen, J. (1988) Biogenesis and membrane Topology of Outer Membrane Proteins in *Escherichia coli. In:* J.A.F. Op den Kamp (ed.): *Membrane Biogenesis,* NATO ASI series, Vol. H16, Springer-Verlag, Berlin, pp. 351–373.

Weiss, M.S., Abele, U., Weckesser, J., Welte, W., Schiltz, E. and Schulz, G.E. (1991a) Molecular architecture and electrostatic properties of a bacterial porin. *Science* 254: 1627–1630.

Weiss, M.S., Kreusch, A., Schiltz, E., Nestel, U., Welte, W., Weckesser, J. and Schulz, G.E. (1991b) The structure of porin from *Rhodobacter capsulatus* at 1.8 Å resolution. *FEBS Lett.* 280: 379–382.

Weiss, M.S. and Schulz, G.E. (1992) Structure of porin refined at 1.8 Å resolution. *J. Mol. Biol.* 227: 493–509.

Biochemistry of Cell Membranes
ed. by S. Papa & J. M. Tager
© 1995 Birkhäuser Verlag Basel/Switzerland

Oxygen damage and mutations in mitochondrial DNA associated with aging and degenerative diseases

T. Ozawa

*Department of Biomedical Chemistry, Faculty of Medicine, University of Nagoya,
65 Tsuruma-cho, Showa-ku, Nagoya 466, Japan*

Summary. This article reviews the theory and molecular genetics of mitochondrial aging and diseases. Mitochondrial DNA (mtDNA) that codes protein subunits essential for the maintenance of the mitochondrial ATP synthesis system located in the inner membrane acquires mutations at a much higher rate than nuclear DNA. Recent study clarifies somatically acquired mutations such as deletions in mtDNA caused by oxygen damage during the life of an individual. Cumulative accumulation of these somatic mutations in postmitotic neuromuscular cells causes bionenergetic deficit leading to age-associated dysfunction of cells and organs. The base-sequencing of the entire mtDNA from individuals revealed that germ-line point mutations transmitted from the ancestor accelerate the somatic oxygen damage and the fragmentation in mtDNA leading to phenotypic expression as premature aging and degenerative diseases.

Introduction

The free radical theory of aging (Harman, 1960), that free radicals cause non-specific damage to macromolecules, such as DNA, lipids, and proteins, has attracted much attention in recent years due to developments in free radical biology. There is enough supporting evidence such as the reciprocal relation between the maximum life span and the specific metabolic rate among animals. However, this theory has not been completely persuasive because biologically active oxygen-damaged macromolecules, such as lipid peroxides or oxidized proteins, usually do not accumulate, but turn over with certain metabolic rate, except for mere inactive degenerative products such as lipofuscin or amyloid. Namely, no adequate bio-marker to support the theory has been established.

In the early 1960s, some people still believed in the *de novo* synthesis of mitochondrion from the cytoplasmic membrane. However, in 1964, the presence of DNA in yeast mitochondria was reported (Shatz et al., 1964), and the existence of mitochondrial DNA (mtDNA), a closed circular duplex DNA, among several biological species was confirmed by extensive studies during the 1970s. In 1981, the entire nucleotide-sequence of human mtDNA of 16 569 base-pairs (bp) was reported

(Anderson et al., 1981). The sequence has been referred to as the Cambridge sequence in the literature. In 1989, we proposed (Linnane et al., 1989) that the somatic accumulation of mitochondrial genome mutations during life is a major cause of human aging and degenerative diseases. This mitochondrial theory of aging is based on the following facts: the high frequency of gene mutation in mtDNA; the small size of the mitochondrial genome and its unknown information content; the lack of a repair mechanism for mtDNA, unlike nuclear DNA; the somatic segregation of individual mtDNA molecules during eukaryotic cell division. After the newly invented polymerase chain reaction (PCR) was put to practical use in 1989, many kinds of mtDNA mutations, such as point mutations and deletions, have been reported to be associated with aging and degenerative diseases. Thus, the theory of mitochondrial aging and diseases caused by mtDNA mutations has now been widely accepted.

Recent findings that the amount of oxygen damage in mtDNA correlates closely with the amount of deletion associated with age (Hayakawa et al., 1991b; Hayakawa et al., 1992) could coordinate the free radical theory and the mitochondrial theory of aging. Namely, it was found that the human mtDNA is a superb bio-marker of cellular oxygen damage for a term of years. In 1990, we reported the presence of mtDNA with a 5 kilo bp deletion in the striatum of patients with Parkinson's disease (PD) as well as in that of aged subjects (Ikebe et al., 1990). The large deletion eliminates several genes encoding protein subunits essential for mitochondrial ATP synthesis, thus resulting in a bioenergetic deficit of the cells. Subsequently, the deletion was found in various human tissues associated with age (Linnane et al., 1990; Yen et al., 1991). Several other kinds of large deletions were reported in patients with premature aging such as those with Kearns-Sayre syndrome (Mita et al., 1990), primary caridomyopathy (Ozawa et al., 1990a), or diabetes with deafness (Ballinger et al., 1992). The underlying mechanism for a large deletion is double-strand separation leading to the generation of long stretches of single-stranded DNA. Accumulation of hydroxyl-radical adducts in mtDNA, such as 8-hydroxy-deoxyguanosine (8-OH-dG) in place of deoxyguanosine (dG), could satisfy this prerequisite to the deletion (Hayakawa et al., 1992). Because of the conformational change, the oxygen-damaged base in mtDNA would play a key role in yielding double strand separation and deletion. Especially in post-mitotic neuromuscular cells, the mitochondrial genome mutations accumulate during life by escaping the excision repair mechanism and the cellular selection by mitosis. Quantitative analyses of oxygen damage in mtDNA by using mass spectrum (MS) as well as of deletions by PCR revealed a synergistic and exponential increase in the somatic oxygen damage and the deletions in normal human subjects as well as in aged experimental animals (Hayakawa et al., 1993). The total base-sequencing of mtDNA among the patients with mitochondrial diseases revealed that even a single germ-line point mutation transmitted from an ancestor extensively accelerates the deletion leading to fragmentation of mtDNA (Katsumata et al., 1994). For a term of years, the difference in the severity of germ-line mutation was demonstrated to be related to the differences in the individual's phenotype, ultimately the patient's life span (cf. Fig. 3).

The accumulated evidence urges one to unify the two ideas of the free radical and mitochondrial theory of aging to obtain "the redox mechanism of aging", that is, that the mtDNA's oxygen damage inevitably associated with cellular respiration results in somatic mutations leading to bioenergetic deficit, aging, and cell death, especially in neuromuscular cells. The germ-line mutations in mtDNA specific for the patients with mitochondrial diseases accelerates these pathological processes leading to their phenotypic expression as premature aging of the patients.

Mitochondrial inner membrane and mtDNA

The basic pathological process in aging and neuromuscular diseases involves mutations in mitochondrial genomes that code essential parts of the energy transducing system. Apart from a minor contribution of anaerobic glycolysis, the ATP essential for the cellular activity is almost exclusively produced by mitochondria. This ATP production depends upon the proper functioning of the mitochondrial electron-transport chain, and of the functionally coupled proton motive ATP-synthase (ATPase). Mitochondria carry out the Krebs cycle and the β-oxidation pathway for fatty acids. These degradative sequences essentially remove hydrogen from metabolic fuels with the release of CO_2 and transfer it through coenzymic carriers to the mitochondrial electron-transport chain. The chain then transfers these reducing equivalents eventually to react with molecular oxygen with the production of H_2O. This latter process is associated with the inner mitochondrial membrane and is achieved in a manner that transduces the redox energy into a proton gradient as the driving force for ATP synthesis carried out by an enzymic apparatus also associated with that membrane. ATP is the direct driving force for virtually all energy-demanding processes of living cells, including neuromuscular activities. Accordingly, deterioration of the mitochondrial electron-transport chain and ATPase leads to cellular dysfunction and cell death, *apoptosis*.

As schematically illustrated in Figure 1A, the mitochondrial electron-transport chain is composed primarily of four complexes (Hatefi, 1985): complex I, NADH-ubiquinone oxidoreductase; complex II, succinate-ubiquinone oxidoreductase; complex III, ubiquinone-cytochrome c oxidoreductase; and complex IV, cytochrome oxidase. These complexes

A   Cytosolic side

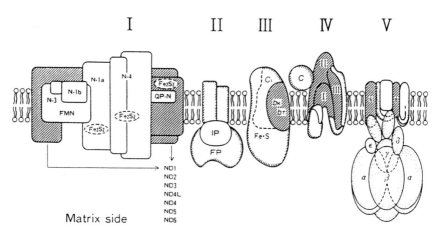

B

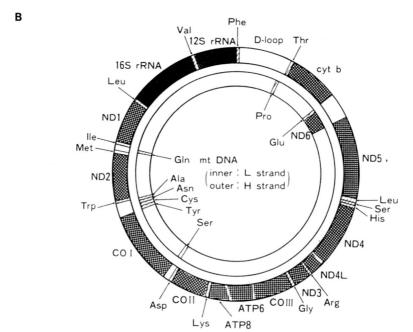

Figure 1. Schematic representation of mitochondrial electron-transport chain and encoding
DNA. (A) The mitochondrial electron-transport chain is composed primarily of four complexes: complex I, NADH-ubiquinone oxidoreductase; complex II, succinate-ubiquinone
oxidoreductase; complex III, ubiquinone-cytochrome *c* oxidoreductase; and complex IV,
cytochrome oxidase. These complexes together with complex V, proton motive ATP synthase,
are all embedded in the mitochondrial inner membrane. The hatched subunits of these
complexes are encoded by mtDNA, as shown below. (B) The human mtDNA is a closed
circular DNA of 16 569 base pairs (bp), which contain 13 protein-coding genes specifying
hydrophobic subunits of mitochondrial electron transport system and ATPase: seven subunits
of complex I, the apocytochrome *b* of complex III, three subunits of complex IV, and two

together with complex V (ATPase) are responsible for the overall process of oxidative phosphorylation, i.e., ATP production. Each complex consists of various numbers of subunits. Some subunits are synthesized in the mitochondrion according to information encoded within mtDNA, and others are specified by nuclear DNA, synthesized in the cytoplasmic compartment, transported into the inner membrane and assembled into functional enzyme complexes of I–V.

Mitochondria are the unique location, in animal cells, of maternally inherited genetic information in the form of extra-nuclear DNA. The determination of the nucleic acid sequence of human mtDNA (Anderson et al., 1981) demonstrated its skeletal organization, as shown in Figure 1B. Each of 16 569 bp of the human mtDNA is numbered according to nucleotide position. The mtDNA contains 13 protein-coding genes specifying hydrophobic subunits of the mitochondrial electron-transport system and ATPase: seven subunits of complex I, the apocytochrome *b* of complex III, three subunits of complex IV, and two subunits of complex V. The rest of the mitochondrial genome contains genetic information essential for the expression of the protein-coding genes; genes specifying two mitochondrial ribosomal RNAs (rRNA), 22 organelle-specific transfer RNAs (tRNA), and genes regulating transcription and replication (D-loop).

Molecular genetics of mtDNA mutations

Recent analyses of human mtDNA mutations demonstrate not only germ-line mutations transmitted from ancestors, but also the somatic mutations acquired during the life of an individual. Classification and characteristics of mtDNA mutations are described below.

Germ-line mutations

The major germ-line mutation is the base substitution causing amino-acid substitution in the protein-coding genes, (mit⁻) mutation. This (mit⁻) genotype is often associated with base-substitutions in the organelle-specific tRNA and rRNA genes essential for the mitochondrial protein synthesis, (syn⁻) mutations, converting the genotype to (mit⁻ + syn⁻). In yeast, (syn⁻) mutants are similar to the strains with

subunits of complex V. The rest of mtDNA contains genetic information essential for the expression of the protein-coding genes; genes specifying two mitochondrial rRNAs, 22 organelle-specific tRNAs and genes regulating the transcription and replication (D-loop).

large deletion, (ρ^-) mutation, in being pleiotropically deficient in the respiratory and ATPase complexes (Tzagoloff, 1982).

One complication in the analysis of germ-line point mutations has been the lack of the normal standard base-sequence of human mtDNA indispensable to survey of (mit$^-$) and/or (syn$^-$) mutation. Although the major part of the Cambridge sequence (Anderson et al., 1981) was derived from a single placenta mtDNA, some regions were derived from HeLa cell's mtDNA. In addition, several ambiguous nucleotides were assumed to be the same as in the bovine mtDNA sequence. Thus, the Cambridge sequence cannot be regarded as the normal standard sequence to the studies on mitochondrial diseases. Wallace et al. (1988) implied that there are some errors in the Cambridge sequence; however, their argument was also not fully documented. On the other hand, basing their theory on the restriction mapping, the Wilson group concluded that all the modern human mtDNAs stem from one woman or one group of women, the mitochondrial Eve, who is postulated to have lived about 200 000 years ago (Cann et al., 1987). However, due to the limitation of restriction enzyme assay, the information derived from the analyses is restricted to only about 9% of the total mtDNA sequence, thus the total base-sequence of the mitochondrial Eve could not be deduced. To overcome the lack of the normal standard mtDNA sequence, we have sequenced the entire mtDNA of 44 individuals of Japanese, American of Irish decent, and Australians of English- or Greek-origin using the fluorescence-based direct sequencing method without using cloning (Tanaka and Ozawa, 1992). The accumulated database of over 750 000 bp permits us to deduce the base-sequence of the mitochondrial Eve and to elucidate the divergence of the Cambridge sequence and those of the patients from Eve. The database identifies the definite genetic pedigrees of the patients with mitochondrial disease (Ozawa et al., 1991a, b) without using an interesting but incomplete statistical method to construct phylogenies from the limited number of base-substitutions obtained by the restriction enzyme assay and/or the partial sequencing of the genome (Felsenstein, 1988).

As shown in Figure 2, a phylogenetic tree was constructed according to the base-substitutions among the patients with mitochondrial cardiomyopthy whose autopsied or biopsied specimens could be obtained clinically (Ozawa, 1994). Some of the base-substitutions are synonymous; however, some are non-synonymous mutations such as (mit$^-$) and/or (syn$^-$) that could produce significant change in the biochemical properties of the gene product and thereby initiate the disease. For example, the base-substitutions of mtDNA of a patient with hypertrophic cardiomyopathy (*HCM, P-3*) demonstrated no (syn$^-$) mutation except for potentially pathogenic mutation (T-to-G transversion at nucleotide position 8993). The mutation causes substitution of a leucine residue conserved among biological species into arginine in a transmembranous helix in the subunit 6 protein, *proton channel* of ATPase (Lewis et al., 1990). Despite no signs of vascular disorder, the *HCM, P*-3 of age 25 showed a negative T-wave in his electrocardiogram that is usually regarded as the sign of ischemic heart in the cardiology clinics. This could be derived from insufficient utilization of oxygen by a defective proton-motive ATP synthase (Fig. 1A). At the early stage of cardiomyopathy, he had shown no signs of heart failure yet, he could be treated with Coenzyme Q_{10} (CoQ) as a sole remedy. After 3 months of administration of CoQ, 300 mg per day, his negative T-wave became positive (Ozawa, 1994). This is the first clinical success in which a negative T-wave became positive among patients with primary cardiomyopathy. This outcome could be interpreted as showing that administration of CoQ, an essential redox carrier in the electron transport chain (Hatefi, 1985) as well as an important proton carrier in the theoretically proposed *proton-CoQ cycle* (Mitchell, 1979), enhanced the respiration rate and the proton gradient, thereby restoring defective ATP synthesis into the near normal level.

There are two distinct features in the phylogenetic tree of the mitochondrial disease patients: (a) There are no predominant point mutations among the patients with PD or with a primary cardiomyopathy, in contrast to those with some neuromuscular diseases; over 80% of the patients with mitochondrial myopathy, encephalopathy with lactic acidosis, and stroke-like episodes (MELAS) had an 3243 A-to-G transition in the $tRNA^{Leu(UUR)}$ gene, and about 50% of the patients with myoclonic epilepsy with ragged red fibers (MERRF) had a 8344 A-to-G transition in the $tRNA^{Lys}$ gene. The sequencing of MELAS and MERRF patients' entire mtDNA in our laboratory has revealed that, besides these (syn⁻) mutations, there are several (mit⁻) mutations that would result in a significant change in the gene products (Ozawa et al. 1991a). (b) Each patient with mitochondrial disease has several (mit⁻) and/or (syn⁻) with potentially serious consequences. These facts suggest that not only a single (mit⁻) or (syn⁻) detected by the restriction enzyme analyses, but also several other serious point mutations, detected by the sequencing of the entire mtDNA, play an important role in the manifestation of mitochondrial disease. Comprehensive analyses of the germ-line mutations among patients with neuromuscular and cardiovascular diseases demonstrated that the effects of these mutations are cumulative. Looking through the whole spectra of the point mutations among the patients with mitochondrial diseases, we have noticed that the patients with (mit⁻ + syn⁻) genotype manifested usually more severe clinical symptoms than patients with (mit⁻) genotype. Figure 3 illustrates the cumulative survival rate of these patients calculated by a statistical method (Kaplan and Meier, 1958). The median survival time of the patients with (mit⁻ + syn⁻) is more than 50 years shorter than that of the patients with only (mit⁻). The clear correlation between the

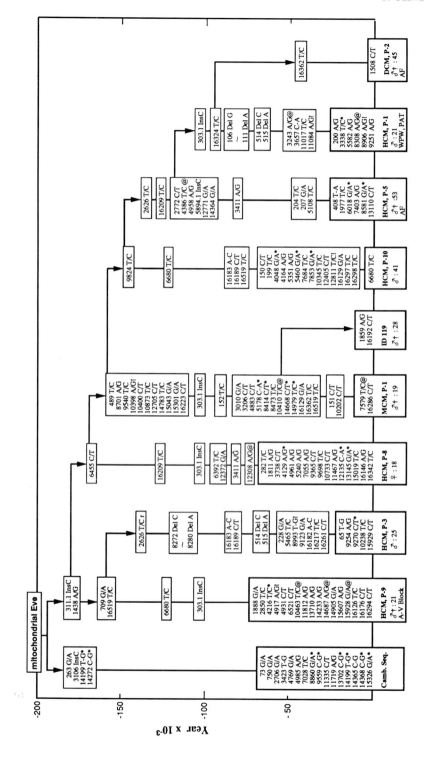

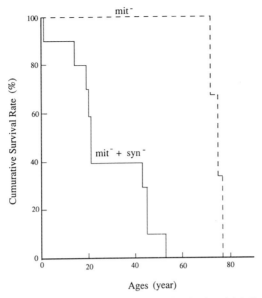

Ages (year)

Figure 3. Cumulative survival rate of the patients with mitochondrial diseases. Cumulative survival rate of the patients with (mit⁻) genotypes (dashed line) and that of the patients with (mit⁻ + syn⁻) genotypes (solid line) are plotted by the method of Kaplan-Meier. All (mit⁻) as well as other base-substitutions of the patients were reported elsewhere (Tanaka et al., 1990; Ozawa et al., 1991a, b; Ozawa, 1994). Among the patients with (mit⁻ + syn⁻), (syn⁻) mutations were detected in the tRNA$^{\text{Leu(UUR)}}$, tRNA^{Trp} and tRNA^{Ile} genes of a patient who died at age 1, in the tRNA$^{\text{Leu(UUR)}}$ and tRNA^{Ala} genes in one who died at 14, in the tRNA^{Asp} and tRNA^{Arg} genes in one who died at 19, in the tRNA$^{\text{Leu(UUR)}}$ gene in one who died at 20, in the tRNA^{Gln}, tRNA$^{\text{Leu(UUR)}}$ and tRNA^{Lys} genes in one who died at 21, in the tRNA^{Arg}, tRNA^{Glu} and tRNA^{Thr} genes in one who died at 21, in the tRNA^{Lys} gene in one who died at 21, in the tRNA^{Cys} and the tRNA^{Thr} genes in one who died at 43, in the tRNA^{Gln} gene in one who died at 45, and in the tRNA^{Gln} gene in one who died at 53.

phenotype and the genotype could be interpreted as showing that (mit⁻ + syn⁻) point mutations, together with provoked somatic mutations, shorten the survival time despite some complementation of the (syn⁻) mutations (Chomyn et al., 1992; Yoneda et al., 1994).

Figure 2. Clustering of point mutations in mtDNA among patients with mitochondrial cardiomyopathy. From the total base-sequencing of 44 individuals including 14 mitochondrial cardiomyopathy patients, the base-sequence of mitochondrial Eve is deduced. Divergence from the mitochondrial Eve of the Cambridge sequence (*Camb. Seq.*) reported by Anderson et al. (Anderson et al., 1981) is also deduced from our data-base. Among the mitochondrial cardiomyopathy patients, subsequently diverged base-substitutions are demonstrated together with their nucleotide positions. Among the base-substitutions, several point mutations are illustrated by abbreviations: /, transition mutation; –. transversion mutation; Ins, insertion of nucleotide; Del, deletion of nucleotide; *, mutation which changes amino acid residue; !, mutation which changes evolutionary conserved amino acid residue; @, mutation in tRNA gene. Patients (*P*) with sex, age of death (†) or of genetic examination and major arrhythmia are illustrated. *HCM:* hypertrophic cardiomyopathy. *DCM:* dilated cardiomyopathy. *MCM:* mitochondrial cardiomyopathy. *ID:* identification number for the legal anatomy.

Another complication in the analysis of germ-line point mutations is that each mitochondrion contains multiple copies of mtDNA. Thus, several thousand copies exist in a single cell. A recent study (Kamiya et al., 1992) on the mutanogenesis of the hydroxyl radical adduct of dG, 8-OH-dG, in mammalian cells clearly demonstrated that a synthetic c-Ha-*ras* protooncogene containing 8-hydroxy-guanine induces random point mutations at the modified site and adjacent positions. As base-sequencing after cloning of mtDNA could pick-up one particular base-substitution out of thousands of copies, the direct sequencing without cloning (Tanaka and Ozawa, 1992) could avoid such complication. In the case of the germ-line point mutations, the mutated bases are *homoplasmic*, except in some cases such as the 3243 A-to-G transition frequently found in the patients with MELAS (Goto et al., 1990; Ino et al., 1991) and some patients with diabetes and deafness (van den Ouweland et al., 1992). The 3243 A-to-G mutation is usually *heteroplas-mic*; namely, the mutated DNA and the wild-type DNA coexist in a tissue. The degree of heteroplasmy could be detected by single-strand conformational polymorphism.

Somatic mutations

A certain percentage of electrons is inevitably eluted from the electron-transfer chain and accepted by molecular oxygen forming oxygen free radicals. In early mitochondrial studies, this phenomenon was observed

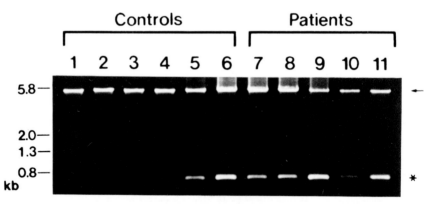

Figure 4. Presence of the deleted mtDNA in the striatum of patients with PD. Detection of the deleted mtDNA in the striatum of patients with PD by the PCR method. The mtDNA fragments were amplified using a primer pair (Ikebe et al., 1990), separated on a 1% agarose gel, and stained with ethidium bromide. Lanes 1–6, the normal controls; and lanes 7–11, the patients with PD. Both the normal controls and the patients with PD are arranged in the order of age. Sizes of amplified fragments are indicated in kbp. The band amplified from the normal mtDNA is indicated by an arrow. The abnormal band of 0.77 kbp amplified from the deleted mtDNA is indicated by an asterisk.

as the antimycin-insensitive respiration that was around 5% of the total mitochondrial oxygen consumption (Brown et al., 1965). Despite the free-radical scavenging activities in mitochondria, certain parts of oxygen free radicals attack macromolecules including mtDNA. Thus, over the years, oxygen free radical adducts of nucleic-acid bases accumulate, leading to double-strand separation and deletion of mtDNA. In the case of deletion, the mutated and the wild-type DNAs are usually *heteroplasmic* as could be detected by PCR amplification. To clarify the role of deletion, it is necessary to locate the deletion span and to quantify the ratio of the deleted mtDNA to the total mtDNA. In our laboratory, several novel methods were devised for a rapid and accurate detection of the somatic mutations in human mtDNA: (a) Non-isotopic Southern blot analysis of deleted mtDNA using PCR amplified probes (Ohno et al., 1991); (b) Primer-shift PCR method for localizing mtDNA deletions (Sato et al., 1989); (c) PCR-S_1 method for localization of deleted mtDNA (Tanaka-Yamamoto et al., 1989); (d) Kinetic PCR to quantify the amount of the deleted mtDNA (Ozawa et al., 1990b); (e) Total detection system for deletion using 180 kinds of primer pairs (Katsumata et al., 1994); (f) Micro high-performance liquid chromatography/mass spectrometry (microHPLC/MS) for the quantification of the oxygen damage in mtDNA (Hayakawa et al., 1991).

The patients with PD had a number of potentially serious germ-line mutations as well as multiple deletions in mtDNA, including a large deletion of 5 kbp that deletes the genes encoding ATPase8 subunit, cytochrome oxidase III subunit, four subunits of NADH: CoQ oxidoreductase, 5 tRNA (Fig. 4). The 5 kbp deletion occurs commonly among mitochondrial disease patients and normal elderly subjects. Quantitative analyses of the 5 kbp deletion in a patient and an elderly patient by the method (d) are shown in Figure 5.

We have detected multiple deletions in mtDNAs of the patients with mitochondrial disease (Ozawa et al., 1990a). Some deletions were maternally inherited (Ozawa et al., 1988), but most were somatically acquired during the life of an individual leading to development of non-atherosclerotic dysfunction of neuromuscular cells. Mitochondrial disease expressed at younger age could be ascribed to premature aging of the neuromuscular cells triggered by endogenous factors such as maternally inherited germ-line mutations and/or exogenous factors such as toxins, drugs, or viral infections (cf. Fig. 9).

Segregation of mutated genomes

The germ-line mutations are maternally inherited. This results in the somatic segregation of individual mtDNA during eukaryotic cell division. It is suggested that the segregation can result in organ-specific

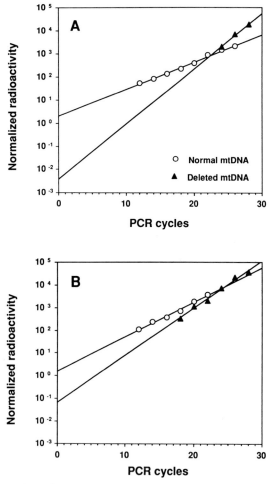

Figure 5. Kinetic analyses of PCR products from deleted mtDNA and normal mtDNA in the Parkinsonian and control striatum. The mtDNA fragments derived from the normal mtDNA and from the deleted mtDNA were amplified by PCR as described in Figure 1, except for the presence of α-[^{32}P]deoxycytosine triphosphate. PCR reactions were terminated after different cycles of amplification as indicated. The fragments were electrophoresed on the gel, transferred onto a nylon membrane, and autoradiographed. The radioactivities of the labeled DNA fragments were quantified by a laser image analyzer. The radioactivity of fragments was normalized by each size. Logarithms of the normalized radioactivities were plotted against the number of PCR amplification cycles. Straight lines with different slopes thus obtained were extrapolated to zero cycle of the amplification. The intercepts of the vertical axis gives the amount of the deleted mtDNA relative to the normal mtDNA that existed in the tissue before PCR amplification. (A) control striatum: (B) Parkinsonian striatum (Ozawa et al., 1990b).

manifestation of mitochondrial cytopathy; e.g., that in brain, skeletal muscle, and myocardium in the case of MELAS, that in skeletal muscle in the case of mitochondrial myopathy, and that in myocardium in the case of mitochondrial cardiomyopathy. Such segregation will also pro-

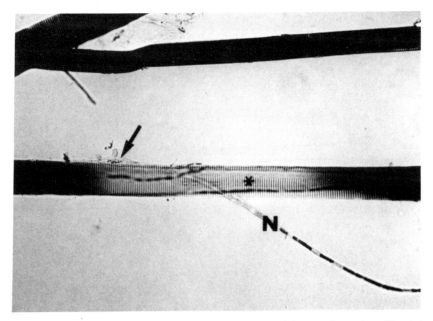

Figure 6. Energy mosaic in a muscle fiber of a mitochondrial myopathy patient. Biopsied muscle fibers from a patient who has a *heteroplasmic* deletion in his mtDNA including cytochrome oxidase subunit 3 were unraveled and subjected to cytochrome oxidase activity staining. Focal destaining (*), deficiency of the activity, near the neuro (N)-muscular junction (arrow) could explain abnormal electromyograph of the patient.

duce an energy mosaic in a tissue whereby some cells of the tissue will be more bioenergetically competent than other cells. In Figure 6, a focal deficit of cytochrome oxidase activity is demonstrated in an unraveled biopsied muscle fiber from a mitochondrial myopathy patient who had a *heteroplasmic* deletion that deletes cytochrome oxidase subunit III gene. Activity of cytochrome oxidase directly reflects the rate of ATP production. Thus, the energy mosaic near the neuromuscular junction could explain the patient's abnormal electromyography. Similarly, the energy mosaic among brain cells could explain the focal low density area in the CT scanning image among mitochondrial disease patients. As the somatic oxygen damage and the deletion in mtDNA increase exponentially with age (Fig. 7), the energy mosaic in the neuromuscular cells would be an important contributor to the senile ataxia and dementia.

Mutational rate

The mutation rate of human mtDNA is many times higher than that of nuclear DNA. The high mutation rate may be based on the following facts. Firstly, in contrast to nuclear DNA, the genes are highly economically packed with no introns, and they are punctuated by tRNA genes. Secondly, expression of the whole genome is essential for the maintenance of mitochondrial bioenergetic function, whereas only about 7% of the nuclear genome are ever expressed at any particular differentiation stage. Thus, by calculation, 10–12 times faster mutation rate than nuclear DNA could be predicted. In addition, the mtDNA is located inside the inner mitochondrial membrane where active oxygen radical species including mtDNA mutation are constantly produced from the normal electron-transport activity. The mtDNA has neither a protective protein, such as histone, nor a repair system. So, the mutated mitochondrial genomes tend to accumulate in a cell, especially in post-mitotic neuromuscular cells. In yeast, the *petite* mutations, a characteristic type of yeast mtDNA mutation, occurs spontaneously in growing cultures at a rate of 10^{-1} to 10^{-3} of total cell population, in contrast to nuclear mutation rate of 10^{-7} to 10^{-8} in *haploid* and 10^{-14} to 10^{-16} in *diploid*. From these facts, we proposed that accumulation of somatic mutations in mtDNA during life is an important contributor to aging and degenerative diseases (Linnane et al., 1989). This proposal was documented by the quantification of mitochondial genome mutations among cardiomyopathy patients and normal aged people by the kinetic PCR method (Ozawa et al., 1990b), which is presented in Figure 7A. As the 7.4 kbp deletion in mtDNA occurs commonly among mitochondrial disease patients, the existence of the deletion among normal aged people was surveyed. The deletion was detected without exception among aged subjects (Hayakawa et al., 1992). In addition, the ratio of mtDNA with a 7.4 kbp deletion to the total DNA increased exponentially with age [log (deleted mtDNA percentage) $= -3.23 + 0.0407 \times$ age, $r = 0.87$, $p < 0.01$] up to 7% at age 97, as shown in Figure 7A. Extrapolation of the exponential curve to 100% mutation of mitochondrial genomes could occur at age 129. The deletion will result in a major defect in the electron-transport chain, leading to energy deficit and cell death, *apoptosis*, in the myocardium. It could be speculated that age 129 could be the maximum life expectancy of humans regarding the myocardial bioenergetic function.

Maternal inheritance

As the mitochondria in the sperm are detached at the time of fertilization, mtDNA is transmitted to a descendant exclusively from an oocyte. So, patients with mitochondrial diseases sometimes have family members with similar symptoms inherited maternally in contrast to familial diseases that show a typical Mendelian pattern of inheritance. Often mitochondrial diseases appear to be sporadically in patients.

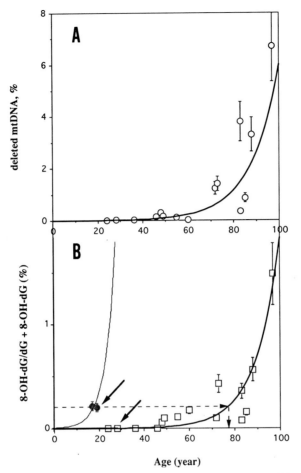

Figure 7. Correlative increase in the deletion and oxygen damage in mtDNA associated with age and mitochondrial cardiomyopathy. (A) The ratio of deleted mtDNA to the total mtDNA was plotted against the age of subjects. The amounts of the total and deleted mtDNA in autopsied myocardium of various ages were determined by the kinetic PCR method as illustrated in Figure 7. The ratio of the deleted mtDNA to the total mtDNA among the specimens increases exponentially with age [$r = 0.87$, $p < 0.01$]. (B) The 8-OH-dG content was plotted against the age of subject. Samples of mtDNA in the rest of sample after the deletion analysis were enzymatically hydrolyzed into nucleosides, subjected to pre-column concentration, and analyzed by the micro HPLC/MS system. Total eluent from the micro column (0.3×150 mm) was directly injected into a mass spectometer. Both selected ion monitoring and ionization mass spectra of 8-OH-dG and dG were recorded. The ratio of 8-OH-dG to (dG + 8-OH-dG) increases exponentially with the ages of subjects [$r = 0.84$, $p < 0.01$]. The 8-OH-dG contents of mitochondrial cardiomyopathy patients who died by heart failure at age 17 (female) and 19 (male, *MCM*, *P-1* in Figure 4, pointed to by an arrow), respectively, are plotted by black circles. The 8-OH-dG content of the genotype control (female, *ID 119* in Figure 4, pointed to by an arrow) is plotted by an open square. Arrows with dashed lines indicate that the 8-OH-dG content of *MCM*, *P-1* is equivalent to that of the normal subjects of age 78.

Mechanism of mitochondrial DNA mutations

Yeast's (mit⁻) mutations are usually associated with a significant rise in
the rate of deletion (Linnane et al., 1989). Sequencing of the deleted
mtDNA revealed that a *pseudo*-recombination occurs between direct
repeats in two genes, as shown in Figure 8. Namely, double-strand
separation leading to the generation of long stretches of single-stranded
DNA is a prerequisite to the occurrence of a large deletion. However,
the mechanism for the occurrence of a large deletion has been unknown
(Grivell, 1989); neither intramolecular homologous recombination nor
slipped mispairing during replication is satisfactory, because there are
no reports of any orthodox recombination mechanisms, and no chance
to expose long stretches of single-stranded DNA at any stage of
mtDNA replication. One possible mechanism for the double-strand
separation is oxygen damage to mtDNA, such as conversion of guanine
to 8-hydroxy-guanine by hydroxy radical, and active oxygen species.
The active-oxygen production is proportional to the respiration rate,

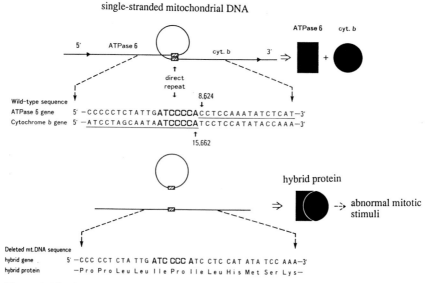

Figure 8. The large deletion of mtDNA in a mitochondrial cardiomyopathy patient. The
cross-over region of a deleted mtDNA from a patient with mitochondrial cardiomyopathy was
directly sequenced. The deletion spanned 7039 bp starting from nucleotide position 8624 on
the 3'-side of the directly repeated (*ATCCCCA*) within the ATPase 6 gene, and ending at
15 662 on the 3'-side of the direct repeat within the cytochrome *b* gene. The sequence of
deleted mtDNA resulting in a frame shift forms an open reading frame that predicts a 12 kDa
hybrid protein composed of 32 amino acid residues from the N-terminal side of the ATPase
sununit 6, *proton channel*, and 75 amino acid residues from C-terminal side of the cytochrome
b apoprotein. Such a hybrid protein would give the cell abnormal mitotic stimuli.

viz., about 5% of the total oxygen uptake, and neuromuscular mitochondria respire at the highest rate among tissues. Both the conformation and the electron negativity of 8-hydroxy-guanine were found to be quite different from those of normal guanine (Aida and Nishimura, 1987), inducing a local DNA structure alternation such as *B-to-Z* transition (Belguise-Valladier and Fuchs, 1991). It could be considered that accumulation of 8-hydroxy-guanine in mtDNA triggers its double-strand separation, thus the amount of 8-hydroxy-guanine and of the deletion in mtDNA should correlate. We have demonstrated that 8-hydroxy-guanine massively accumulated in mtDNA of the azidothymidine treated mice, an experimental model of mitochondrial myopathy, associating with deletions, and that hydroxy radicals cause oxygen damage *in vitro* and *in vivo* to mtDNA, converting guanine to 8-hydroxy-guanine (Hayakawa et al., 1991a). Using the same human specimen as for the quantification of the deletion in Figure 7A, the ratio of 8-OH-dG to (dG + 8-OH-dG) in mtDNA hydrolysate was determined using the microHPLC/MS system (Hayakawa et al., 1991a; Hayakawa et al., 1992). Even among normal aged people, oxygen damage increase exponentially with age [$\log (\text{8-OH-dG \%}) = -3.82 + 0.0401 \times \text{age}$, $r = 0.84$, $p < 0.01$] as shown in Figure 7B. Accumulation of 8-OH-dG in mtDNA up to 1.5% of dG (one per 150 bp of mtDNA $\approx$ 15 turns of double helix) could trigger the double-strand separation that is the prerequisite for a large deletion. The analyses established a clear correlation between the amount of the deleted mtDNA and the 8-OH-dG content [(deleted mtDNA percentage) $= 0.0715 + 4.734 \times (\text{8-OH-dG \%})$, $r = 0.93$, $p < 0.01$]. Abnormal ion accumulation into mitochondrial matrix due to energy deficit of the cell would also facilitate the double-strand separation of mtDNA. Consequently, the oxygen damage in mtDNA induces a large deletion. Conversely, a defective mitochondrial respiratory chain encoded by the deleted mtDNA would enhance oxygen free radical formation, resulting in more accumulation of the oxygen damage. Such a vicious cycle of the oxygen damage and deletion, schematically illustrated in Figure 9, would result in those changes being synergistic and exponential with a significant correlation, $r = 0.93$, as shown in Figure 7. These deteriorative sequences finally lead to cellular energy deficit and death. Thus, the development of non-atherosclerotic neuromuscular dysfunction seems to be an inevitable progressive change in human life (Hayakawa et al., 1992).

The mitochondrial diseases starting at a young age could be regarded as premature aging; namely, premature phenotypic expression of maternally inherited point mutations. Contents of 8-OH-dG in the mtDNA hydrolysates from young patients with mitochondrial diseases are shown in Figure 7B. Two patients, one female who died of heart failure at age 17, and one male who died at age 19 (*MCM, P-1* in Fig. 2),

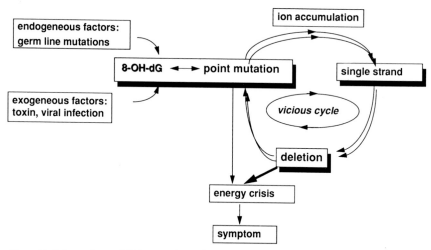

Figure 9. Scheme for mtDNA mutations. Active oxygen species production is enhanced by both endogenous factors such as maternally inherited point mutations (cf. Fig. 4) and exogenous factors such as toxins, drugs or viral infections. Hydroxy-radical adduct of guanosine, 8-OH-dG, resulted in random point mutations and the double-strand separation leading to a long stretch of single-stranded DNA, then deletion. Defective mitochondrial electron-transport chain encoded by the deleted mtDNA enhances the hydroxy-radical forma- tion resulting in more accumulation of 8-OH-dG and ions in mitochondrial matrix. High concentration of ions also facilitates the double-strand separation. Such a vicious cycle of oxygen damage and deletion in mtDNA seems to result in those changes being synergistic and exponential, as demonstrated in Figure 8.

accumulated 8-OH-dG equivalent to the normal control subjects of age 78, as shown in Figure 7B. By the total detection system for deletion (Katsumata et al., 1994), it was demonstrated that myocardial mito- chondria of *MCM, P-1* accumulated over 325 kinds of multiple dele- tions in mtDNA with abnormal mitochondrial morphology. Deleted mtDNA accounted for 84% of the total indicating extensive fragmenta- tion of his mtDNA. Four point mutations (a 10398 A-to-G transition in the ND3 gene, a 14797 T-to-C transition in the cytochrome *b* gene, a 7579 T-to-C transition in the tRNA^{Asp} gene and a 10410 T-to-C transition in the tRNA^{Arg} gene) were detected in his mtDNA. In a woman who died from an accident at age 28 (*ID 119* in Fig. 2), having almost the same mtDNA genotype except the 7579 (syn⁻) mutation, 67 types of deleted mtDNA, accounting for 23% of the total, were detected in her heart mtDNA. In the case of a baby of age 3 with ventricular septal defect, only five types of deleted mtDNA accounting for less than 1% of the total were detected in her biopsied myocardium. Thus, it could be concluded that the 7579 (syn⁻) mutation with combination of other (mit⁻) mutations of *MCM, P-1* is a major initiator of his mtDNA deterioration. As such, germ-line mutations are usually *homo-*

plasmic among tissues and among patients' maternal relatives; the differences in phenotypic expression could be due to additional somatic mutations such as the amount of deletions, and/or to difference in the expression of *apoptosis* regulating protooncogene, such as Bcl-2/Bax/Ced-9 gene (Haecker and Vaux, 1994).

Clinical symptoms as phenotypic expression

The accumulation of bioenergetically defective cells with time is a key factor in the process of aging and mitochondrial disease. Thus, a progressive decrease in the efficiency of the oxidative phosphorylation process occurs in the neuromuscular cell. Normal individuals naturally adjust their physical activities to compensate for this loss and it does not necessarily lead to any clinical treatment. However, when there is an excessive increase in the cell's energy demand under stress, there may be sudden death or severe impairment of the neuromuscular cell with pathological consequences. Idiopathic PD is a very common degenerative brain disease causing movement disorder. By calculation, it was reported that about 1 person in 40 will become symptomatic with PD during normal lifetime (Kurland, 1958). As the disorder has a slight familial connection not often following any clear Mendelian pattern of inheritance, a polygenic mode of inheritance and exposure to environmental factors have been proposed as the etiological factor. As introduced here, the underlying mechanism of PD is progressive accumulation of polygenic deletion due to oxygen, a prominent environmental factor. A similar exponential increase in the oxygen damage and the extent of deletions was detected in the mtDNA from diaphragm muscle (Hayakawa et al., 1991b; Torii et al., 1992). These mutations would account for a progressive reduction of vital capacity and pulmonary compliance with an increase of residual volume in the elderly lung. Together with clinical and histochemical data relevant to the genetic mutations, such as mild left ventricular hypertrophy associated with age (Gestenblith et al., 1977) and focal decrease of cytochrome oxidase activity with age (Müller-Höcker, 1989), it could be concluded that every human will suffer from neuromuscular mitochondrial disease to some extent with aging.

In the case of mitochondrial diseases, there is tissue-specific phenotypic expression. Patients with mitochondrial cardiomyopathy often show left ventricular hyperplasia and negative T-wave, which is a sign of ventricular ischemia, despite of vascular disorder. In the case of *MCM P-1*, his heart size did not increase, but heart failure developed over a 10-year period with proliferation of abnormal mitochondria and atrophy of muscle fibers. In the terminal stage of disease, his heart failure was of the NYHA Class IV. Due to the bioenergetic mosaic in myocardium, patients with mitochondrial cardiomyopathy suffered from arrhythmias as shown in Figure 2. Some cases of HCM died of premature ventricular contraction. Cardiac output of DCM patients tended to decrease in association with the course of the disease, and the patients died of cardiac failure (*DCM, P-2*). Some patients showed a transition from HCM to DCM (e.g., *HCM, P-5*). *HCM, P-5* died of cerebral infarction with atrial fibrillation. Another patient showing the transition from HCM to DCM died of heart failure and renal failure at age 43 (Hattori et al., 1991). Examination of the mtDNA from his autopsied heart demonstrated several serious germ-line point mutations similar to *MCM, P-1*, and multiple deletions.

Treatment

The accumulation of somatic mitochondrial gene mutations will lead to a mosaic of cells ranging in their bionergetic capacity from normal to partially or grossly defective (Fig. 6). It is possible to attenuate the pathological consequences by circumventing particular defective enzyme complexes that are part of the normal bioenergetic pathway. The pharmacological addition of a suitable redox substance could potentially restore the rate of electron-transport with an increase in chemical energy available to the cells sufficient to allow the cells to respond to stress which would otherwise result in neuromuscular damage (Linnane et al., 1989). Several oxidation-reduction carriers are routinely used in mitochondrial studies to bypass specific blocks in the electron-transport chain *in vitro*; some have already been used clinically as redox-therapy for human mitochondrial disorders (Eleff et al., 1984; Ozawa et al., 1987; Jinnai et al., 1990). These redox substances include ubiquinol, menadione, and ascorbic acid. The efficacy of CoQ administration to *HCM, P-3* was mentioned in the above section.

A fundamental therapy for serious mitochondrial aging and diseases would be a gene therapy to replace mutated mtDNA. However, technical difficulties, at present formidable, must await new developments. Meanwhile, the redox therapy holds out a good possibility for immediate therapy.

References

Aida, M. and Nishimura, S. (1987) An *ab initio* molecular orbital study on the characteristics of 8-hydroxyguanine. *Mutat. Res.* 192: 83–89.
Anderson, S., Bankier, A.T., Barrell, B.G., de Bruijn, M.H.L., Coulson, A.R., Drouin, J., Eperson, I.C., Nierlich, D.P., Roe, B.A. Sanger, F., Schreier, P.H., Smith, A.J.H., Staden, R. and Young, I.G. (1981) Sequence and organization of the human mitochondrial genome. *Nature* 290: 457–465.

Ballinger, S.W., Shoffner, J.M., Hedaya, E.V., Trounce, I., Polak, M.A., Koontz, D.A. and Wallace, D.C. (1992) Maternally transmitted diabetes and deafness associated with a 10.4 kb mitochondrial DNA deletion. *Nature Genetics* 1: 11–15.

Belguise-Valladier, P. and Fuchs, R.P.P. (1991) Strong sequence-dependent polymorphism in adduct-induced DNA structure: analysis of single *N*-2-acetyl-aminofluorene residues bound within the *Nar*I mutation hot spot. *Biochemistry* 30: 10091–10100.

Brown, C.B., Russel, J.R. and Howland, J.L. (1965) Antimycin-insensitive respiration in beef heart mitochondria. *Biochim. Biophys. Acta* 110: 640–642.

Cann, R.L., Stoneking, M. and Wilson, A.C. (1987) Mitochondrial DNA and human evolution. *Nature* 325: 31–36.

Chomyn, A., Martinuzzi, A., Yoneda, M., Daga, A., Hurko, O., Jhons, D., Lai, S.T., Nonaka, I., Angelini, C. and Attardi, G. (1992) MELAS mutation in mtDNA binding site for transcription termination factor causes defects in protein synthesis and in respiration but no change in levels of upstream and downstream mature transcript. *Proc. Natl. Acad. Sci. USA* 89: 4221–4225.

Eleff, S., Kennaway, N.G., Buist, N.R.M., Darley-Usmer, V.M., Capaldi, R.A. and Chance, B. (1984) ^{31}P NMR study of improvement in oxidative phosphorylation by vitamins K_3 and C in a patient with a defect in electron transport at complex III in skeletal muscle. *Proc. Natl. Acad. Sci. USA* 81: 3529–3533.

Felenstein, J. (1988) Phylogenies from molecular sequences: influences and reliability. *Ann. Rev. Genet.* 22: 521–565.

Gestenblith, G., Frederiksen, J., Yin, F.C.P., Fortuin, N.J., Lakatta, E.G. and Weisfeld, M.L. (1977) Echocardiographic assessment of a normal adult aging population. *Circulation* 56: 273–278.

Goto, Y., Nonaka, I. and Horai, S. (1990) A mutation in the transfer RNA$^{Leu(UUR)}$ gene associated with the MELAS subgroup of mitochondrial encephalomyopathy. *Nature* 348: 651–653.

Grivell, L.A. (1989) Mitochondrial DNA: Small, beautiful and essential. *Nature* 341: 569–571.

Haecker, G. and Vaux, D.L. (1994) Viral, worm and radical implications for apoptosis. *TIBS* 19: 99–100.

Harman, D. (1960) The free radical theory of aging: The effect of age on serum mercaptan levels. *J. Gerontol.* 15: 38–40.

Hatefi, Y. (1985) The mitochondrial electron transport and oxidative phosphorylation system. *Ann. Rev. Biochem.* 54: 1015–1069.

Hattori, K., Ogawa, T., Kondo, T., Mochizuki, M., Tanaka, M., Sugiyama, S., Ito, T., Satake, T. and Ozawa, T. (1991) Cardiomyopathy with mitochondrial DNA mutations. *Am. Heart J.* 122: 866–869.

Hayakawa, M., Ogawa, T., Tanaka, M., Sugiyama, S. and Ozawa, T. (1991a) Massive conversion of deoxy-guanosine to 8-hydroxy-guanosine in mouse liver mitochondrial DNA by administration of aziodothymidine. *Biochem. Biophys. Res. Commun.* 176: 87–93.

Hayakawa, M., Torii, K., Sugiyama, S., Tanaka, M. and Ozawa, T. (1991b) Age-associated accumulation of 8-hydroxydeoxyguanosine in mitochondrial DNA of human diaphragm. *Biochem. Biophys. Res. Commun.* 179: 1023–1029.

Hayakawa, M., Hattori, K., Sugiyama, S. and Ozawa, T. (1992) Age-associated oxygen damage and mutations in mitochondrial DNA in human hearts. *Biochem. Biophys. Res. Commum.* 189: 979–985.

Hayakawa, M., Sugiyama, S., Hattori, K., Takasawa, M. and Ozawa, T. (1993) Age-associated damage in mitochondrial DNA in human hearts. *Mol. Cell. Biochem.* 119: 95–103.

Ikebe, S., Tanaka, M., Ohno, K., Sato, W., Hattori, K., Kondo, T., Mizuno, Y. and Ozawa, T. (1990) Increase of deleted mitochondrial DNA in the striatum in Parkinson's disease and senescence. *Biochem. Biophys. Res. Commun.* 170: 1044–1048.

Ino, H., Tanaka, M., Ohno, K., Hattori, K., Ikebe, S., Ozawa, T., Ichiki, T., Kobayashi, M. and Wada, Y. (1991) Mitochondrial leucine tRNA mutation in a mitochondrial encephalomyopathy. *Lancet* 337: 234–235.

Jinnai, K., Yamada, H., Kanda, F., Masui, Y., Tanaka, M., Ozawa, T. and Fujita, T. (1990) A case of mitochondrial myopathy, encephalopathy and lactic acidosis due to cytochrome c oxidase deficiency with neurogenic muscular changes. *Eur. Neurol.* 30: 56–60.

Kamiya, H., Miura, K., Ishikawa, H., Inoue, H., Nishimura, S. and Ohtsuka, E. (1992) c-Ha-*ras* containing 8-Hydroxyguanine at codon 12 induces point mutations at the modified and adjacent positions. *Cancer Res.* 52: 3483–3485.

Kaplan, E.L. and Meier, P. (1958) Nonparametric estimation for incomplete observations. *J. Am. Stat. Assoc.* 52: 457–481.

Katsumata, K., Hayakawa, M., Tanaka, M., Sugiyama, S. and Ozawa, T. (1994) Fragmentation of human heart mitochondrial DNA associated with premature aging. *Biochem. Biophys. Res. Commun.* 201: 1158–1164.

Kurland, L.T. (1958) Epidemiology: Incidence, Geographic Distribution and Genetic Considerations. *In:* W. Fields (ed.): *Pathogenesis and Treatment of Parkinsonism.* Thomas, Springfield, IL, pp. 5–49.

Lewis, M.L., Chang, J.A. and Simoni, R.D. (1990) A topological analysis of subunit a form *Escherichia coli* F1F0-ATP synthase predicts eight transmembrane segments. *J. Biol. Chem.* 265: 10541–10550.

Linnane, A.W., Marzuki, S., Ozawa, T. and Tanaka, M. (1990) Mitochondrial DNA mutations as an important contributor to ageing and degenerative diseases. *Lancet* i: 642–645.

Linnane, A.W., Baumer, A., Maxwell, R.J., Preston, H., Zhang, C.F. and Marzuki, S. (1990) Mitochondrial gene mutation: the ageing process and degenerative diseases. *Biochem. Int.* 22: 1067–1076.

Mita, S., Rizzuto, R., Moraes, C.T., Shanske, S., Arnaudo, E., Fabrizi, G.M., Koga Y., DiMauro, S. and Schon, E.A. (1990) Recombination via flanking direct repeats is a major cause of large-scale deletions of human mitochondrial DNA. *Nuceic Acids Res.* 18: 561–567.

Mitchell, P. (1979) *David Keilin's respiratory chain concept and its chemiosmotic consequences.* The Noble Foundation, Stockholm.

Müller-Höcker, J. (1989) Cytochrome-*c*-oxidase deficient cardiomyocytes in the human heart. An age related phenomenon. *Am. J. Pathol.* 134: 1167–1173.

Ohno, K., Tanaka, M., Sahashi, K., Ibi, T., Sato, W., Takahashi, A. and Ozawa, T. (1991) Mitochondrial DNA deletions in inherited recurrent myoglobinuria. *Ann. Neurol.* 29: 364–369.

Ozawa, T., Tanaka, M., Suzuki, H. and Nishikimi, M. (1987) Structure and function of mitochondria: Their organization and disorders. *Brain Dev.* 9: 76–81.

Ozawa, T., Yoneda, M., Tanaka, M., Ohno, K., Sato, W., Suzuki, H., Nishikimi, M., Yamamoto, M., Nonaka, I. and Horai, S. (1988) Maternal inheritance of deleted mitochondrial DNA in a family with mitochondrial myopathy. *Biochem. Biophys. Res. Commun.* 154: 1240–1247.

Ozawa, T., Tanaka, M., Sugiyama, S., Hattori, K., Ito, T., Ohno, K., Takahashi, A., Sato, W., Takada, G., Mayumi, B., Yamamoto, K., Adachi, K., Koga, T. and Toshima, H. (1990a) Multiple mitochondrial DNA deletions exist in cardiomyoctyes of patients with hypertrophic or dilated cardiomyopathy. *Biochem. Biophys. Res. Commun.* 170: 830–836.

Ozawa, T., Tanaka, M., Ikebe, S., Ohno, K., Kondo, T. and Mizuno, Y. (1990b) Quantitive determination of deleted mitochondrial DNA relative to normal DNA in Parkinsonian striatum by a kinetic PCR analysis. *Biochem. Biophys. Res. Commun.* 172: 483–489.

Ozawa, T., Tanaka, M., Ino, H., Ohno, K., Sano, T., Wada, Y., Yoneda, M., Tanno, Y., Miyatake, T., Tanaka, T., Itoyama, S., Ikebe, S., Kondo, T. and Mizuno, Y. (1991a) Distinct clustering of point mutations in mitochondrial DNA among patients with mitochondrial encephalomyopathies and with Parkinson's disease. *Biochem. Biophys. Res. Commun.* 176: 938–946.

Ozawa, T., Tanaka, M., Sugiyama, S., Ino, H., Ohno, K., Hattori, K., Ohbayashi, T., Ito, T., Deguchi, H., Kawamura, K., Nakane, Y. and Hashiba, K. (1991b) Patients with idiopathic cardiomyopathy belong to the same mitochondrial DNA gene family of Parkinson's disease and mitochondrial encephalomyopathy. *Biochem. Biophys. Res. Commun.* 177: 518–525.

Ozawa, T. (1994) Mitochondrial cardiomyopathy. *Herz* 19: 105–118.

Sato, W., Tanaka, M., Ohno, K., Yamamoto, T., Takada, G. and Ozawa, T. (1989) Multiple populations of deleted mitochondrial DNA detected by a novel gene amplification method. *Biochem. Biophys. Res. Commun.* 162: 664–672.

Schatz, G., Haslbrunner, E. and Tuppy, H. (1964) Deoxyribonucleic acid associated with yeast mitochondria. *Biochem. Biophys. Res. Commun.* 15: 127–132.

Tanaka, M., Ino, H., Ohno, K., Hattori, K., Sato, W. and Ozawa, T. (1990) Mitochondrial tRNAIle mutation in fatal infantile cardiomyopathy. *Lancet* 336: 1452.

Tanaka, M. and Ozawa, T. (1992) Analysis of Mitochondrial DNA Mutations. *In*: A. Longstaff, and P. Revest (eds): *Protocols in Molecular Neurobiology*. The Humana Press, Totawa, pp. 25–53.

Tanaka-Yamamoto, T., Tanaka, M., Ohno, K., Sato, W., Horai, S. and Ozawa, T. (1989) Specific amplification of deleted mitochondrial DNA from a myopathic patient and analysis of deleted region with S1 nuclease. *Biochim. Biophys. Acta* 1009: 151–155.

Torii, K., Sugiyama, S., Tanaka, M., Takagi, K., Hanaki, Y., Iida, K., Matsuyama, M., Hirabayashi, N., Uno, Y. and Ozawa, T. (1992) Aging-associated deletions of human diaphragmatic mitochondrial DNA. *Am. J. Respir. Cell. Mol. Biol.* 6: 543–549.

Tzagoloff, A. (1982) Mitochondrial Genetics. *In: Mitochondria*. Plenum Press, New York and London, pp. 267–322.

van den Ouweland, J.M.W., Lemkes, H.H.P., Ruitenbeek, W., Sandkuijl, L.A., de Vijlder, M.F., Struyvenberg, P.A.A., van de Kamp, J.J.P. and Maassen, J.A. (1992) Mutation in mitochondrial tRNA$^{Leu(UUR)}$ gene in a large pedigree with maternally transmitted type II diabetes mellitus and deafness. *Nature Genetics* 1: 368–371.

Wallace, D.C., Singh, G., Lott, M.T., Hodge, J.A., Schurr, T.G., Lezza, A.M.S., Eisas, L.J. and Nikoskelainen, E.K. (1988) Mitochondrial DNA mutation associated with Leber's hereditary optic neuropathy.

Yen, T.C., Su, J.H., King, K.L. and Wei, Y.H. (1991) Ageing-associated 5 kb deletion in human liver mitochondrial DNA. *Biochem. Biophys. Res. Commun.* 178: 124–31.

Yoneda, M., Miyatake, T. and Attardi, G. (1994) Complementation of mutant and wild type human mitochondrial DNAs coexisting since the mutation event and lack of complementation of DNAs introduced separately into a cell within distinct organelles. *Mol. Cell. Biol.* 14: 2699–2712.

Subject index

(The page number refers to the first page of the chapter in which the keyword occurs)

Tissue Culture Techniques

An Introduction

Edited by
B. Martin, *Boston University School of Medicine, Boston, MA, USA*

1994. 248 pages. Hardcover
ISBN 3-7643-3718-4

Also available in Softcover:
ISBN 3-7643-3643-9

Designed to teach the fundamentals of tissue culture to readers ranging from undergraduate students taking their first courses, to veteran scientists needing a quick review of tissue culture, it is the most up-to-date laboratory course on the subject. Written by an experienced teacher with student-tested techniques and protocols, this volume carefully covers the field, with safety and sterility key underlying themes. The ultimate goal is to help the reader think like a tissue culturist.

Extensively illustrated and written in an informal easy-to-read style, this book progresses with a logical and coherent teaching method. The many troubleshooting tips and problem sets ensure that the novice will be up and running quickly, with a firm grasp of the tissue culture and manipulations. The book leads right up to molecular biology experiments like cell transformation, transfection, and isolation.

Contents
Preface • Introduction • Sterility: Aseptic Technique • Sterilization Methods • Quality Control of Sterilization • Routine Cell Culture: Feeding Schedules and Media Components • Subcultivation • Cell Enumeration and Cell Viability • Putting Routine Methods to Work • Detecting and Disposing of Contamination • Troubleshooting • Experiments in Culture: Alterations of the Media • Substrata • Altering the Environment • Primary Cell Culture: Isolation • Characterization • Cell Preservation • Cell Cloning • Culture Changes: The Loss of Differentiated Phenotype • Immortality and Aging • Cell Transformation • Transfection (Gene Transfer) • Information for New Cell Studies: Using Advertising and Catalogues • Appendices: The Cell Cycle, Solutions: Media and Salt Solutions, Vendors, Glossary, Answers to Problem Sets and Exercises

Birkhäuser